James Challis

Lectures on Practical Astronomy And Astronomical Instruments

James Challis

Lectures on Practical Astronomy And Astronomical Instruments

ISBN/EAN: 9783744743709

Printed in Europe, USA, Canada, Australia, Japan

Cover: Foto ©berggeist007 / pixelio.de

More available books at **www.hansebooks.com**

LECTURES

ON

PRACTICAL ASTRONOMY

AND

ASTRONOMICAL INSTRUMENTS

BY THE

Rev. JAMES CHALLIS, M.A., F.R.S., F.R.A.S.
PLUMIAN PROFESSOR OF ASTRONOMY AND EXPERIMENTAL PHILOSOPHY IN THE
UNIVERSITY OF CAMBRIDGE, AND FELLOW OF TRINITY COLLEGE.

CAMBRIDGE:
DEIGHTON, BELL AND CO.
LONDON: GEORGE BELL AND SONS.
1879

Cambridge:
PRINTED BY C. J. CLAY, M.A.
AT THE UNIVERSITY PRESS.

PREFACE.

I BEGAN in the year 1843 a Course of Lectures on Practical Astronomy and Astronomical Instruments, having been at that time seven years Director of the Cambridge Observatory. Lectures on these subjects had already been given by Dr Peacock as Lowndean Professor, which he discontinued on the understanding that I would undertake to carry them on. The circumstances that an Astronomical Observatory so well appointed as that of Cambridge was near at hand, and was provided with various instruments of first-rate quality, appeared to me to give facilities for lecturing on Practical Astronomy which ought to be taken advantage of; and accordingly I commenced lecturing in the above-named year, having previously produced a 'Syllabus' of the subjects of the lectures, and procured a considerable number of wooden models, which, together with some apparatus that had been collected by Professor Peacock, I made use of in conducting and illustrating the Lectures. I had also the advantage of having at command several portable instruments pertaining to the Observatory. At the end of the Syllabus a list of formulæ applicable to the reduction of astronomical observations was introduced, accompanied by brief demonstrations. Also in giving the lectures orally, I adopted the plan of exhibiting in ink-writing on large sheets of white paper descriptions of the instruments, and investigations of formulæ, with the requisite illustrations by Figures, all being of such size as to be readily seen by the students from their seats. I had recourse to this method of lecturing partly because I thought it would serve to convey adequate information with little expenditure of the students' time, and partly because I was unable to meet with a text-book on this department of Astronomy which I could regard as sufficiently accurate and complete.

The contents of the Syllabus, together with the explanations inscribed on the above mentioned papers, have formed the ground-work of the present publication. In fact, as far as regards the divisions of the subjects treated of under the head of each instrument, and the order of their treatment, the Syllabus has been closely followed; but in the course of writing the Treatise, the composition of the Lectures, as originally conceived, has been in various respects modified and added to. The subjects which make up the additional matter are the following:—A method of correcting the errors of a transit-instrument for deviation of the pivots from the cylindrical form; a detailed description of the construction and applications of the collimating eye-piece; the chronographical method of registering transit-observations; a discussion respecting personal equation in taking eye-and-ear transits; an experimental investigation of the effect of the flexure of a Mural Circle on the mean of its Microscope-readings; a description of the construction of the new Transit-Circle of the Cambridge Observatory and the mode of using it in taking observations (inserted with the consent of Professor Adams); the method employed at Greenwich for making observations of the Moon with an Altazimuth Instrument (not previously introduced, as far as I am aware, in an Elementary Treatise); and the process of taking observations with the Greenwich Reflex Zenith Tube. The treatment of these subjects has given to the work an extension much beyond what I at first contemplated, but at the same time its usefulness as an astronomical manual may be considered to be much increased by being thus made to exhibit with a great degree of completeness the actual state of observational astronomy.

Although the instruments of the Cambridge Observatory, and processes of observation I adopted in the use of them, have been more especially described, and the Treatise consequently partakes somewhat of a local and personal character, I may venture, I think, to say that as having been written after twenty-five years of continuous labors in astronomical observations and calculations, and containing what may have occurred to me in the course of that experience as contributory to the advancement or improvement of practical astronomy, it will be found of some general utility as respects the work carried on in an Astronomical Observatory.

A particular account has been given of the investigation of formulæ for the calculation of the mean places and annual variations of the fixed stars, and care has been taken to derive the numerical coefficients from the most accurate data procurable. The values assigned to the coefficients were calculated for the year 1879, and generally may be used without important error for epochs less than ten years before and after that date.

I take this opportunity for saying a few words respecting Practical Astronomy considered as a subject included in the course of the Mathematical Studies of the University of Cambridge. It seems to me that the tendency of our mathematical instruction and examinations has been of late years to promote the acquisition of a knowledge of formal relations of symbols and the power of readily producing them, apart from a distinct exercise of the reasoning faculty. Now as far as regards Practical Astronomy and Astronomical Instruments, it may be asserted that it would be impossible for the student to make himself acquainted with this department of science without understanding the *reasons* of the mutual relations of the different parts of the instruments, and the processes of observation by which the intended purposes are effected. He must be able to see *how* the many ingenious mechanical contrivances which the wants of astronomy have called forth, contribute to facility and precision in making and recording observations, and although this accomplishment may not demand a very high order of intelligence, it is still a mental exercise of much educational value, inasmuch as it is altogether unlike any process of reasoning by abstract symbols, and may serve as a corrective of the effect of too exclusive an attention to reasoning of that kind. The amount of mathematical knowledge which a complete understanding of Practical Astronomy requires is not more than what every candidate for Mathematical Honors is expected to possess, and accordingly the subject may be regarded as being within the reach of candidates of all degrees of proficiency. For these reasons I think it is much to be desired that Lectures on Practical Astronomy should always form a part of University teaching, and that the subject should at the same time receive due recognition in the Senate House Examination.

I am of opinion that a Professor of Astronomy, or Lecturer, who may undertake to lecture in this University on Practical Astronomy, if he should recommend selected portions of a Treatise like the present for the students' *reading*, and give only a *few* Lectures, would be able to convey sufficient instruction to candidates for Mathematical Honors, without requiring a disproportionate expenditure of time on this one branch of only one of the many subjects that engage their attention. The Lectures might be given partly in the Astronomical Lecture-Room in the Museum Buildings, to take advantage of the collections of instruments and models contained either in the Cabinet of the Plumian Professor's private room or in the Astronomical Apparatus Room, and partly at the Observatory for the purpose of pointing out the construction and mounting of the several instruments, and the methods of taking observations.

The contents of this Volume to page 337 are limited to the consideration of what is done with fixed instruments in a fixed observatory, and as having, consequently, exclusive relation to the *foundations* of exact Astronomical Science, might be appropriately separated from all other parts of Practical Astronomy, and form a distinct Volume. But because in the before mentioned Syllabus additional subjects were inserted which might be ranged under the heads of *Observations with Transportable Instruments*, and *Miscellaneous Astronomical Information* not given in the previous part of the Treatise, which also in delivering the Lectures formed the concluding portion of the Course, I have included them in the last two Sections of the present Volume, although on account of failing health and strength, I have not been able to treat of them so fully and in as much detail as the subjects comprised in the fundamental portion. It will, however, I think, be found that *The Theodolite* and *The Sextant*, two portable instruments of essential use in the Sciences respectively of *Geography* and *Navigation* which are of so much national importance at the present time, have been adequately handled. Also due consideration has been given to the methods recently employed for determining the Solar Parallax, and, in particular, inferences deduced from the observations of the Transit of Venus across the Sun's disk, on December 8, 1874, have been brought into notice.

CAMBRIDGE,
May 8, 1879.

CONTENTS.

N. B. The numbers refer to *Articles*, except where *Pages* are expressly signified.

INTRODUCTION . . . . Pages 1—19.

Art. 1, division of Astronomy into Practical, Spherical, and Gravitational: the Lectures embrace only the first division. 2, data obtained by observations with fixed instruments in a fixed Observatory constitute the *Foundation* of astronomical science. 3, two kinds of astronomical observation; how related to the contents of the *Nautical Almanac*, and the Sciences of *Geography* and *Navigation*. 4, antecedent knowledge required for understanding the Lectures. 5, general statement of the immediate purpose of observations in a permanent Observatory, namely, to determine celestial positions by means of spherical coordinates in Right Ascension and Declination. 6, names of the principal fixed instruments employed for this purpose, and general account of what they respectively effect. 7, reasons for treating of the Transit-instrument, the Mural Circle, and the Transit-Circle separately. 8—9, remarks on the Cambridge Observatory and its site; processes for drawing a meridian line, fixing the positions of the piers of a transit-instrument, and placing a meridian mark.

Art. 11, optical conditions to be satisfied by an Astronomical Telescope. 12—13, definition of the optical centre of a single, or compound, lens. 14, analogous optical centre of a circular spherical mirror. 15—16, optical axes of pencils of light refracted by a simple or compound lens, or reflected by a mirror. 17—19, method of adjusting and centering the two lenses of an achromatic object-glass; exemplified in the instance of the object-glass of the Northumberland Telescope. 20—23, The principal eye-pieces of an Astronomical Telescope; Huyghens's, Ramsden's, the four-glass erecting eye-piece, and the diagonal eye-piece.

Art. 24 (1)—(15), short notices of modern works relating to Practical Astronomy; authors:—Woodhouse, Pearson, Coddington, Airy, A. R. Grant (Cambridge), R. Grant, Loomis (New York), Brünnow (translation in part by R. Main), R. Main (Spherical Astronomy), P. T. Main (Plane Astronomy), Brünnow (the whole translated by the Author), Godfray, Chambers (two works), Chauvenet (Philadelphia and London), F. W. Simms (published by Troughton and Simms). 24 (16), additional sources of information; Herschel's "Introduction to Astronomy" and "Outlines of Astronomy"; the Articles "Mural Circle and "Equatorial" by Sheepshanks in the Penny Encyclopædia; Introductions to the volumes of Cambridge Observations of 1836, 1837, and 1838; the volumes of Greenwich Observations from 1811, by Pond and Airy; references to those volumes (1842—1871) which contain particular descriptions of Greenwich instruments. 25, general relation between Instrumental Observation and Practical Astronomy; treatment of each instrument under four Divisions.

THE TRANSIT INSTRUMENT . . . . . P<small>AGES</small> 20—148.

I. *Description of the Parts and Mounting of the Instrument* . P<small>AGES</small> 20—23.

Art. 27, the Piers; the pivots of the transverse axis, and the supporting Y's. 28—29, test of the cylindrical form of the pivots; effect of wear of the pivots by friction, requiring correction; counterpoises to diminish the friction. 30, description of Ramsden's eye-piece, and adaptation for use in taking transit-observations. 31—32, the single-wire and double-wire Micrometers. 33, the field of view of the Telescope, as seen with a Ramsden's eye-piece; the vertical transit-wires and the horizontal wires; reading of the Micrometer. 34—36, the external parts of the eye-end of the Telescope; the setting-circles, their uses, and correction of their index-errors; the action of the tangent-screw and clamp of a setting-circle, distinguished from other applications; adjustments of the wire-frame by angular motion about the telescope-axis, shifting it horizontally, and fixing it at the geometrical focus of the object-glass. 37—38, methods of illuminating the field of view for seeing the wires on a bright ground, and of illuminating the wires for seeing them on a dark ground.

II. *The mechanical adjustments of a Transit-Instrument and corrections of Errors by calculation* . . . . . P<small>AGES</small> 33—102.

Art. 40, definitions of axis of motion, line of collimation, plane of collimation, and optical axis, the pivots being cylindrical. 41—42, the plane of collimation to be adjusted to the meridian either mechanically, or virtually by calculation; the latter method preferable. 43, preliminary adjustments of the middle wire and the micrometer-wire. 44, definitions of collimation, level, and azimuth errors. 45—47, arrangements in the case of the Cambridge Transit-instrument for correcting the three errors; special means of correcting the level and azimuth errors of the Greenwich and Cambridge Transit-Circles. 48—50, mechanical correction of collimation error by a meridian mark and reversing the instrument. 51—52, calculation of uncorrected collimation error; effect of employing two marks, one northward and the other southward. 53—56, description and applications of collimators; substitution of a collimator for a meridian mark; Gauss's method of obtaining collimation error by two horizontal collimators without reversion. 57—58, determination of the sign of the collimation-correction. 59, calculation of the value in arc of the micrometer-revolution. 60—61, observations and calculations for ascertaining the equatorial time-intervals of the several wires from the mid-wire and from the mean of all. 62, example of calculating collimation-error from bisections of the two marks in reverse positions of the instrument. 63, error of pointing due to diurnal aberration; formula for calculating the correction; application to the latitude of the Cambridge Observatory. 65—66, method of obtaining collimation-error by transits of Polaris and reversing the instrument; investigation of a formula for calculating the correction.

Art. 67, construction of the spirit-level; radius of curvature of its axis, and measure of its sensibility; case of the Cambridge Level. 68—73, adjustment of a spirit-level; levelling by adjusted marks; by fixed adjusted scales; by movable scales unadjusted; calculation of level-error in scale-intervals, the pivots being cylindrical and equal. 74—76, correction for inequality of cylindrical pivots; determination of the value in arc of the scale-interval by a mural circle, or by a bubble-trier; sign of the correction for unequal cylindrical pivots. 78—79, example of the calculation of the level-correction, and determination of its sign; list of corrections for apparent inequality of the pivots obtained in 1828—1854, giving evidence of wear of the pivots; discontinuance of the use of the spirit-level.

Arts. 80—83, the collimating eye-piece, its optical construction, and mode of its application to the Telescope; supersedes the plumb-line, or spirit-level, for determining the vertical direction; its use for placing the wire-frame at the geometrical focus of the object-glass. 84—86, simultaneous determinations of collimation and level errors by the collimating eye-piece and either reversing the instrument, or employing north and south collimators; the latter method practised with the Greenwich and Cambridge Transit-Circles. 87, level error independently obtainable by direct and reflection observations of transits of Polaris.

Arts. 88—90, azimuth error; necessarily obtained by astronomical observation; the process of observing, and general formula of calculation; modifications of the formula in four cases. 91, mechanical correction of azimuth error, and determination of the distance of a fixed mark from the meridian. 92, formulæ for calculating the time-corrections of the three errors of collimation, level, and azimuth. 93—97, definition of the total error of collimation (see errata), the pivots being equal and cylindrical; special means of obtaining its value when the pivots are unequal and of any irregular form nearly cylindrical; investigation of a formula for calculating the change of collimation-error due to deviation from the cylindrical form; the time-correction for such deviation; determinations of the constants involved in the formula; example of the calculation of formulæ for both positions of the illumination-end of the axis. 102, formation of a mean table of pivot-corrections from special measures taken in three years, the argument being P.D.; inference of the correction for any P.D. by interpolation; an account of the method of obtaining the pivot-corrections of the Cambridge Transit-Circle (in a Note to Art. 102).

Arts. 103—106, the Astronomical Clock, with mercurial pendulum; mode of altering the clock's rate; correction of imperfect compensation by adding or subtracting mercury; formula for calculating the required amount.

III. *Methods of taking Transit Observations, and calculations of Corrections* [*of certain transits as observed*]. . . . . PAGES 102—126.

Art. 107, adjustment of the setting circles by the collimating eye-piece, or by a known star; setting for an object of given P.D. 108—112, methods of observing transits of the Sun, the Moon full or nearly full, Jupiter, Mars, the Ring or Body of Saturn, Mercury, Venus, and very small planets; transits of stars very near the Pole, taken either at the several wires, or at the micrometer-wire in arbitrary positions, or at an additional set of wires separated by small intervals (the case of the Greenwich and Cambridge Transit-circles). 113—115, deduction of time of transit of Centre from the observed time of transit of Limb; correction of transit-time of a Limb for defect of illumination, in observations of the Moon, Venus (horned and gibbous), Mars, and Jupiter; investigations of formulæ for calculating the corrections. 116—126, reduction of the mean of transits across some of the wires to transit across the mean of all the wires; previous determination of the interval of each wire from the mean of all for any star or moving body; auxiliary table of intervals for an equatorial star, and for Polaris and δ Ursæ Minoris; reduction to the mean of the wires of transit-times taken at dark bars for observing faint objects; investigation of factors for obtaining the correction of broken transits of the Sun, Moon, and Planets from that for a star of the same P.D.; additional factor for the Moon to include the effect of parallax; (the measure of the Moon's diameter by transits not affected by parallax); rule for deriving from the intervals calculated for omitted wires the reduction of the mean of the observed times to the mean of all the wires. 127—128, investigation of a formula for calculating level-correction from transits of Polaris taken at part of the wires directly, and at part by reflection; the result independent of collimation and azimuth errors; deduction thence of collimation-error from measures by the collimating eye-piece; the method by horizontal collimators preferable.

Arts. 129—131, registering of transit-observations by means of a galvanic circuit; description of the apparatus, and of the chronographic operation; advantages of this method as respects accuracy, and diminution of difference of personal equation. 132—135, discussion relative to personal equation both in galvanic and in eye-and-ear observations; notice of personal discordance in taking transits of first and second limbs.

IV. *Calculation of Apparent Right Ascensions from the transit-observations, and of the Mean Right Ascensions of Stars* . . . . PAGES 126—148.

Arts. 136—147, historical dissertation on the determination of Right Ascensions reckoned from the First Point of Aries; Flamsteed's method, and formation of his *Historia Cœlestis*; Bradley's more perfect method through discovery of the effects of Aberration and Nutation; formulæ for reducing Apparent to Mean places of Stars, and for calculating their Annual Variations, deduced by Bessel from researches in his *Fundamenta Astronomiæ* and *Tabulæ Regiomontanæ*; corrections of the constants of Bessel's formulæ by O. Struve, Dr Peters, W. Struve, and Leverrier; Peters's formulæ for calculating the Annual Precessions in R.A. and P.D.; the formulæ for reduction to Mean R.A. and P.D. adopted in the Naut. Alm.; Airy's modification of the same for avoiding negative signs.

Arts. 148—150, determination of the error of the clock on true sidereal time of the place of observation; effected by means of a catalogue of fundamental stars; origination of the catalogues of such stars contained in the Naut. Alm.; deductions therefrom of the R.A. of the fundamental stars used for the reduction of the Cambridge transit-observations. 151—154, calculation of the clock's daily rate and the clock-error; rules for finding the correction for clock-error to be applied to the clock-time of any meridian transit; allowance for difference of personal equations when the transits are not all taken by the same observer; inference of the mean R.A. of a star from all its apparent R.A. obtained in the course of a year; calculation of the star's mean R.A. at a given epoch from the results of several years' observations; formation of a catalogue of the mean R.A. of the stars, with the annual variations and proper motions; an enumeration of the principal Catalogues of Fixed Stars, in a Note to Art. 154 (page 146).

Art. 155, final determination of the mean error of the assumed R.A. of the fundamental stars; only obtainable by combining Circle and Transit observations of the Sun on the meridian. 156, example of the complete reduction of the transit-observation of a star.

Description of the Cambridge Transit-Instrument, as represented in Plate I. . . . . PAGE 149.

THE MURAL CIRCLE . . . . . . PAGES 148—233.

I. *Description of the Parts and Mounting of the Instrument* . PAGES 151—159.

Art. 158, special description of the Mural Circle of the Cambridge Observatory. 159, protection from the Sun's heat by a screen. 160—161, details and advantages of the tangent-screw and clamp attached to the Cambridge Circle. 162, construction and mounting of a micrometer-microscope for reading off the Circle graduation; its field of view, cross-wires, toothed comb, and zero-indicator. 163, the field of view of the Telescope, the five vertical wires, fixed horizontal wire and parallel micrometer-wire, toothed comb, and zero-indicator coinciding with the fixed wire; external parts of the eye-end, namely, means of adjusting the fixed wire, or micrometer-wire horizontally, the micrometer-head and opposite spring, apparatus for attaching the Telescope to the limb of the Circle and shifting the position, screws for correction of collimation-error, and screw-adjustment of the wire-frame to the geometrical focus of the object-glass. 164, illumination of the field of view, setting-circle, pointer for setting by a calculated Circle-reading. 165, the pivots of the Cambridge Circle, counterpoise action by friction-wheels, means of correcting mechanically level and azimuth errors.

II. *The adjustments of a Mural Circle, and corrections of instrumental errors* . . . . PAGES 160—187.

Arts. 166—167, mechanical adjustments to the plane of the meridian sufficiently accurate; processes of fixing the position of the wall, and of mounting the instrument, for the application of such adjustments. 168--170, description of Ramsden's Ghost Apparatus, and method of using it for levelling the axis of motion of a mural circle. 171, mechanical correction of a level-error obtained by direct and reflection transits of Polaris. 172, correction of collimation-error by the collimating eye-piece and screw-adjustment, after correcting the level-error; corrections of collimation and level errors by means of two horizontal Collimators and the collimating eye-piece; mechanical correction of azimuth error found by transits of stars. 173—174, temporary use of the Circle Telescope for transit-observations, after numerical determinations of the three errors; usual mode of testing the adjustment of its plane of collimation to the meridian. 175—179, adjustments of the pointings of the axes, and positions of the combs, of the micrometer-microscopes; setting of the microscopes; definition of the error of Runs; the microscope reading of a Circle-graduation; calculation of the correction, positive or negative, of an error of Run. 180—182, the Circle-reading and application of total correction for Runs; the Pointer-reading; concluded Circle-reading; proof that with the use of six microscopes the concluded reading is not affected by eccentricity of the graduation, and irregularity of the forms of the pivots.

Arts. 183—184, equatorial adjustment of the fixed horizontal wire for reference-position of the observations of P. D.; use of the micrometer-wire alone for measures of P. D.; readings for coincidence of the micrometer-wire with the fixed wire; determination of the arc-value of the micrometer revolution, for reducing bisections with the micrometer-wire to the reference position; elimination, by alternate measures, of the disturbing effect of atmospheric refraction in making the determination by a meridian mark; general formula applicable to alternate measures (in a Note to Art. 185, page 178). 187, discussion of residual errors by which Circle-readings may be affected.

Arts. 188—191, definition of Zenith Point, or Index Error; inference therefrom of Zenith Distances; method of finding the Zenith Point by a vertical, or a horizontal, floating collimator; applications of the collimating eye-piece for the same purpose, with and without a fixed wire; formula for coincidence of the micrometer-wire with its image when a fixed wire is used.

DESCRIPTIONS, ACCOMPANYING PLATE II., OF THE GREENWICH AND CAMBRIDGE MURAL CIRCLES . . . . PAGES 186, 187.

III. *Methods of observing with the Mural Circle, and Calculation of Apparent Zenith Distances* . . . . PAGES 187—225

Arts. 192—196, use of a setting-circle with spirit-level; adjustment of the index for setting in a direct observation; additional index for setting in a reflection observation; uses of two setting-circles of a Transit-circle which is not reversed. 197—201, the direct observation of a star; note of the place of bisection in the field, the star not being near the Pole; note of the clock-time of bisection for a star near the Pole; formulæ for reduction to the meridian for curvature of path in the two cases; determination of the sign of the correction. 202—203, the observation of a star by reflection; process of taking the double observation; placing a mercury-trough, and clearing the surface of the mercury, for the reflection-observation; correction for curvature of path, the same as that for the direct observation with opposite sign. 204, additional correction for reduction to the meridian in observations of the P.D. of moving bodies; determinations of its amount and sign in the cases of the Sun and Planets. 205—206, method of observing both Limbs of the Sun at the same transit with the Mural Circle; the practice with the Cambridge and Greenwich Transit-circles.

Arts. 207—208, mode of observing the P.D. of a Limb of the Moon; reduction to the meridian, requiring two additional factors on account of the Moon's orbital motion and parallax; observations for P.D. of a N. or S. Limb, and R.A. of a first or second Limb, by one observer at the same meridian passage; biséctions, or transits, of opposite limbs nearly full by a Transit-circle with assistance. 209—213, Circle-observations of Planets having sensible disks; Jupiter, the globe of Saturn, Mars, Venus; bisections of the estimated centres of other planets; investigation of corrections for defect of illumination of gibbous and horned disks; inference of apparent semi-diameters; amount and sign of the correction for the Moon when nearly full; general rule for determining which limb is defective. 214, bisection, at the edge of a broad micrometer-bar, of an object too faint for illumination of the field, and reduction of the micrometer-reading to the reference-position.

Arts. 215—218, determination of apparent Zenith Distances; correction for Refraction by Bessel's Tables; local Zenith and Polar Distances; discordance of zenith points deduced from reflection and direct observations of stars; method of correcting apparent Z. D. for discordance of zenith points. 219—220, calculation of the colatitude of an Observatory from corrected Z. D. of the same star above and below Pole; estimation of the weights of results given by observations of different stars; concluded colatitude of the Cambridge Observatory from observations in 1833, 1836, 1837, and 1838. 221—228, particulars of an experiment for determining the effect of mechanical flexure on the circle-readings, made by two collimators mounted for collimating with each other and with the circle Telescope at all Zenith Distances. 229, result of the experiment, shewing that mechanical flexure does not account for the discordance of zenith points exhibited by the Cambridge Mural Circle; conjectural reasons for ascribing it to the large size of the Circle. 230, errors of graduation; evidence that they do not sensibly affect the indications of the Cambridge Circle.

IV. *Calculations of the Mean Polar Distances of Stars, and of the Geocentric Polar Distances of the Sun, Moon, and Planets.* . . PAGES 225—233

Art. 231, the apparent P.D. of a Star, and calculation of its mean P.D. at the beginning of the year; application of corrections for error of adopted

C. *b*

xviii   CONTENTS.

colatitude and discordance of zenith points to the mean of all such mean P.D. of the star obtained in the year; the mean P.D. concluded from observations in several years; complete formation of a Catalogue of mean places of stars for a given epoch.

Art. 232, calculated local P.D. of a bisected N. or S. Limb of the Sun, Moon, or a Planet; deduction of the Geocentric P.D. by correcting for Parallax; special formula for calculating the Moon's parallax; formula for any other body. 233—235, inference of geocentric P.D. of centre, and of geocentric measures of diameter, from bisections of opposite limbs; calculation of the Moon's geocentric diameter from transits of both limbs taken on the meridian, or out of the meridian with an Equatorial; correction thereby of the value assumed in the Naut. Alm.; confirmation of the tabular value of the Sun's diameter by Transit and Circle observations; derivation of the semidiameter of Venus from the micrometer measures of 1838 and 1839, by an investigation which eliminates the effect of irradiation; consequent correction of the value adopted in the Naut. Alm. 236, correction of the observed P.D. of the Moon for error of position of the plane of collimation.

Art. 237 (pages 230—233), examples of the complete reduction of Circle observations of a Star, and of the Sun.

THE TRANSIT CIRCLE   .   .   .   .   .   .   PAGES 233—256.

Arts. 238—239, notices respecting the Greenwich and Cambridge Transit Circles; statements relative to the latter (see Note in page 234), and extracts from the account of it given in the M.N. of the R.A.S. 240—245, particular descriptions, accompanying Plate III., of the various parts and appendages of the Instrument, the sets of wires, the two collimators, the piers for the mounting, the graduation and microscopes, the clamping and slow-motion apparatus, the illuminations, and generally of the manipulation and arrangements for taking simultaneous observations of R.A. and P.D. 246, special means employed for obtaining corrections of R.A. observations for deviations of the pivots from the cylindrical form.

Arts. 247—252, calculations for deducing from the observations of the Sun made in the course of a year with a Transit-instrument and Mural Circle, or with a Transit Circle, together with data from Solar Tables, the exact position of the First Point of Aries, the Obliquity of the Ecliptic, and the mean error of the assumed R.A. of the fundamental stars; conversion of errors in R.A. and P.D. into errors in Longitude and Ecliptic Polar Distance. 253—258, derivation of data for exactly measuring time-intervals and determining epochs from meridional observations; definitions of a sidereal day and a mean solar day; formula for computing the Sun's Mean Longitude at the Greenwich mean noon of January 1, $1800+t$, derived from results obtained by Bessel by comparing meridian observations of the Sun separated by a long interval; consequent formula for computing the sidereal time of mean noon; analogous formula for the epoch of Greenwich mean noon of January 1, $1850+t$, from Leverrier's Solar Tables; example of calculating by each set of formulæ the sidereal time of a given mean noon; rule for computing the mean time of transit of the first point of Aries. 259—261, rules for converting the sidereal time of a place into its mean time, and the reverse operation; conversion of local sidereal time into Greenwich mean time, the Longitude of the place being known. 262—264, principle of determining terrestrial Longitude by astronomical means; comparison for that purpose of simultaneous sidereal times at two localities by transfer of chronometers; by galvanic signals; determination by both methods of the Longitude of the Cambridge Observatory. 265, concluding remarks on the relation of the results of meridian observations to Physical Astronomy.

Statement of the subjects treated of in the subsequent portion of the Lectures
PAGES 256—257.

THE EQUATORIAL . . . . PAGES 257—316.

Art. 266, general statement of the purposes to which an Equatorial is adapted; its uses in taking observations out of the meridian, especially differential measures of R.A. and P.D.

I. *Description of the parts and mode of mounting* . . PAGES 258—262.

Arts. 267—271, the Five-feet Equatorial of the Cambridge Observatory; indication of its parts and mounting by reference to Figure 40; the piers supporting the instrument, place and support of the clock, the upper and lower pivots and Y's, the frame bearing the Telescope, the polar axis, the declination-circle and two microscopes, the declination-axis, the hour-circle and two microscopes, the graduations and slow-motion apparatus of the two circles; the wires and illumination of the field, and the Finder. 272—273, construction of a diagonal eye-piece for conveniently observing objects near the Pole.

II. *The Adjustments of an Equatorial* . . . . PAGES 263—276.

Arts. 274—275, mechanical corrections of instrumental errors; determinations of the errors by observations of stars; the errors arranged in the logical order of the corrections, namely, index error of the declination-circle, error of the angular elevation of the polar axis, the deviation of that axis from the plane of the meridian, the collimation error, the error of position of the declination-axis, and the index error of the hour-circle. 276, adjustments that may be made mechanically without observations of stars, placing the declination-axis at right angles to the polar axis by a swinging level (the construction and mode of application of which are shewn by Fig. 44); correcting collimation error by a mark; correcting the index error of the hour-circle by the swinging level. 277, test of the stability of the polar axis by a swinging level with its axis of motion parallel to the telescope-axis. 278, method of adjusting the position of the polar axis of an Equatorial which has no declination-circle; formulæ for calculating the deviation of the instrumental Pole southward and westward. 279, the other adjustments of an Equatorial for taking observations of R.A. and P.D. the same as those of the Transit and Mural Circle, or Transit-circle.

III. *Methods of observing with an Equatorial and elimination of instrumental errors* . . . . Pages 276—280.

Arts. 280—282, setting for observing an object; the process of observing, the same as that with a Transit-circle; correction for Runs of the declination-microscopes; investigation of formulæ for correcting, if required, for error of position of the polar axis; advantage of observing in reverse positions of the declination-circle. 283—285, independent determination of celestial places by an Equatorial; not as accurate as the differential method by comparisons with known stars; details of the differential observation when the declination-circle is read, and when the objects are bisected by a micrometer-wire.

IV. *Final calculation of R.A. and P.D. from Equatorial Observations*
Pages 280—288.

Arts. 286—290, formulæ for calculating corrections for refractions in absolute determinations of R.A. and P.D. with an Equatorial; deductions therefrom of formulæ applicable to differential observations; description of Ramsden's Refraction-piece, and its use in shortening the calculations of refraction-corrections. 291, the reductions of Equatorial observations of Limbs to observations of Centres, the same as for meridian observations. 292—294, investigation of formulæ for calculating the corrections of Equatorial observations of R.A. and P.D. for Parallax; special investigation of formulæ for calculating exactly the Moon's parallax in R.A. and P.D.; formulæ expressing the ratio of the Moon's geocentric semidiameter to an apparent or measured semidiameter.

THE NORTHUMBERLAND EQUATORIAL OF THE CAMBRIDGE OBSERVATORY (REPRESENTED IN PLATES IV. AND V.) . . . . . Pages 288—296.

Arts. 295—300, historical notices respecting the completion of the Instrument and erection of the Dome; descriptions of the Telescope, and the construction of the Polar Frame for supporting it; the attachment of a Declination-sector for measuring small differences of P.D.; the Hour-Circle and connected Clock; method of measuring differences of R.A. with the instrument carried by clock-movement. 301—302, more particular account of the parts and appendages of the Telescope and Dome, by reference to Plates IV. and V., the former of which represents the eye-end of the Telescope and Double-wire-micrometer, and generally the observing-apparatus within reach of the observer, and the other a north-and-south section of the Dome and Instrument, with the movable chair, the observer in position, and the apparatus for slow-motion in P.D. and hour-angle. 303, determinations of the values of the revolution of the Telescope-micrometer, the Sector microscope-micrometer, and the interval between the graduations of the Sector.

The counterpoise mounting of an Equatorial . . . Pages 296—301.

Arts. 304—307, the instrumental arrangements, advantages, and disadvantages of this form of mounting; a mode of counterpoise mounting

CONTENTS. xxi

(shewn by Figure 50) suitable for out-door sweeping for objects; processes of sweeping when the object is recognisable by its physical aspect, and when it appears as a star.

Other observations made with an Equatorial . . . PAGES 301—316.

Arts. 308—315, observations for parallax of Mars in opposition, at two places or at one; elongations of a satellite from its primary; measures of the Moon's diameter; eclipses of a star, or a Planet, or the Sun, by the Moon; calculations of results from the several eclipses. 316—322, measures of relative positions of the components of double, triple and multiple stars; use of the double-wire micrometer; two examples given of measures by the double-wire micrometer; method of measuring by the double-image micrometer (represented in Plate IV., figs. 2 and 3). 323—325, measures of diameters of Planets by the double-wire and double-image micrometers. 326, equatorial observations for discovering comets and small planets, and for furnishing data for calculating ephemerides after discovery.

OTHER FIXED INSTRUMENTS OF AN OBSERVATORY . PAGES 316—337.

THE ALTITUDE AND AZIMUTH INSTRUMENT . . . . PAGES 317—329.

Arts. 327—328, the principle of the Mounting (shewn by Fig. 53); the use of the Altazimuth of the Greenwich Observatory for observations of the Moon. 329—338, methods of observing separately azimuths and zenith distances, and applying instrumental corrections; dependence of the lunar observations on catalogued places of stars. 339—342, reduction of the observations for comparison with theoretical places; corrections for semi-diameter and parallax; conversion of tabular errors in azimuth and altitude into errors in R. A. and P. D.

THE FIXED ZENITH SECTOR . . . . PAGES 329—333.

Arts. 343—346, adaptation of the instrument for measuring small zenith distances; applied at Greenwich in measuring the zenith distances of γ Draconis for determining the constants of aberration and nutation; Bradley's Zenith Sector; Pond's; Airy's Reflex Zenith Tube; particular description of the process of observing with the last-named instrument.

THE TRANSIT IN THE PRIME VERTICAL . . . . . PAGES 333—337.

Arts. 347—350, Repsold's instrument; its construction, and the process of using it employed by Professor W. Struve for obtaining the constants of aberration and nutation. 351, Bessel's method of determining the latitude of an Observatory by a Transit in the prime vertical.

OBSERVATIONS WITH TRANSPORTABLE INSTRUMENTS
PAGES 337—366.

Arts. 352—359, Ramsden's Zenith Sector; Airy's Zenith Sector; the portable Altazimuth; the portable Transit-instrument. 360—369, the Theodolite; description accompanying Figure 55; its adjustments; use of a compass-needle; application of the Theodolite in large trigonometrical surveys; exemplification in measuring by base-line and triangulation the differences in Longitude and Latitude between the transit-instrument of the Cambridge Observatory and a clock in the Museum building. 370—376, Ramsden's and Troughton's Sextants; description by reference to Figure 56; the graduation and the adjustments; the use of the Sextant at sea for obtaining the time of day by equal altitudes of the Sun, the latitude of the place, and measures of Lunar Distances for calculating the longitude; observations with a sextant on land by means of an artificial horizon. 377—379, constructions of, and methods of observing with, Troughton's Reflecting Circle and Borda's Repeating Circle.

MISCELLANEOUS ADDITIONAL SUBJECTS . . . Pages 366—394

Art. 381, different methods of determining terrestrial longitude; by Transfer of Chronometers; by Galvanic Signals, applied in finding the longitudes of Cambridge and of Paris from Greenwich; by Lunar Distances (formula for calculating to second differences tabular Lunar Distances, with an example of the calculation); by observations of occultations of stars and planets by the Moon; by transit-observations of the Moon and Moon-culminating Stars; by observations of Eclipses of Jupiter's Satellites; and by Trigonometrical Surveys; comparison of the chronometric and geodetic determinations of the longitude of Valentia. 382—385, Solar Parallax; method of finding it by Transits of Venus across the Sun's disk; allowance for the black drop; investigation of formulæ for calculation in the cases of Delisle's method and Halley's method; the results deduced from the processes hitherto employed for determining Solar Parallax.

Arts. 386—390, Reticles; differential observations of R.A. and P.D. made with Lacaille's Reticle; with Valtz's Reticle; with a Ring Micrometer. Boguslawski's method by transits of two known stars and an unknown object across a single wire. 391—399, the Vernier; the Transit-reducer, for the time-reduction of a transit-observation to the meridian; the use of Huyghens's eye-piece for observing physical phenomena; magnifying powers; the Dynameter, described by reference to Figures 61 and 62; interpolations to second differences and to fourth differences.

APPENDIX containing investigations of formulæ for ascertaining the magnification and brightness of an image produced by any combination of lenses and mirrors . . . . Pages 395—400.

ERRATA.

Page 84, beginning of Art. 93, *for* This angle *read* The bracketed angle.

,, 115, line 6 from the bottom, *omit*, In this calculation the arcs p and P should not be substituted for their sines. See the foot-note in page 201.

,, 118, line 6 from the bottom, *for* azimuth *read* level.

,, 131, line 1, the Note proposed to be introduced was omitted, as being of too theoretical a character. The theory referred to is given in the Philosophical Magazine for April 1872, pp. 289—295.

,, 201, line 3, *for* $z - \epsilon$ *read* $Z - \epsilon$.

LECTURES

ON

PRACTICAL ASTRONOMY.

INTRODUCTION.

1. THE science of Astronomy may be regarded as composed of three distinct parts: Practical Astronomy, relating to the construction and use of Astronomical Instruments; Spherical Astronomy, consisting for the most part of deductions, according to the rules of Spherical Trigonometry, from data furnished by Practical Astronomy, apart from any dynamical considerations; Physical Astronomy, or calculations of the motions of the Sun, Moon, and Planets, depending essentially on the hypothesis of Universal Gravitation. The Lectures will embrace only subjects which may be considered to pertain to the first part.

2. Practical Astronomy, while it is distinguished from the other two parts by being immediately concerned with the means and methods of taking observations of celestial objects, is at the same time related to them as the *foundation*[1] on which they rest. Spherical Astronomy and Physical Astronomy admit of actual application only so far as they depend on results obtained by observations made with *Astronomical Instruments*.

3. Astronomical observations are of two kinds: (1) observations made at a permanent *Observatory* with fixed instru-

---

[1] The work entitled *Fundamenta Astronomiæ*, produced in the year 1818 by Bessel, the eminent astronomer of Königsberg, consists mainly of deductions from observations of Stars and the Sun, made by Bradley at Greenwich, in the years 1750—1762.

ments, for the purposes, generally, of determining and arranging in *Catalogues* exact positions of fixed stars, for use as points of reference in other observations, or of ascertaining at recorded times the positions of moving bodies in order thereby to test and correct their physical theories; (2) observations made either with instruments placed in temporary observatories, and with theodolites, for geodetical surveys and mapping the earth's surface, or with portable instruments for nautical purposes. The sciences of Geography and Navigation are respectively dependent on the two kinds of observation in class (2), conjoined with data furnished by means of observations that come under class (1), such data, for instance, as those contained in the *Nautical Almanac*. We may also consider to be included in class (2) observations of eclipses and occultations of celestial objects, measures of the diameters of Sun, Moon, and Planets, and measures of the relative positions of the components of double and multiple stars. Although the observations of the first class, as being fundamental, will be chiefly treated of in these Lectures, all those of the other class will receive some share of attention.

4. For understanding the Lectures some previous knowledge, of no great amount, is required. I shall suppose the student to be acquainted with the terms by which certain points and great circles of the celestial sphere are designated, as given, for instance, in the introduction to Hymers's Astronomy, or in Chapters I. and II. of Godfray's Treatise on Astronomy. A few propositions of Spherical Trigonometry and Analytical Geometry will have to be employed, and in explaining the modes of using telescopes and microscopes for astronomical observation certain propositions of Geometrical Optics will be taken for granted. The proofs of two optical propositions of practical importance, one relating to the magnifying power of a telescope, and the other to the brightness of the image formed in its field of view, will be given in an Appendix.

5. It may be asserted generally that what is effected in Practical Astronomy is the measurement of celestial arcs by means of clocks, or graduated instruments, or by micrometers. The measurement of an arc by the intervention of an astro-

nomical clock depends on the assumed uniformity of the rotation of the earth about its axis. A position on the celestial sphere is assigned by the measurement of two arcs, or spherical co-ordinates, one of which is reckoned from an assumed origin on a great circle of the sphere to the point of intersection of this great circle with another passing through its pole and through the position itself, and the other co-ordinate is the arc of the second great circle intercepted between the position and the aforesaid point of intersection. The co-ordinates are Right Ascension and Declination if the pole be that of the equator, and Longitude and Latitude if it be the pole of the ecliptic. Instead of Declination and Latitude, their complements, called respectively North Polar Distance and Ecliptic Polar Distance, are frequently employed as co-ordinates of position.

6. The principal instruments used in a permanent Observatory for determining exact celestial positions are the Transit Instrument, the Mural Circle, the Transit-Circle (a combination of the Transit Instrument and Mural Circle), the Equatorial, and the Altitude and Azimuth Instrument. The parts and appendages of these instruments are appropriate to the purposes they are designed to answer. The Transit Instrument measures Right Ascension by intervals of time, and serves to determine epochs of time; the Mural Circle measures by its graduation Declination, or North Polar Distance; the Transit Circle is capable of giving at the same time both Right Ascension and Declination, and of effecting all that is done by the separate use of the Transit Instrument and the Mural Circle. The observations with these three instruments are made in the meridian of the place of observation. The Equatorial is adapted to ascertain simultaneously both the Right Ascension and the Declination of an object out of the meridian. The same assertion is true of the Alt-azimuth Instrument, the Altitude and Azimuth immediately observed being converted by calculation into Right Ascension and Declination. The Longitude and Latitude of an object are not directly observed, but are deduced by calculating according

to rules of Spherical Trigonometry from its Right Ascension and Declination obtained by means of instrumental observation.

7. The use of the transit circle is preferable to observing separately with a transit instrument and a mural circle, because the latter method requires two observers, and circumstances may occur under which only one of the co-ordinates of position is obtained, whereas with the transit circle it is generally possible for a single observer to obtain both co-ordinates. But although the transit circle has this advantage in practice, I propose to treat of the transit instrument and mural circle, at first, as separate instruments, and afterwards to consider them as combined to form the transit circle. This course is adopted for the sake of simplicity and distinctness in exhibiting the principles of the *Adjustments* and *Corrections* which are required for ensuring accuracy in taking meridional observations.

8. It is not necessary to give in detail the description of an observatory to those who have the opportunity any day of inspecting the Cambridge Observatory. This institution, although not to be classed with national observatories like that at Greenwich, fulfils the main requisites of an observatory as respects site, rooms appropriated to instruments and the work of calculation, and residences for observers, and in being furnished with powerful instruments of the best quality. The situation of an observatory ought to be such as to present no obstacle to pointing a telescope to a celestial object whatever be its position in the heavens[1]. It is, however, very rarely possible to observe satisfactorily an object at an altitude less than two

---

[1] The Cambridge Observatory was erected in 1822-4, one-third of the cost being obtained by subscriptions, and the rest granted from the University Chest. The chief promoters of the undertaking were Professor Woodhouse, Mr Catton of St John's College, and Mr Peacock, afterwards Lowndean Professor. The first Director, Professor Woodhouse, was succeeded in 1828 by Mr Airy, now Astronomer Royal, whom I succeeded in 1836. The present Director, Professor Adams, was appointed in 1861. The site of the Observatory sufficiently fulfils the condition of commanding an unobstructed view of the heavens; but occasionally I found it necessary to cut off the tops of trees in the surrounding plantations to secure observations of certain comets.

degrees, on account of the disturbing effects of refraction, and the prevalence of mists, near the horizon.

9. When the Cambridge Observatory was erected it was thought necessary to arrange that the telescope of the transit instrument should be capable of pointing to a fixed meridian mark, and accordingly the position of the instrument was determined by satisfying this condition with respect to a mark on the tower of Grantchester church, situated to the south of the observatory at the distance of about $2\frac{1}{2}$ miles. This was done for a first approximation by bisecting the spire of the tower by the middle wire of the telescope of a small transit instrument, and comparing the Sun's transit across the wire with true time of apparent noon brought up by the chronometer used for taking the transit from Mr Catton's observatory at St John's College. Nearer approximations were made by transit observations of *high and low stars* with the same small instrument[1]. Probably the simplest method of determining the position of the piers of a transit instrument, the telescope of which is required to point to a mark in a given position, would be to employ a theodolite, placed a little to the north or south of the mark, as an altazimuth instrument for finding by the method of equal altitudes of the Sun or a star the direction of the meridian passing through the mark. Then by pointing the telescope in this direction, and depressing it to the horizon, a position might be fixed upon, such that if the piers be placed equally distant from it and at a distance from each other suitable for mounting the instrument, and if their azimuths be adjusted till the image of the mark is seen at the middle of the field of view of the mounted telescope, the required conditions will be satisfied.

In the actual state of Practical Astronomy meridian marks for adjusting instruments in the meridian can be dispensed with. There are, however, certain purposes (which will be subsequently indicated), for which they may occasionally be

[1] This statement is derived from the "account of the transit instrument made by Mr Dollond, and lately put up at the Cambridge Observatory," given by Professor Woodhouse in the Philosophical Transactions for 1825 (p. 419).

made use of. The use of the Greenwich marks placed by Pond at Blackwall and at Chingford was discontinued by the present Astronomer Royal in 1836[1].

10. The account given in Art. 9 of methods of directing the telescope of a transit instrument to a meridian mark was scarcely logical, inasmuch as these methods depend on employing astronomical instruments the descriptions of which, and of the modes of observing with them, properly belong to a subsequent stage of the Lectures. If it were required to draw, *ab initio*, a meridian line through a given spot, for the purpose of marking out positions for the piers of a transit instrument, this might be done by setting up a pole vertically at the spot, drawing around it several horizontal circles, and noting the points of coincidence of the extremity of the shadow of the pole with these circles before and after noon. Then if the two points on each circle be joined by a chord, the mean of the directions of the middle points of the chords from the pole will be approximately the direction of the meridian line. The experiment would be best made about midsummer, when the Sun's diurnal path is high, and the change of Declination small. This process would probably be sufficiently exact for a first determination of the azimuths of the piers; and also for fixing upon the position of a *near* meridian mark which, by being put in coincidence with the focus of an achromatic compound lens, might by the intervention of this lens be distinctly seen in the transit-telescope. A meridian mark might thus be set up without having recourse to an astronomical instrument[2].

[1] These marks are both on the north side of the Observatory, that at Blackwall at a distance of about two miles, and that at Chingford "on the Essex hills" about eleven miles. The latter on account of its distance could rarely be seen, and the sight of the other was often prevented by ships on the Thames. For a south mark Pond had recourse to a small transit-telescope used as a Collimator on the principle, now very generally adopted, of collimating with two telescopes pointed towards each other. What is said in Woodhouse's *Astronomy* (Vol. I. p. 84) respecting the Greenwich meridian marks is altogether inaccurate. (See Pond's Introduction to the Greenwich Observations of 1834.)

[2] Admiral Smyth joined to this method of viewing a meridian mark apparatus for adjusting the position of the mark by screws. After adjusting it to the

11. We may now proceed to the consideration of the *optical conditions* which must be satisfied by the telescope and eye-pieces of an astronomical instrument which is to be employed for the accurate measurement of celestial arcs. The remarks will apply, at first, only to refracting telescopes. The object-glass of a Refractor is composed of two lenses, the outer one a double convex lens, and the inner one a double concave lens. The adjacent surfaces of the two lenses have exactly equal and opposite curvatures and generally are in actual contact, and the curvature of the second surface of the concave lens is small compared to that of the first surface. The outer lens is of crown-glass, the other of flint-glass. The forms and refractive powers of the glasses are such as satisfy, as far as may be, the conditions of correcting both spherical aberration and colour. It is not possible with two lenses completely to correct colour; but it is of importance to remark that the residual colour is least injurious when it has a purplish tint, being composed of rays lying towards the violet end of the spectrum, because the light of these is much less intense than that of those towards the red end.

12. We have next to define exactly the point which is called *the optical centre* of the object-glass, on the existence and position of which the purposes to which the astronomical telescope is applied essentially depend. First, it must be understood that the *axis* of a lens is the straight line which, passing through its middle point, cuts the two surfaces at right angles, and that the *axis* of a system of lenses is the straight line which similarly cuts at right angles the two surfaces of each of the components of the system. Now by Geometrical Optics it is shewn that a *beam* of light diverging from a point and incident on the whole of the surface of any lens (i.e. a *centrical beam*), on emerging from the lens very approximately converges towards, or diverges from, a point so situated that the following rule holds good: The straight line joining this point

meridian in the first instance, and fastening it in its position, to prevent disturbance the screws were withdrawn. (Smyth's *Celestial Cycle*, Vol. I., p. 331.)

and the point of original divergence (which may be called the *axis* of the beam) cuts the axis of the lens at a certain point which is very approximately the same for all small inclinations of the former axis to the other. This point is the optical centre of the lens.

13. If instead of being divergent the incident centrical beam had been convergent, its axis would still have passed through an optical centre. Hence in the case of the combination of two lenses in juxtaposition, such as those which form the object-glass of an astronomical telescope, there will be an optical centre of each of the lenses, the centrical beam emerging from the first lens being an incident centrical beam with respect to the other. But these centres will be points on the common axis of the lenses so near each other, that for small inclinations we may suppose without sensible error that the axis of the beam incident on the first lens, and that of the beam emergent from the second, are coincident in direction, and that consequently the compound glass has an invariable optical centre. This supposition has not been found by experience to be in any sensible degree erroneous.

14. Analogous considerations are applicable when a centrical beam is incident on a spherical *mirror*, whether convex or concave, the contour of which is a circle. For a mirror of this form, the centre of its curvature takes the place of the optical centre defined in Art. 12, and the straight line joining the focus of an incident beam and the focus of the same beam after reflection passes *quam proxime* through that centre of curvature.

15. From what has been said (Arts. 12 and 13) respecting the incidence of a diverging or converging beam on a lens, it is evident that if the beam be conceived to be composed of indefinitely small *pencils* of light and we select one of them, the straight line joining the focus, actual or virtual, of this pencil before incidence, and the focus, actual or virtual, of the same pencil after emergence, will equally pass through the optical centre of the lens. This is true also for a compound lens like the object-glass of an astronomical telescope.

16. Similarly in the case of a spherical mirror, the straight line joining the focus of an incident pencil and the focus of the same pencil after reflection passes through the centre of the spherical surface.

17. Generally the two lenses of an object-glass are joined together, and the compound glass is put in position in the tube of the telescope, by the optician once for all, and the observer has not the means of altering the adjustments. Professor Airy, who, as my predecessor at the Cambridge Observatory, designed and superintended the mounting of the telescope of the Northumberland Equatorial, provided apparatus whereby the observer could himself execute the adjustments of the object-glass. This I succeeded in doing in the first instance by myself, as mentioned in p. 34 of the Astronomer Royal's "Account of the Northumberland Equatoreal and Dome," which forms Appendix II. of the Volume of Cambridge Observations for the year 1843 (Vol. XV). Subsequently (in 1848), having occasion to take out the two glasses for the purpose of cleaning them (which the above-mentioned apparatus allowed of doing), I called in the assistance of Mr Simms for replacing them, and had the opportunity of witnessing, and taking part in, the whole operation. Considering that an account of the process might be generally useful to practical astronomers, as indicating means of testing, or judging of, the performance of their telescopes, I thought it might be appropriately given in detail in this Introduction.

18. The mechanism of the apparatus was arranged so as to enable the observer to execute *three* movements by the intervention of iron rods extending from the object-glass to the eye-end of the telescope, giving the means, at the same time that the movements were produced, of seeing the effects they had on the image of a bright star. These movements were (1) a rotation of the crown-glass lens about the axis of the telescope for the purpose of observing the effects of different relative positions of the surfaces of the two lenses, and choosing that which appeared to give the best effect as to definition; (2) a *tilting* of the whole object-glass in any required direction, by

turning three screws tapped in projections from the object-glass cell and holding it in position; (3) motions of translation, in two directions at right angles to each other, of the crown-glass lens relatively to the flint-glass lens for the purpose of *centering* the lenses. The mechanical means of producing these movements being provided, the process of adjusting the object-glass is such as follows.

19. First the image of a bright star (as Arcturus) is looked at with the eye-piece well adjusted to focus in order to see whether there is a determination of the light of the image towards one direction. If so, the appearance (technically called a *tail*) has to be corrected tentatively by tilting the object-glass. The rule is, that if the tail is towards the left hand, the object-glass requires to be thrown off on the right hand. The tail being thus corrected, the image is next looked at much out of focus by pushing in, or drawing out, the eye-piece a considerable distance from its normal position. There will then be seen in the midst of a confused coloured image a line of light of the nature of a caustic formed by the intersection of rays of different refrangibilities. This caustic is required to be an exact circle, and if it be of any other form, it must be brought, as nearly as may be possible by the judgment of the eye, to the circular form by the rectangular transverse movements of the crown-lens upon the flint-lens. After this has been done and the image is looked at in focus, there should be a well-defined point from which the light of the image is radiated in all directions, or about which concentric bright *rings* are formed. The rings are more conspicuous the smaller the telescope, and in one so large as the Northumberland telescope (of $11\frac{3}{4}$ inches aperture) they are scarcely noticeable, while the radiations in the case of bright stars are very intense. These spurious optical effects, the theory of which does not appear to be well understood, do not sensibly interfere with fixing upon the central point of the image for bisection. It is advisable, after proceeding so far, to turn again the crown-glass about the axis of the telescope in order to choose the relative position of the glasses most favourable for good definition, and then with the

glasses in the selected position to repeat the processes of tilting and centering for a final adjustment. This, if I remember rightly, was the course we adopted in adjusting the object-glass of the Northumberland telescope. Very little optical effect was discernible by causing the crown-glass to revolve relatively to the other.

20. The principal eye-pieces of the astronomical telescope are Huyghens's, or the Negative eye-piece, and Ramsden's, or the Positive eye-piece. Occasionally use is made of an Erecting eye-piece, and a Diagonal eye-piece. I shall not consider it necessary to give in these Lectures mathematical details respecting the forms and arrangement of the lenses of eye-pieces, this part of the subject being fully treated of in standard works on Geometrical Optics[1]. It will suffice for my purpose to make only the following remarks.

21. Huyghens's eye-piece is sometimes called "the Achromatic eye-piece," because the condition of achromatism is very approximately fulfilled when, as is practically the case, the directions of the pencils are nearly parallel to the common axis of the two lenses. This eye-piece is, consequently, suitable for examining and recording the physical features of celestial objects, and, in fact, is generally employed for that purpose; but because the image of the object is formed between the lenses, it is not one that can be adapted to taking observations which require *measures* of time or space. For such observations the observer must be able to see distinctly, in coincidence with the optical image of the celestial object, fine wires, or threads, stretched across a stop fixed in the telescope-tube and limiting the field of view, and also the edges of the stop which form the boundary of the field. This condition can be satisfied by a Ramsden's eye-piece, the image formed by the telescope's object-glass being situated close to the inner glass of the eye-piece on the side towards the object-glass, and being distinctly visible in

---

[1] The requisite mathematical investigations relating to lenses and mirrors are well given in Coddington's Treatise on Optics, Part I., and the parts and uses of various kinds of telescopes and eye-pieces are fully described, with accompanying figures, in Chapter IX., Part II., of the same work.

that position by adjusting the eye-piece to focus. Accordingly this arrangement allows of seeing the image of the celestial object in coincidence with wires or diaphragms, at the geometrical focus of the object-glass, adapted for taking the desired astronomical measures. Ramsden's eye-piece is not achromatic like Huyghens's; if, however, care be taken to look at the image or wire as nearly as may be in the direction of the optical axis, all injurious appearance of colour due to the eye-piece is got rid of. To enable the observer to do this the eye-piece is carried by a slide movable in grooves laterally by hand, and can thus be placed exactly opposite the wire or image looked at. By this contrivance the extent of the field of view is in effect enlarged, and the image can be seen distinctly during the whole of its transit across the field defined by the stop.

22. The erecting eye-piece, consisting of four glasses, is generally used with telescopes intended chiefly for looking at terrestrial objects. The way in which it has been modified for converting a telescope into a double-image micrometer will be taken notice of in the course of these Lectures.

23. A diagonal eye-piece is furnished with a plane metallic reflector inclined at an angle of 45° to the optical axis of the telescope for the purpose of changing the direction of vision through 90°. The two lenses of this eye-piece are placed at the ends of a broken tube consisting of two straight parts at right angles to each other, and the reflector is at the angle of their junction. By these means the observer is able to get sight of an object under circumstances in which the parts of the instrument may not allow of his looking directly along the axis of the telescope. An eye-piece of this kind is mostly required for small portable instruments. In some cases of the mounting of an equatorial it is needed for looking at the Pole star.

24. I propose to conclude this Introduction with giving some account of works on Practical Astronomy to which the student might have recourse for information on the subject as a whole, or on particular parts of it. It is hoped that the list, although it has no pretension to be considered

complete, may, with the accompanying remarks, be found to be useful.

(1) Woodhouse's Treatise on Astronomy Theoretical and Practical. A new Edition. Cambridge: Vol. I. Part I. (1821) and Part II. (1824). Part I. treats of the Fixed Stars, and Part II. gives the Theories of the Sun, Moon, and Planets, as founded on Kepler's Laws. [Vol. II., published in 1818, contains exclusively Physical Astronomy.] The principles involved in making Astronomical Observations, and in reducing them for subsequent use, are clearly stated in this work; but the accounts of Astronomical Instruments, and of the methods of observing, fall much behind the improved state of Practical Astronomy at the present time. In Chapter XXXVIII. (p. 754) there is given an investigation of formulæ for calculating the Sun's Parallax from observations, according to Halley's method, of a transit of Venus over the Sun's disk, with application to that of 1769.

(2) Pearson's Introduction to Practical Astronomy. London: printed for the Author. Vol. I. (1824) and Vol. II. (1829), of quarto size, with a volume of Plates. The first volume consists of Tables computed for facilitating the Reduction of Celestial Observations, with explanations of their construction and use. The second volume contains "descriptions of the various Instruments that have been usefully employed in determining the places of the heavenly bodies, with an account of the methods of adjusting them and using them." The descriptions of Instruments, fixed, portable, and auxiliary, are illustrated by numerous Plates, and, regard being had to the date of the publication, are very complete.

(3) Coddington's System of Optics. Cambridge: Part I. (1829) and Part II. (1830). The first Part is entitled "A Treatise on the Reflexion and Refraction of Light," and, as the author acknowledges, "contains the substance of Professor Airy's Papers *on the Achromatism* and *the Spherical Aberration of Eye-pieces*, published in the Cambridge Philosophical Transactions." Part II., which consists for the most part of applications of mathematical results obtained in the first part, and is illus-

trated with excellent engravings, gives much information useful to the practical astronomer. (See foot-note to Art. 20.)

(4) "Six Lectures on Astronomy, delivered at the Temperance Hall, Ipswich, in the month of March, 1848," by George Biddell Airy, Astronomer Royal. London: 1849, 2nd Edition. These Lectures, remarkable as respects the method and clearness of the explanations, seem to prove the possibility of conveying a large amount of astronomical information, practical, theoretical, and physical, to an audience not possessing the advantage of a mathematical education.

(5) "Plane Astronomy, including explanations of Celestial Phenomena and descriptions of the principal Astronomical Instruments," Part I. By the Rev. A. R. Grant, Fellow and Assistant Tutor of Trinity College (Cambridge: 1850). This work, of which only Part I. was published, is mentioned here as containing a good account of the striding *spirit-level*, and of the principles on which its indications of level depend.

(6) "History of Physical Astronomy from the earliest ages to the middle of the nineteenth century." By Robert Grant, F.R.A.S. London: 1852. Although the History is devoted mainly to Physical Astronomy, some portions of the work, especially Chapters XVIII—XXI., consist of historical notices respecting successive advances made in the construction and use of Astronomical Instruments, and the additions to Astronomical Science thereby effected. These notices, which I shall take occasion to refer to in the course of the Lectures, are particularly interesting as indicating how the foundation of Physical Astronomy has been laid in the results obtained by Observational Astronomy, and by what means, in proportion as theoretical calculations made on the hypothesis of universal gravitation called for observations in greater number and of greater accuracy, the demand was met by improved instruments and modes of observing, and by a diligent and intelligent use of the apparatus provided by the skill of the instrument-maker for taking exact observations.

(7) "An Introduction to Practical Astronomy, with a collection of Astronomical Tables." By Elias Loomis, LL.D., Pro-

fessor of Mathematics and Natural Philosophy in the University of the city of New York. New York: 1855. In Chapters X. and XI. of this treatise are given at considerable length investigations of rules for calculating Eclipses of the Moon, Eclipses of the Sun, and Occultations of Stars by the Moon, accompanied by examples of the application of the rules. In an example of the calculation of the beginning and end of a Solar Eclipse, and in two examples of the calculation of the time of an occultation, Bessel's method of computation is adopted, and the arithmetical details are fully exhibited.

(8) Brünnow's Spherical Astronomy, translated by the Rev. Robert Main, M.A., F.R.S., Radcliffe Observer at Oxford. Part I., including the Chapters on Parallax, Refraction, Aberration, Precession and Nutation. Cambridge: 1860. Mr Main did not produce a translation of the remainder of the work. The whole, as translated into English by the author himself, will presently be remarked upon.

(9) "Practical and Spherical Astronomy, for the use chiefly of Students in the Universities." By the Rev. Robert Main. Cambridge: 1863. Chapter II. of this publication treats of Instruments and corrections of instrumental errors, and having been composed by an author who has had large experience in Practical Astronomy, may be depended upon as being free from inaccuracies and defects which would seem to be inevitable when the writer has had no practice in the use of astronomical instruments. The major part of the work is devoted to Spherical Astronomy and deductions from Kepler's Laws of Planetary Motion. The department of Practical Astronomy might, as it seemed to me, as respects both the actual state of the science, and the reading of Cambridge mathematical students, receive with advantage fuller and more extensive treatment.

(10) "An Introduction to Plane Astronomy, for the use of Colleges and Schools." By Philip Thomas Main, M.A., Fellow of St John's College, Cambridge. Cambridge: 1865. This elementary work, which, as to arrangement, follows the above-mentioned treatise on *Practical and Spherical Astronomy* by Mr Main's father, seems well adapted to answer the purpose, intended

by the author, "of a text-book for that part of the subject which is required in the first three days of the Examination for Mathematical Honours."

(11) Spherical Astronomy. By F. Brünnow, Ph. Dr., translated by the author from the second German edition. London, Asher and Co., 1865. The first edition was the one translated in part by Mr Main, as mentioned above (No. 8). Previous to Dr Brünnow's translation of the second edition, the work had been "considerably enlarged." As its title implies, it treats of Spherical Astronomy apart from physical considerations, and makes no reference to Kepler's Laws. In the seventh and last section, Practical Astronomy is included under the head of "Theory of the Astronomical Instruments." Although, as being a German work, it differs in composition and details from English treatises on the same subjects, it contains much that the English astronomer might derive advantage from.

(12) "A Treatise on Astronomy for the use of Colleges and Schools." By Hugh Godfray, M.A., of St John's College. Cambridge and London: 1866. The purpose of this publication is of the same kind as that already mentioned relatively to No. 10; but as it takes in a wider range of subjects, "embracing all those branches of astronomy which have, from time to time, been recommended by the Board of Mathematical Studies," it is suitable for the reading of Candidates for Mathematical Honours who are preparing for the Examination in the last five days. The descriptions in Chapter II. of the construction of clocks for astronomical use, and the account given at the end of Chapter VI. of the principle of Rochon's Double Image Crystal Micrometer, may be mentioned as conveying information not usually met with in elementary treatises on Astronomy.

(13) "A Hand-book of Descriptive and Practical Astronomy," London: 1861, and "Descriptive Astronomy," Oxford, at the Clarendon Press: 1867. By George F. Chambers, F.R.A.S. These two works, the contents of which are partly identical, the second being mainly an expansion of the first, convey a very large amount of miscellaneous astronomical information. In the first, the subject of Book VIII. is "Astronomical Instru-

ments," and in the other the subject of Book VII., consisting of nine Chapters, is "Practical Astronomy." These two volumes are remarkable for containing a very large number of carefully executed diagrams, and illustrative representations of celestial objects, amounting in the first to 160, and in the later volume to 224.

(14) "A Manual of Spherical and Practical Astronomy: embracing the general problems of Spherical Astronomy, the special applications to Nautical Astronomy, and the theory and use of fixed and portable Astronomical Instruments; with an appendix on the method of least squares." By William Chauvenet, Professor of Mathematics and Astronomy in Washington University, Saint Louis. Volumes I. and II. Philadelphia and London: 1863. The first volume contains "Spherical Astronomy;" the other "the theory and use of Astronomical Instruments" and "the method of least squares." The contents of the two volumes form a complete Treatise on Observational Astronomy, properly making no mention of Kepler's Laws. The American method of recording transits by the intervention of a galvanic circuit is described in Arts. 71—77 of Vol. II. The appendix to Vol. II., consisting wholly of a very elaborate discussion of the method of least squares, occupies not less than ninety pages.

(15) "A Treatise on the Principal Mathematical Instruments employed in Surveying, Levelling, and Astronomy, explaining their construction, adjustments, and use; with an Appendix and Tables." By Frederick W. Simms, F.R.A.S., F.G.S., Civil Engineer. 8th edition. Troughton and Simms, 138 Fleet Street, London: 1850. This publication is entirely of a practical character, giving very useful information respecting the construction and manipulation of the various instruments employed in surveying and levelling. It does not, however, describe the form of theodolite which, as constructed by Mr James Simms of Fleet Street, is now in general use for Trigonometrical Surveys. I shall have occasion to speak of this serviceable instrument in the course of the Lectures. The instruments referred to in the title of the work as being

employed in Astronomy are those of a portable kind, as the sextant, the reflecting circle, the portable transit, and the portable altitude and azimuth instrument. The subject of the Appendix is, Protracting and Plotting, &c., in surveys, and the instruments employed for such purposes.

(16) In addition to the sources of information indicated in the foregoing list, there are some others which may properly be adverted to here. Herschel's "Outlines of Astronomy" is principally devoted to Physical Astronomy; but Chapters I., II., III., of this work, as well as "The Introduction to Astronomy" by the same author, forming a Volume of Lardner's Cabinet Cyclopædia (1833), treat of Practical Astronomy, and may be read with advantage. The Articles "Mural Circle" and "Equatoreal" in the *Penny Encyclopædia*, written by Richard Sheepshanks, an experienced astronomer, although they are brief, are intelligible and much to the purpose, especially that on the Equatorial. The Introductions to the Cambridge Observations of 1836 and 1837 describe in detail the methods of observing with the meridian instruments of the Cambridge Observatory, and the Introduction to the Volume for 1838 contains an account of the adjustments of the Northumberland Equatorial and the uses of its mechanical appliances. Information derivable from all the different sources that have now been mentioned would be amply sufficient for the Cambridge student, as far as regards the educational effect of a knowledge of this department of science; but if he desires to pursue the subject beyond the requirements of a university examination, and to acquire a thorough acquaintance with the successive modern improvements in the construction of instruments, the modes of observing, and the reductions of the observations, he cannot do better than have recourse to the Volumes of Greenwich Observations published in the interval, beginning at 1811, during which the Royal Observatory has been under the direction of Pond and his immediate successor, the present Astronomer Royal. Of the principal instruments in actual use at Greenwich, descriptions, copiously illustrated by plates, are given in several of the Annual Volumes, as here indicated:—

INTRODUCTION.

The Altitude and Azimuth Instrument is described in pp. iv—xvi of the Introduction to the Volume of Observations of 1847, the plates being at the end of the Introduction, and a perspective view of the instrument (taken from the "Illustrated News" for the week ending Oct. 2, 1847, p. 221) forming a frontispiece to the Volume.

The descriptions of the Transit Circle, the Reflex Zenith Tube, and the Great Equatorial, with the plates, constitute respectively Appendix I. to the Volume for 1852, Appendix I. to that for 1854, and Appendix III. to the one for 1868.

A history and description, with figures, of the Water Telescope (which there will subsequently be occasion to speak of) are given in the Introduction to the Volume for 1871, pp. cxix—cxxxii. This instrument, mounted for a special purpose, is no longer required for use [1].

From the consideration I have given to the contents of the elementary works on Practical Astronomy used as text-books, I have come to the conclusion that, while other parts of the subject have been satisfactorily handled, there is generally a want of precision and completeness as respects the principle and uses of the collimating eye-piece, and the adjustments and calculations relating to equatorial observations. These faults I shall endeavour to rectify by explanations given in the course of the Lectures.

---

25. PRACTICAL ASTRONOMY may be regarded as including whatever is directly effected by means of INSTRUMENTS adapted for observations of celestial objects. Lectures on this subject ought therefore to give an account (1) of the instruments, as to the modes of mounting them, their component parts, and the purposes for which the different parts were designed; (2) of the adjustments that are required to be either actually or virtually

[1] A plan of the buildings and grounds of the Greenwich Observatory (1863, August), with explanation and history, forms Appendix II. to the Observations of 1862.

effected before the instruments are used for taking observations; (3) of the methods of exact observation; (4) of the calculations that are proper for making the immediate results of observation available for the intended purposes. The Lectures on each principal instrument will be given, both as to order and matter, in conformity with these four divisions.

THE TRANSIT INSTRUMENT.

26. A Transit Instrument is specially employed for recording the times, as shewn by a regulated Astronomical Clock, of the passage of a celestial object across the meridian of a given spot, and its construction and mounting are arranged so as to be suitable for fulfilling this purpose.

I. DESCRIPTION OF THE PARTS AND MOUNTING OF THE INSTRUMENT.

27. The telescope is furnished with an achromatic[1] object-glass, and a set of Ramsden's eye-pieces, and attached to the middle part of its tube there is a transverse axis, the ends of which are formed into cylindrical pivots. When the telescope is mounted these pivots rest in two grooves, usually called Y's, formed by two intersecting planes, and inserted, with their angles vertically downwards, into the two piers which support the instrument. In the model represented by Figure 1, a, a are the piers, b, b the two arms of the transverse axis, c, c the positions of the pivots resting in their Y's, d the object-end and e the eye-end of the telescope tube, and f, f are the positions of two microscopes pointing towards the ends of the pivots, to be

---

[1] In many of Dollond's telescopes, the achromatic object-glass was composed of three lenses. The improvement as to achromatism effected by the third lens being found not to compensate for the additional loss of light, object-glasses of two lenses are now universally adopted. The object-glass of the Dollond telescope of the Cambridge Transit Instrument consists of two lenses; that of the equatorially mounted 46-in. Dollond had originally three lenses, but in consequence of two of the lenses having been accidentally broken, it was replaced by a two-lens object-glass.

used for a purpose which will be indicated under the head of *Division* III. The piers are usually of massive stone, resting on a firm foundation in the ground, and sometimes on a common

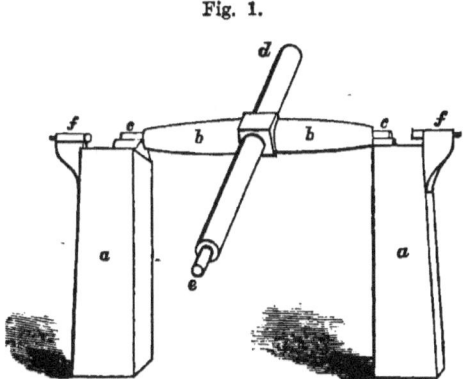

Fig. 1.

slab. They are found to be steadier when the sub-stratum is sand than when it consists of clay or rock.

28. Supposing the pivots, as first turned, to be exactly cylindrical, whatever material they are made of the friction between them and the Y's must in some degree produce continual deviation from that form, the use of the instrument being such that the surfaces of the pivots are not equally worn at all points. The observer ought, therefore, to have the means of testing from time to time the cylindrical form of the pivots, and of correcting any error by which the observations may be affected in consequence of change of form. Unless this can be done the *logic* of the use of the transit instrument is not perfect. The form of the pivots may to a certain extent be tested by making use of a striding spirit-level, and applying the Y's of its feet to the pivots in a manner that will be indicated under *Division* II. If on applying the level, its readings undergo no change while the instrument is turned through a complete revolution about its axis, we may conclude that both the pivots are cylindrical, whether or not they are equal, excepting in the very improbable

case of the alterations of form being so exactly alike that their effect is merely a change of position of the level without change of its inclination to the horizon. This experiment, however, cannot be completely made, because there is a considerable angular range of the pointing of the telescope through which its tube prevents application of the striding level to the pivots. The level-readings may still be adequate to determine whether the pivots change form, and to some extent may furnish data for calculating the amount of change. But in *Division* III. it will be shewn how this source of error may be completely got rid of by *optical* means. At present it is sufficient to remark that under all circumstances it is desirable to diminish the friction between the pivots and the Y's by taking off some of the weight of the instrument.

29. This is done by means of *Counterpoises*, acting either by leverage or by steel springs. The counterpoises of the Cambridge Transit Instrument were originally of the former kind, one end of the lever, in the form of a Y, pressing upwards against a collar adjacent to the pivot, and the other end carrying a box containing shot to allow of regulating the weight. Under this arrangement it was possible the instrument might happen to be sustained by resting partly on a face of the Y of the counterpoise, and partly on a face of a Y of the pier; which I once discovered to be actually the case[1]. On account of the unsatisfactory action of these counterpoises I afterwards substituted for them spring-counterpoises fastened to the inner faces of the piers, and pressing upwards on collars near the pivots by means of friction-wheels. The Transit Circle of the Cambridge Observatory (which occupies the place of the original Transit Instrument) has lever-counterpoises acting similarly by means of friction-wheels, but in place of springs the wheels are sustained by rods resting on arms of the levers, the opposite arms carrying

---

[1] The lever-counterpoises might have the sustaining Y at the lower extremity of a bar hanging from the end of one arm of the lever, the weight being adjustible on the other arm. (See the Figure in p. 45 of Mr Godfray's *Treatise on Astronomy*.)

THE TRANSIT INSTRUMENT. 23

the weights, and the levers passing through holes in the piers nearly on a level with the floor of the room.

30. Figure 2 represents any section of a Ramsden's eye-piece by a plane through its axis, together with the course in that plane of a pencil of rays converging from the object-glass to the focus r. If this focus be the position of any point of a

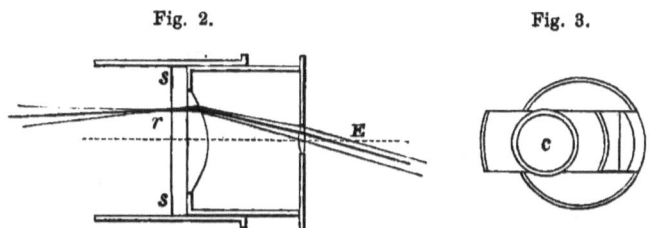

Fig. 2. Fig. 3.

wire, or diaphragm, situated in the transverse plane coincident with the stops ss which limit the field of view, the figure shews how by means of the two lenses the eye at E sees distinctly the point of the wire at r, and in coincidence with it the image of the celestial object from which the pencil of rays originally proceeded. Figure 3 exhibits the sliding-piece in grooves, mentioned in Art. 21, which carries the eye-piece, and by being shifted horizontally, enables the observer to see distinctly through an aperture at c the image of a star or other celestial object in successive positions during its transit across the field of view of the telescope, and also a series of fixed wires at the instants the image crosses them.

31. Before proceeding farther in the description of auxiliary apparatus, it will be proper to introduce an account of the principle and construction of the Filar Micrometer[1], which has

---

[1] The first invention and use of this Micrometer may certainly be ascribed to William Gascoigne, who, "while only eighteen years of age appears to have been actively engaged in observing the celestial bodies, and in advancing the state of optics and practical astronomy." He was killed at the battle of Marston Moor on July 2, 1644, in the twenty-fourth year of his age. (See Grant's *History of Physical Astronomy*, pp. 449—453.)

been applied in an important manner not only in observations taken with the Transit Instrument, but also, under different forms, in the observational use of other astronomical instruments. The principle of the Micrometer is to measure very small spaces by the intervention of much larger spaces, which by mechanical construction are made to have to the first a fixed ratio, and are capable of being easily read off from a graduated scale. Figure 4 represents a double-wire Micrometer as constructed by Troughton. *A* is one of the micrometer-heads, the

Fig. 4.

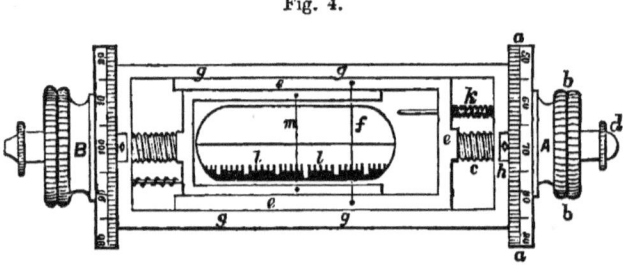

form of which is symmetrical with respect to an axis; *aa* is an attached circle graduated from 0 to 100 on its rim, and movable about the axis by turning the milled head *bb*; *c* is the micrometer-screw working in the micrometer-head, and passing into the hollow terminal *d*; the screw is fastened to the fork *eee*, which slides within a box of which *gg, gg* are projected sides, and carries with it the micrometer-wire *f* crossing from arm to arm transversely; *h* is the place of the index by which the readings of the graduation are taken: the screw being of the form called *right-handed*, which is that universally adopted by mechanists, if the circle be turned in the direction for which the index-readings of the graduation increase—the opposite to that in which the graduation-numbers in the figure increase—the fork and wire will be drawn *towards* the micrometer-head, so that the distance of the wire from the micrometer-head is less the greater the micrometer-reading. At *k* there is an antagonistic compressed spiral spring, which, when the graduation is turned in the direction of *decreasing* index-readings, pushes the fork and wire *from* the micrometer-head. This construction is consequently adapted to measuring any small space through which the wire is moved by turning the milled-head, by recording the larger space simultaneously described by a given line of the graduation; and it is evident that the ratio of the former space to the other is equal to that of the interval between the threads of the screw to the circumference of the graduated circle. The comb ll serves to indicate the integral number of revolutions, the fraction of a revolution is read off in graduation intervals, and the deepest indenture in the comb shews approximately the position of the wire at the zero reading. The exact performance of a screw depends essentially on the equality of the intervals between consecutive threads. A "drunken" screw is one defective in this respect. Although this condition is usually fulfilled by the mechanist with wonderful accuracy, as a matter of precaution the micrometer-measurement of spaces requiring a large number of turns of the screw should, as far as possible, be avoided.

32. The foregoing description may be sufficient as respects the single-wire micrometer. When there is a second wire (as m in Fig. 4), this is drawn towards the micrometer-head B in exactly the same way as the other is drawn towards A, the second graduated circle being turned in the direction of increasing index-readings of the scale marked on its rim. The fork of this micrometer slides within the fork of the first, and the two wires can pass each other in close proximity, so as to be both sufficiently in focus at the same time.

33. The field of view of the Transit Telescope, as seen with a Ramsden's eye-piece, is shewn in Figure 5, in which are exhibited the vertical wires (usually of spider's web) at which the Transit observations are taken, and crossing them horizontally, two wires mid-way between which the image of the object is made to traverse the field by adjusting the pointing of the telescope, it being of some consequence that the transits should all be taken at the same parts of the wires and field. The number of vertical wires (seven in the figure) is always uneven in order that the mean of the times of transit across all the

wires may be nearly the same as that across the middle wire. The additional wire mn is a micrometer-wire, movable, as explained in Art. 31, by turning the milled-head a of the micro-

Fig. 5.

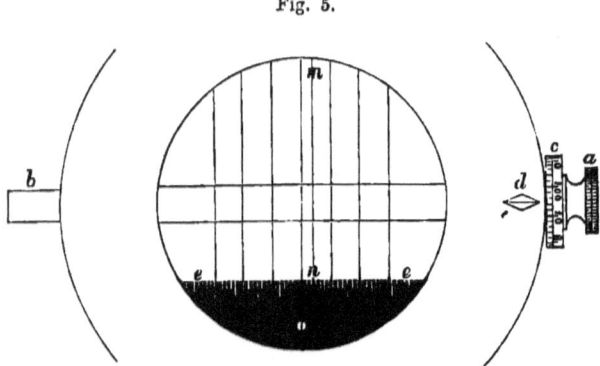

meter-screw, and by the action of an antagonist spring at b. The fractional part of a reading of micrometer-revolutions is taken from the graduated circle c by means of the index d, and the integral number of revolutions is shewn by the indents of the comb $e\ e$, the place of the wire for zero reading being approximately marked by the centre of a small circular hole in the comb.

34. Figure 6 represents those parts of the apparatus pertaining to the eye-end of the Transit-telescope which are seen *externally*. I proceed now to describe these in detail and to mention their respective uses. The two small circles ab, $a'b'$, attached to the telescope tube near the eye-end, are *setting-circles*. Their use is principally to enable the observer to point the telescope in readiness for observing the transit of an object of which he already knows approximately the Right Ascension and clock-time of Transit, and the Zenith Distance, or North Polar Distance. Each circle is graduated expressly for this purpose, and is provided with a small spirit-level movable about an axis through the centre of the graduation, and carrying at one end an index and vernier by means of which the angular position

of the level may be set to any required graduation-reading. In the figure, cd is the spirit-level of one of the setting-circles, the vernier being attached to the end 'd. The position of the zero

Fig. 6.

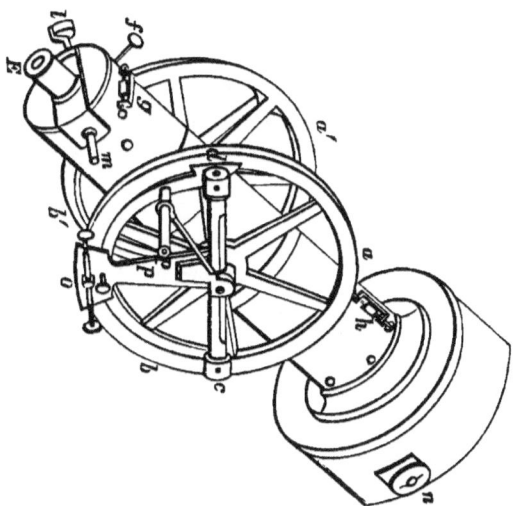

reading of the setting-circle is so adjusted that when the telescope points to the zenith, and the bubble of the spirit-level is in its middle position, the index reading is equal to the co-latitude of the Observatory. Hence for any other pointing of the telescope the index-reading for bubble in mid-position is North Polar Distance. This arrangement for setting for an object in North Polar Distance is made suitable to the graduation of either circle when on the West side, the graduations being such that when by reversing the instrument that on the East side is transferred to the West side, the process of setting is precisely the same as before. For this reason the setting is always done on the *West* side[1]. Occasionally, when the interval between the transits of two objects put down for observation allows of little time for setting for the second after the transit of the first, the reading of the *East* circle should, to save time, be prepared beforehand for setting for the second observation. This may be conveniently done by first setting for the second observation with the West circle, and then, without moving the telescope, clamping the East level with bubble in mid-position; after which the setting for the first observation may be performed in the usual manner, and immediately after the transit across the seven wires the telescope may be pointed for the second observation by turning it till the bubble of the East circle is in mid-position. Before making the above-mentioned uses of the setting-circles, their spirit-levels require to be adjusted by a process which will come under consideration in *Division* III.

Figure 7 is an enlarged representation of the setting-circle

Fig. 7.

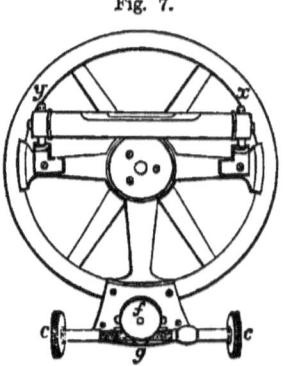

[1] Formerly, each setting-circle of the Cambridge Transit Instrument had two verniers, one for setting in zenith distance, and the other for setting in north polar distance, and the graduations were suitable for setting on the west side. As one of these verniers might be dispensed with, the second one was removed, and in order that the other might point to north polar distance, each circle was re-fixed, after being shifted about its centre through an angle equal to the co-latitude, and that on the west side was used as before, for setting. The setting-circles of the Cambridge Transit-circle also have each one vernier for setting in N. P. D., but as this instrument is not used in reverse positions, their graduations are exactly alike, and either circle may be chosen for setting.

THE TRANSIT INSTRUMENT.

of a meridian Instrument: x and y are screws with capstan-heads for adjusting the spirit-level. The action of the tangent-screw cgc and clamp f is explained in the next article.

35. To make the index-reading of the setting-circle equal to a given angle, which it is necessary to do before setting the telescope for observing a selected object, the observer makes use of a *tangent-screw and clamp* (the position of which in Figure 6 is at o), which serves to give slow motion to the vernier for exactly setting the index to the given angle. As this apparatus for slow motion is frequently applied in the use of astronomical instruments, I propose to state here the principle of its construction and the mode of its action. Its essential parts may be generally described as follows by reference to Figure 8. A fine screw aa, with milled-heads bb at the two

Fig. 8.

ends, is supported by a nut c, in which its stem, detained by flanges, is caused to revolve by turning either milled-head, as may happen to be most convenient. Another nut o is tapped for the screw to work in, and is fixed to a plate which overlaps a little the limb ff of the instrument on the farther side. At the same time another plate gg is in contact, partly with the first plate, and partly with the limb on the nearer side, overlapping to the same extent as the first plate, and to allow of both contacts a portion of the thickness of the second plate equal to the thickness of the limb is cut away. The farther plate is tapped for the action of a screw which passes through a hole in the plate gg, and has its head and flange in the position h. By turning this screw the surface of the second plate is brought up to that of the first in such manner that the limb is nipped tightly between the two overlapping parts. When the pressure is taken off by turning back the clamping-screw h, the two plates are separated by springs and at the same time are kept in position by suitable detents, in order to allow of large movements either of the tangent-screw relative to the limb, or of the limb relative to the tangent-screw. Supposing now the clamping to have been effected by means of the screw-head h, by turning one of the heads bb the nut o will approach or recede from the nut c, according to the direction of the turning. These arrangements being understood, it will be seen that there are *two* cases of the action of a tangent-screw and clamp. In one case the supporting nut c is fixed to a plate in rigid connection with a vernier which is to be used either for reading off from the graduated limb, or for setting the reading to a given angle. This is the usual case of the tangent-screw of the setting circle of a meridian instrument. Since the limb of the circle is fixed relatively to the telescope-tube, the nut o and the plates, when clamped to the limb, are also fixed, and consequently turning either head b causes the nut c to move relatively to the limb. In the instance represented by Fig. 6 this nut is rigidly connected with the tongue p, which consequently moves with it, carrying the spirit-level cd and the vernier d. In the other case, the supporting nut c is fastened to a wall, or immovable plate, and turning bb gives the slow motion to the nut o, the clamping plates, and the clamped limb ff. This is the kind of action of the tangent-screw of a Mural Circle, or Transit Circle. What is said above may suffice for explaining generally the principle and use of the tangent-screw and clamp. The details of the construction differ according to the purpose for which the instrument is designed. The tangent-screw of the Mural Circle, for instance, is constructed very differently from that of a Ramsden's Sextant, as will be shewn when those instruments are under consideration.

36. Again in Figure 6, the screw-head seen projecting at f belongs to the tangent-screw of the setting-circle on the farther side of the eye-end. At g, two screws acting oppositely on a projecting piece of the frame which carries the micrometer-wire, and thereby causing angular motion of the frame about the axis of the telescope, serve for fixing this wire in the position of

exact parallelism to the middle wire. At *h* there is an adjusting apparatus, like that at *g*, by the action of which on a piece projecting from an interior tube which carries the two wire-frames and the eye-piece, this tube can be shifted longitudinally, and the

Fig. 6.

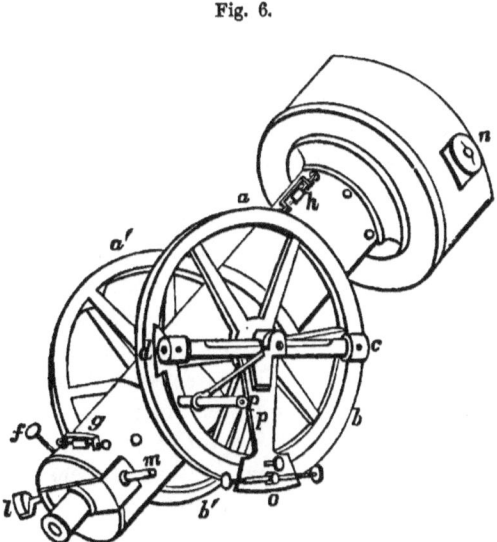

wires be placed at the distance from the object-glass of its geometrical focus. The particular reason that the observer should have the means of making this adjustment will be explained in *Division* II. At the extremity of the eye-end the figure represents the eye-piece *E* inserted in a plate movable in grooves (as mentioned in Arts. 21 and 30), together with the places of the micrometer-head *l* and the opposite spring *m*. In addition to the parts above mentioned there are two opposite screws with capstan-heads (not exhibited in the figure), which work in the frame to which the transit-wires are attached, and answer the purpose of shifting it horizontally and fixing it in any required position.

37. In order to see the wires distinctly at night-time for observing transits, it is necessary to *illumine the field of view*, that the wires may appear dark on a bright ground. The illumination may be effected and its amount regulated in the following manner. The light of a lamp, put opposite the end of one of the pivots through which a perforation has been made, falls upon a reflector which is situated where the axis of motion intersects the axis of the telescope, and, being inclined to the latter at an angle of 45°, sends the reflected light to the eye-end, and thus illumines the field. The reflector is ring-shaped in order to allow of free passage of the pencils of light from the object-glass, and its surface is gilded for the purpose of increasing the intensity of the scattered light reflected from it. The degree of illumination is regulated by varying the size of the aperture at the end of the pivot, which is done by means of rack-work and wheels, moved by turning the milled-head at n, which is within reach of the observer when he is looking through the eye-piece. In the Greenwich Transit Circle, as also in that of the Cambridge Observatory, the illumination can be gradually diminished to the point of extinction by means of apparatus provided by Mr James Simms for enabling the observer to vary the inclination of the reflector.

38. When the light of the object to be observed is very feeble, it may happen that the least amount of illumination required for seeing the wires with sufficient distinctness for the observation is too strong for distinguishing the object. In that case recourse is had to *illumining the wires* on a dark field. To do this the illuminating light has to pass in the direction from the eye-glass to the wires. The Cambridge Transit Instrument was furnished with the means of suspending a small lamp in such manner that its light fell in a *slanting* direction on the wires so as to make them visible on a dark ground. This method is liable to the objection that as one side of a wire is illumined more than the other, the axis of the line of illumination does not exactly coincide with the axis of the wire. This fault might be got rid of by having two lamps, one on each side of the system of wires: but at best this method is practi-

cally inconvenient. It has been proposed to substitute for the spider-lines fine lines engraven on plate glass. These lines would ordinarily appear dark on a bright ground; but on applying at night-time the light of a lamp to the edge of the piece of glass, they would be seen, by the light scattered from innumerable facets, as bright lines on a dark ground. I do not know that this ingenious method has been put in practice. I have used with success for the same purpose the collimating eye-piece (of which a description will subsequently be given), by holding in the hand a lamp so that its light reflected from the plate-glass, situated between the eye and the wires and inclined to the axis at an angle of 45°, is incident on the wires. This and the preceding method are objectionable on account of the loss of light from the faint object caused by its passage through the plate-glass. The most complete method appears to be one invented by Mr Simms, according to which by continuing to vary the inclination of the reflector beyond the point of extinction mentioned in Art. 37, the light of the lamp is thrown on the wires, the field remaining dark. The apparatus for this purpose has been supplemented by furnishing the observer with means of regulating by turning a milled-head the amount of the illumination of the wires.

39. From the foregoing descriptions of the mounting and the parts of a Transit Instrument we may proceed to treat of it under the *second* Division; that is, to consider what preliminary arrangements have to be made in order that it may effect the immediate purpose of determining the clock-time of transit of a celestial body across the meridian passing through its centre of figure, which meridian will be taken to be the reference meridian of the Observatory.

II. THE ADJUSTMENTS OF A TRANSIT INSTRUMENT AND CORRECTION OF ITS ERRORS BY CALCULATION.

40. It will at first be supposed that the pivots are exactly equal and cylindrical, and that their axes are in the same

straight line (see Art. 27). It is evident that on this supposition *the axis of motion* coincides with the common axis of the pivots, and that it will have the same position in space in reverse positions of the instrument, if only the Y's in which the pivots rest retain the same position. It has been stated (Arts. 12, 13, 15) that the axes of all pencils of rays that go to form the image looked at with the eye-piece pass through the optical centre of the object-glass, so that a plane through this point cutting the axis of motion at right angles has not only a fixed position relative to the instrument, but is also related in a definite manner to the directions of the pencils of rays that come from a celestial object. *A line of collimation* may be defined to be the axis of any pencil of convergent rays which forms the image of a point of the object at the geometrical focus of the object-glass, and its direction, as far as concerns transit observations, is determined by its inclination to the above-mentioned plane. This, in fact, is the plane to which collimation is referred, and, for brevity, will be called *the plane of collimation.* There is no line which can properly be named the optical axis of collimation; but a straight line drawn from the optical centre of the object-glass to the point of the plane of collimation which is mid-way between the two horizontal lines represented in Figure 5, is in some sense an optical axis, because, for the sake of uniformity and exactness, the images whose transits are taken should all be made to cross the field nearly through this point. If, however, the wire-frames be accurately adjusted, crossing a little higher or lower will be of no consequence.

41. That a Transit Instrument may answer the purpose to which it is applied, the optical axis (defined as in the preceding article) must either be made *by mechanical adjustments* to move in the plane of the meridian, or its small deviation from that plane must be ascertained for the pointing of the telescope in each observation, in order that the error of the observed time of transit thence arising may be *corrected by calculation* in the reduction of the observations, which is *virtually* to correct the errors of an unadjusted instrument.

42. The method of correcting for instrumental errors by calculation is preferable for two reasons to that by mechanical adjustment: (1) because it dispenses with frequent turning of screws, which, as tending to unsteadiness, should as far as possible be avoided; (2) because the data for the calculation are generally obtainable with great precision by the use of optical instruments for magnifying.

43. In certain circumstances mechanical adjustment is always employed; namely, when it is sufficiently exact, and not liable to be disturbed, and at the same time the method by calculation is not readily available. This is the case with respect to adjusting the middle transit-wire in order to make it coincident with a plane perpendicular to the axis of motion, and the micrometer-wire, so that it shall be exactly parallel to this fixed wire. These two adjustments, which it will be proper to consider first of all, may be performed by the following processes. For the first adjustment there should be added to the apparatus at h provided for shifting longitudinally the tube which carries the eye-piece and wire-frames (see Fig. 6 and Art. 35), two opposite screws which, by acting *transversely* on the same projecting piece, would give the means of turning that tube about its axis, and fixing it in any required angular position. Then, assuming that the axis of motion has been roughly adjusted to horizontality, and the wire-frames have been approximately put into position, that carrying the system of seven wires being *fixed* by opposite screws, the middle wire is to be brought to bisect (by turning the screw for azimuthal adjustment of the axis of motion, as hereafter spoken of) a well-defined fixed point, which may be some point of a terrestrial object, or a meridian mark, or, what is still better, the intersection of crossing wires at the focus of a collimator. Now by means of the horizontal adjustment by the opposite screws referred to at the end of Art. 36, and the angular adjustment just mentioned, it is always possible to make the bisected point run exactly along the middle-wire from end to end, while the telescope is shifted about the axis of motion. The position of the wire when this condition is fulfilled is coincident with a

plane perpendicular to that axis. (I have found that the Polestar, by reason of its slow diurnal motion, may be used when on the meridian for testing the accuracy of this adjustment.) This being done, the micrometer-wire is to be placed exactly parallel to the middle wire, partly by its micrometer-movement, and partly by employing the apparatus at g (Fig. 6) for shifting the wire about the axis of the telescope and fixing it in the required position[1]. The parallelism is effected when by the two movements a fine line of light of uniform breadth is formed between the two wires.

44. When these adjustments of the wires have been effected, three other adjustments, necessary and sufficient for the exact performance of a Transit Instrument, having, as supposed in Art. 40, equal cylindrical pivots, may be proceeded with. (1) The middle wire is required to be coincident with the plane of collimation defined in Art. 40; (2) the plane of collimation must be vertical, and therefore pass through the zenith of the point of its intersection with the axis of the instrument; (3) this vertical plane must also pass through the pole of the heavens in order that it may be coincident with the meridian of that central point. The first of these conditions has reference to the instrument, the second to the earth, and the third to the heavens. Deviations from fulfilment of these conditions are *instrumental errors*, which have either to be corrected by mechanical means, or to be allowed for by calculation (see Art. 41). As it is proposed to indicate both methods of correction, it will be proper to introduce here a description of the apparatus employed for correcting the three errors mechanically.

45. For correcting *the error of collimation*, that is, the

---

[1] In the Cambridge Transit Instrument no apparatus was provided for the rotational movement of the tube which carries the wire-frame, in consequence of which, having to take out the tube for the purpose of inserting the apparatus at h, I was obliged, in replacing it, to make the angular adjustment by hand in the best way I could. The Transit Circle now in use is furnished with means of making both the longitudinal and the rotational adjustment, two sets of antagonist screws acting for that purpose upon the projecting piece at h.

deviation of the middle wire from the plane of collimation, the movement of the wire-frame by capstan-headed screws, mentioned before in Art. 36, is provided. For correction of *the error of level*, or deviation of the axis of motion from a horizontal plane, the Y in which one of the pivots rests is movable vertically, and can be fixed in position, by adjusting screws; and for correction of *the error of azimuth*, or deviation of the plane of collimation, corrected for level error, from the plane of the meridian, the other Y is movable horizontally, and can be fixed in position, by antagonist adjusting screws. The mechanical arrangements made for these two movements in the mounting of the Cambridge Transit Instrument are exhibited in Figures 9 and 10, as seen when the brass plates that concealed them were removed[1].

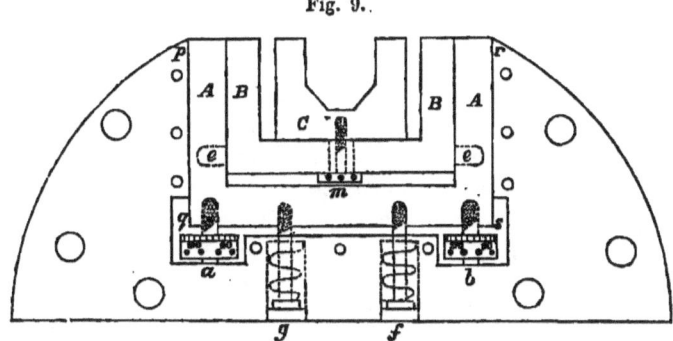

Fig. 9.

46. The vertical adjustments were made by turning with spikes the capstan-headed screws at a and b (Fig. 9), and the azimuth adjustments by turning the antagonist screws at c and

---

[1] As this apparatus has been dismounted, I had the opportunity of examining it before drawing the Figures 9 and 10. These are, for the most part, copied from figures contained in the second plate to Woodhouse's account of the instrument (before referred to in a note to Art. 9), which, as no explanation accompanied them, were scarcely intelligible without inspection of the apparatus.

38   PRACTICAL ASTRONOMY.

d (Fig 10). The former adjusting screws were maintained in bearing by the spiral springs at g and f. The capstan-heads were graduated in such manner that after completing the

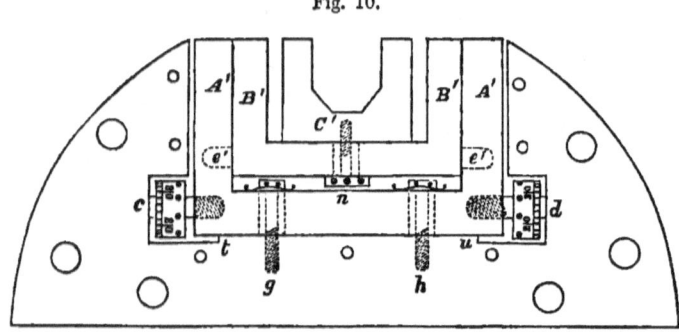

Fig. 10.

adjustment of the axis, the index-readings of the two heads for movement in altitude should be the same, as also those of the two heads for movement in azimuth. This appears to have been arranged in order to avoid the *strain* which such action of two screws as that which the above figures indicate is evidently liable to produce if they should be turned unequally. Accordingly after turning the screws for either the altitude or the azimuth adjustment, the index readings of both screws were always left the same. (It would have been better if the altitude adjustment had been effected by a *single* capstan-headed screw placed between the spiral springs at g and f (Fig. 9), supposing these to be of considerable strength, and if for the azimuth adjustment one of the screws at c and d, instead of working in the movable plate $A'A'$, had simply abutted on this plate for fixing it, the movement being effected by the screw at the other end.) The vertical movement of AA took place along the fixed edges pq and rs, and the horizontal movement of $A'A'$ along the fixed edge tu, on which it was made to bear by the steel springs and screws g and h (Fig. 10). I found that the plate BB was made to be movable in contact with AA about a horizontal axis ee supported by the latter, and that the like was the case relatively to the plates $B'B'$ and $A'A'$. This was done in order that by the weight of the instrument the pivots might be made to rest in perfect contact with the Y's; and for the same reason the pieces C and C', in which the Y's are cut, admitted of being shifted horizontally till the pivots obtained their proper bearing, after which they were fixed in position by the capstan-headed screws m and n. Since, however, with all the above-mentioned precautions it was not certain that strain, or unsteadiness, was wholly prevented, I consider the mode of adjustment described in the next paragraph, which dispenses with screw-movement, to be much preferable.

47. The foregoing explanations may suffice to indicate generally what is required to be done for correcting the three errors, whether the Transit Instrument be small or large, portable or fixed. But the details of the mechanical means may differ according to the size or purpose of the instrument. In the Transit Circles of the Greenwich and Cambridge Observatories, the wire-frame is not fixed in position by capstan-headed screws, but is movable by a micrometer-screw with graduated head and antagonist spring, and the use of the single micrometer-wire is dispensed with. Also the pivots are made to rest in portions of cylindrical surfaces exactly fitting them, and instead of employing adjusting screws for correcting the errors of level and azimuth when they have become large, this is done by simply *scraping* these surfaces. The fitting portions extend on each side through 15°, and their centres are about 120° apart.

Correction of Collimation-Error by mechanical adjustment.

48. We may now proceed to consider particular processes by which the three errors may be *corrected*, beginning with *the mechanical correction of the error of collimation*. The method for a long time generally adopted for making this correction was to set up a fixed terrestrial mark very near the meridian, and at a distance from the telescope so great that the rays from any point of it entering the object-glass would be very nearly parallel (see Art. 9). The Cambridge mark, which was on the

40   PRACTICAL ASTRONOMY.

tower of Grantchester church (distant 2½ miles), was a cross in the form of X[1], and its angular distance West from the meridian was about 14″. Under these circumstances the middle wire could be brought to bisect the image of the cross by means of the screws for azimuth adjustment (Art. 46). Then on reversing the instrument[2] by exchanging the positions of the pivots in the

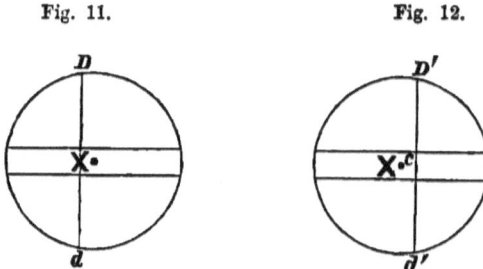

Fig. 11.   Fig. 12.

Y's, the wire would usually be found to be separated from the cross.

49. Figures 11 and 12 represent what is seen before and

---

[1] For this form I substituted a small circular disk, held by fine iron wires at the centre of a wooden circular frame which was supported by two legs equally inclined to the vertical through the disk. This mark was also used for *horizontal* bisection.

[2] The Cambridge transit instrument was reversed by being lifted from the Y's by a system of pulleys hanging from the end of an iron bar, which was drawn out for the occasion so far as to bring the point of suspension vertically over the centre of the instrument. Leathern straps which passed under the transverse arms being attached to the pulley-frame, the lifting was effected by pulling the rope by hand. This method I found to be attended with some risk, having discovered (in good time) that the check which limited the drawing out of the bar required to be made fast, having gradually given way. A better method is adopted for lifting the present Transit-circle, a much heavier instrument, this being done by a jack-screw machine, which stands usually at a corner of the room, and by means of rails and a turn-table can be brought into a position proper for acting from below on the transverse arms. There is rarely occasion for this operation, the observations being all taken, not necessarily, but for greater convenience, in the same position of the instrument, and the error of collimation being found, without reversing, by processes which will shortly be described.

after the reversion. As the line of collimation (Art. 40) of any point of the wire makes equal angles with the plane of collimation before and after reversion on opposite sides, it follows that this plane passes through a point c midway between the bisected cross and the wire in its new position $D'd'$. Hence to correct the error of collimation of the middle wire, it is only required to move the wire after the reversion half-way towards the image of the cross by means of the screws for collimation-adjustment (Art. 45). If this movement is effected by capstan-headed screws, as is usually the case in small and portable instruments, it can only be done tentatively, and after a first trial the same process should be repeated till the wire appears to bisect the cross both before and after the reversion. When this is the case the correction has been made with as much accuracy as is attainable by mechanical means.

50. If the telescope has a micrometer-wire (as mn in Figure 5), the mechanical correction of collimation-error can be more readily effected and with greater precision. In that case the mark is bisected by the micrometer-wire before reversing, and the micrometer-reading is recorded; and the same steps are taken after the reversion. If in Figures 11 and 12 Dd and $D'd'$ represent the micrometer-wire in the two positions, the difference of the two readings will evidently measure the distance of $D'd'$ from the cross, or twice the collimation error. The mean of the readings will be the micrometer-reading for the mid-position c, and if the micrometer-head be set to this reading, the micrometer-wire will actually be placed in the plane of collimation. Consequently the mid-wire will be corrected for collimation-error by being brought by the screws for horizontal adjustment of the wire-frame into coincidence with the micrometer-wire according to the judgment of the eye, which may be done with very considerable accuracy. If, however, the wire-frame be movable by a micrometer (as mentioned in Art. 47), the mid-wire is at once placed in the plane of collimation by making the micrometer-reading equal to the mean of the two recorded readings.

Calculation of uncorrected Collimation-Error.

51. In the use of large meridian instruments the processes of mechanical correction are only employed for securing that the errors (which from one cause or another are always changing) shall not exceed very small amounts, no attempt being made to maintain them at zero by adjustments. Hence it is necessary from time to time to calculate the actual amounts, for the purpose of getting rid of their effects on the results of the observations. Proceeding, accordingly, to indicate how collimation-error may be *calculated*, I shall, at first, suppose that the measured quantities are expressed in micrometer-revolutions, and afterwards shew how to determine the value in celestial arc of the space the wire is carried over by one revolution of the micrometer-head. It has been ascertained (Art. 50) that the micrometer-reading for coincidence with the plane of collimation is the mean of the micrometer-readings for bisection of the mark before and after reversion. (For the sake of precision each adopted micrometer-reading is usually the mean of the readings for, at least, six bisections.) The coincidence of the micrometer-wire with the mid-wire is not judged of in the way mentioned in Art. 50, but is determined by just excluding the line of light between the two wires three or more times on one side and an equal number of times on the other side, and taking the mean of the micrometer-readings for all the positions for the adopted coincidence-reading. The collimation-error of the mid-wire expressed in micrometer-revolutions is the difference between this coincidence-reading and that for coincidence with the plane of collimation. In the case of a Transit Circle which has no micrometer-wire, and is not reversed, the collimation-error is calculated by means of collimators, as will be presently shewn.

52. On the supposition that the pivots are cylindrical and equal, the determination of collimation-error by this method is complete unless the operation of reversing the instrument produces horizontal displacement of the Y's. As it is possible this might happen, especially with such apparatus for supporting the Y's as that described in Art. 46, the observer ought always to employ *two* marks, one southward and the other northward, because the effect of horizontal angular displacement of the axis is thereby eliminated, as is proved by the following reasoning. The telescope being horizontal and pointing southward, suppose the eye-end to be shifted, by the displacement, towards the east. Then the image of the south mark will be shifted, relatively to the mid-wire, towards the *West* through a space which measures the angular displacement. So when the telescope is turned to point northwards, and the eye-end is consequently shifted, by the displacement, westward, the image of the north mark is shifted, relatively to the mid-wire, through an *equal* space towards the *East*. Hence if r_1 and r_2 be the micrometer-readings for bisection of the marks before the reversion, and r_1' and r_2' those for bisection after reversion, we may infer from what has just been said that the latter would have been $r_1' \pm a$ and $r_2' \mp a$ if there had been no displacement. Hence the reading for plane of collimation is $\frac{1}{2}(r_1 + r_1' \pm a)$ by one mark, and $\frac{1}{2}(r_2 + r_2' \mp a)$ by the other, and the mean of the two determinations is $\frac{1}{4}(r_1 + r_1' + r_2 + r_2')$, which is the mean of the four readings actually recorded. This mean, therefore, is the true reading for coincidence with the plane of collimation, notwithstanding any small angular displacement of the axis. The same result would evidently have been obtained if the shifting of the eye-end, the telescope pointing southward, had been towards the West.

53. Instead of a second distant mark on the opposite side of the instrument (an arrangement which it would rarely be possible to make), at the Cambridge Observatory, as well as at Greenwich (see Note to Art. 9), a small transit telescope was formerly set up near the telescope of the transit instrument, to serve for a *Collimator*. The two telescopes being so placed that they pointed directly towards each other, an image of the wires of the small telescope, which are supposed to be accurately

44   PRACTICAL ASTRONOMY.

placed at its geometrical focus, was formed at the geometrical focus of the large telescope, and a definite point of this image was selected for bisection. The selected point, as being a focus of convergence of parallel rays incident on the object-glass, was equivalent to the image of a mark at an infinite distance. It is obvious that if *two* collimators, one towards the north and the other towards the south, were employed in this manner, a fixed terrestrial mark would not be required. Gauss introduced the improvement, now generally adopted, of setting up two collimators so that each could be a collimator to the other as well as to the intermediate telescope, by which means the error of collimation can be calculated without reversing the instrument, as may be shewn by the following argument.

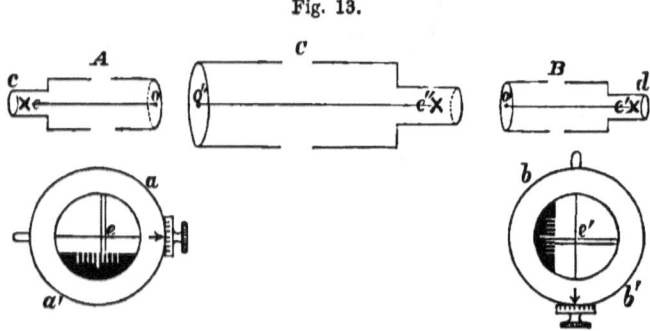

Fig. 13.

54. In Figure 13 A and B represent the two collimators, and C the transit telescope between them. The optical centres of the three object-glasses are respectively at o, o', and o'', and the positions e, e', e'', marked by the intersections of crossing lines, are their geometrical foci. It will be supposed that the system of wires of the telescope C is placed exactly at the focal distance $o''e''$, so that rays proceeding from a point of the wires situated on or very near the optical axis will emerge from the object-glass as parallel rays. This adjustment, usually performed by the instrument-maker, might be effected by the observer himself by using the screw-adjustment at h mentioned in Art. 36,

the wire-frame being thereby placed in such a position that the mid-wire, where it is intercepted between the horizontal wires, and the image of a star or planet crossing it, are both seen distinctly. (A better method of making this adjustment will be indicated when the collimating eye-piece is under consideration.) The additional figures aa' and bb' represent, within the interior circles, the fields of view of A and B as seen by eye-pieces at c and d. The wires thus exhibited are used for collimating, and it is necessary that they should be exactly at the geometrical foci of the collimators. This condition may be satisfied as follows. After adjusting, as above stated, the wire-frame of C, A and C' are made to point towards each other, with their optical axes approximately in the same straight line, by screw-movements provided for roughly adjusting the optical axis of A both vertically and horizontally. An image of the wires of C is thus formed at the geometrical focus of A, being due to parallel rays issuing from the object-glass of C and incident on that of A. The wires of A are then shifted (by an external ring-screw, or other apparatus) till they are seen distinctly through the eye-piece at c coincidently with the above-named image, in which case the required adjustment is effected. The wires of B have to be adjusted to focus by a like process. The mid-wire of C being assumed to be accurately parallel to the plane of collimation (see Art. 43), the vertical wires of A and B have to be adjusted so as to be parallel to that wire. This may be done by means of the horizontal and vertical adjustments of their optical axes above spoken of, and by unclamping, moving rotationally by hand, and then clamping, the tubes in which the wires are fixed. No other means of making the rotational adjustment is required, this being sufficiently exact if the collimating operation be performed according to the following process.

55. The wires and comb in the two fields of view aa', bb' (Fig. 13) are exactly alike; but the micrometer-movement in the former is horizontal and is used for collimating in right ascension, whilst that in the other is vertical and might be used for collimating in declination, and for determining, in the case

of a Mural Circle or a Transit Circle, the effect of its flexure. It would suffice for all purposes to furnish only one of the collimators with a micrometer, if means were provided of rotating its wires through 90°. I adopted an arrangement by which this could be done by hand with sufficient exactness, and was thus able, by means of two collimators, only one of which had a micrometer, to correct the collimation-error of the Mural Circle, and also to calculate corrections for the effect of its horizontal flexure. It would, however, be preferable to furnish each collimator with a micrometer in the manner exhibited in Fig. 13. (This is the case in the collimators of the Cambridge Transit-circle.) In the field of each collimator are represented two parallel wires, separated by a small interval, and movable together by turning the micrometer-head. Supposing the parallel wires of B to be put nearly in the middle of the field, as indicated by the zero of the comb, the point of the vertical wire which bisects the interval between them may be taken for e'. If B has no micrometer the point e' is marked by the intersection of the vertical wire by a horizontal wire through the middle of the field. By preliminary operations the optical axes of A, C, and B should be placed nearly in the same straight line, and the optical axes of A and B nearly in the plane of collimation of C. These arrangements, which may be made by means of the vertical and horizontal adjustments of the optical axes of A and B, require no great precision. Also in the tube of the telescope C an aperture is required to be made through which, when the axis of the telescope is vertical, the collimators can point directly towards each other. By this means an image of e' is formed at the geometrical focus of A, and by the adjustments of its axis may be brought nearly to the centre of its field. These precautions prevent any appreciable error arising from defect of parallelism of the vertical wires of A and B to the mid-wire of C. The parallel wires of A may now be moved till the image of e' exactly bisects the interval between them. This may be done with great precision by the judgment of the eye. For greater certainty it is usual to make six or more such trials, to record the micrometer-reading after each, and finally to set the micrometer to the mean of all the readings. The point e in A may then be taken to be that point of the horizontal wire which is midway between the parallel wires. The telescope C being next pointed to A, let the image of e be formed at e'' on or near the optical axis. This image is to be bisected six or more times by the micrometer-wire of C (if the system of wires be not movable by a micrometer), and the mean of the readings taken. Exactly the same process is to be gone through with respect to the point e' of the collimator B. Now although the collimation between A and B does not place eo and $e'o'$ in the same straight line, it secures that their directions shall be *parallel*. Hence if the line $e''o''$ determined by the collimation between A and C be inclined to the plane of collimation of C, that determined by collimating between B and C will be *equally* inclined on the opposite side. Hence the mean between the two readings for bisection of the two points e'' will be the micrometer-reading for the plane of collimation, and the difference between this reading and that for coincidence with the mid-wire (found by the method indicated in Art. 51), will be the measure of the collimation-error of the mid-wire in micrometer-revolutions.

56. If the wire-frame be movable by a micrometer, the process for finding the micrometer-reading for plane of collimation is the same as before, the mid-wire being used in place of the separate micrometer-wire. It is the practice at Greenwich to obtain the micrometer-reading for collimation-plane every day, after each operation to set the micrometer to a certain convenient reading that may be presumed to be little different from the true reading, and to adopt for true reading the mean of several determinations, usually those of a week. Then the adopted error of collimation is the difference between this adopted reading and that for the arbitrary position occupied by the wires when the transits are taken. The error of collimation for the transit observations made with the Cambridge Transit Circle is determined on the same principle.

57. The amount of collimation-error having been found, it is next required to determine the *sign* with which it is to be

applied for correcting its effect on the observed times of transit. If the instrument be reversed from time to time, as was formerly the practice at Cambridge, the sign of the collimation-correction depends on the position of the transverse axis as to East or West, which, therefore, has to be noted. This was done by recording after each reversion 'Illumination East,' or 'Illumination West,' according as the pivot which is perforated for admitting the light for illuminating the field of view was left eastward or westward. (It would have been simpler to indicate the position by the eastward or westward direction of the micrometer-head, which happened to be opposite to that of the perforated pivot.) Let us suppose the micrometer-head to be *westward*. Then taking into account that a micrometer-reading is always greater the nearer the micrometer-wire is to the head, it will be seen that the mid-wire is to the east or west of the plane of collimation according as the reading for coincidence with it is less or greater than that for plane of collimation. Hence since the image of a star above pole crosses the field from west to east, in the case of the less coincidence reading it crosses the mid-wire after crossing the plane of collimation, that is, too late, and the correction of the time of transit is *negative*. In the contrary case of the coincidence-reading being *greater* than that for plane of collimation, the correction is *positive*. The signs of the correction for a star below pole are opposite to these, because its image crosses the field in the opposite direction. The signs of the corrections for micrometer-head eastward are evidently contrary, under the same circumstances, to those for micrometer-head westward.

58. The micrometer-head of the Greenwich Transit Circle is *eastward*, and the instrument is not reversed. Hence for a star above pole the sign of the collimation-correction is always the same as that of the algebraic excess of the reading for collimation-plane above that for the position assigned to the wire-frame (see Art. 56). In the case of the Cambridge Transit Circle the micrometer-head is always *westward*, and consequently the signs of the collimation-corrections, under the same circumstances, are opposite to those for the Greenwich instrument.

59. It is now required to express the amount of the error, as obtained by the foregoing processes, in terms of the arc of a great circle ; that is, *to determine the value in arc of one revolution of the micrometer.* For this purpose it will be necessary to anticipate what has to be said under *Division III.* so far as to assume that the observer is able to record with great precision the clock-time of the instant of the passage of a star across the micrometer-wire. In Figure 14, P is the pole of the heavens,

Fig. 14.

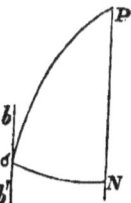

and PN is a circle of north polar distance in or very near the meridian, and coincident with the micrometer-wire. A transit of Polaris (the pole star) is taken at the micrometer-wire in this position, and after moving the wire to the position $b\sigma b'$ by turning the micrometer-head through an integral number (n) of revolutions, the instant of transit at σ is in like manner taken. (Any other star the north polar distance of which is approximately known and sufficiently small to allow by its slow movement of time for this operation, might be selected.) $P\sigma$ is the star's polar distance, and σN is an arc of a great circle at right angles to PN. Let x be the value of one revolution of the micrometer in seconds of space. Then $\sigma N = nx$. Hence if $P\sigma = \delta$, and the angle $\sigma PN = 15t$, t being the interval between the transits in seconds of time, we have by the right-angled spherical triangle σPN, $\sin nx = \sin \delta \sin 15t$. Let R be put for 206265″, the number of seconds in an arc equal to radius. Then, nx being small, $R \sin nx = nx$ nearly, and hence

$$x = \frac{R}{n} \sin \delta \sin 15t.$$

If the arc $15t$ be not larger than it is required to be for making an exact determination, we should have $R \sin 15t = 15t$ nearly, and the formula becomes

$$x = \frac{15t}{n} \sin \delta,$$

which would generally be sufficiently approximate. The value of δ, if not known by a contemporaneous observation of the star's polar distance, might for certain stars be taken from the Table of Apparent Places of circumpolar stars in the Nautical Almanac. The adopted value of x, if used for arcs exceeding that for one micrometer-revolution, should be obtained from a formula in which n is a considerable integer[1].

60. The foregoing processes only determine the error of collimation of the middle wire, whereas the transits are generally taken at all the wires. We must, therefore, find the collimation-correction for the position corresponding to the mean of all the clock-times of transit, which is, in fact, to take into account the collimation error of each of the wires. This mean correction may be inferred from that for the middle wire as follows. Recurring to Figure 14, suppose PN to be coincident with the mid-wire, $b\sigma b'$ now to be the position of any other wire, and transits of Polaris, or some other known star near the pole, to be taken at both wires. Then, σ being the point of transit of $b\sigma b'$ by the star, the arc σN of a great circle perpendicular to PN measures the horizontal, or equatorial, interval between the wires, because it is the projection on the heavens of the rectilinear perpendicular let fall from the point σ on the mid-wire. Hence putting $15a$ for the arc σN, δ for the star's polar distance $P\sigma$, and t for the interval between the transits in seconds of time, we have by Spherical Trigonometry,

$$\sin 15a = \sin \delta \sin 15t,$$

from which formula a, the interval between the wires in seconds

---

[1] The value $17''{,}06$, used in all instances of the reduction of micrometer-measures taken with the Cambridge transit instrument, was adopted by Mr Airy in 1829, and subsequently proved by myself to be sufficiently exact. In no case is it used for arcs that are not very small.

of time for an equatorial star, may be calculated. The same calculation is required for each of the wires, and as accuracy in these determinations is important, they should be mean results from repeated observations.

61. Naming the wires $A, B, C, D, E, F, G,$ and supposing a star above pole to cross them in this order when the micrometer-head is *westward*, let $-a, -b, -c,$ be the calculated equatorial intervals of the first three, and e, f, g those of the last three, from the middle wire $D,$ and let x be the unknown interval of D from the mean of all. Then

$$(x-a)+(x-b)+(x-c)+x+(x+e)+(x+f)+(x+g)=0;$$

$$\therefore x = \frac{1}{7}(a+b+c-e-f-g).$$

According as this result is *positive* or *negative*, the star passes the mean of the wires *after* or *before* passing the mid-wire, and the sign of the correction to be added to the collimation-correction for mid-wire is *plus* or *minus*. For the transit of a star below pole the signs have to be changed; and when the micrometer-head is *eastward*, the sign is the opposite to that under the same circumstances for micrometer-head westward.

After determining the intervals $-a, -b, -c, e, f, g$ by a large number of observations, and calculating the value of $x,$ the equatorial intervals $(x-a), (x-b),$ &c. of the several wires from the mean of all may be readily calculated.

62. I proceed next to give an instance of the numerical calculation of the correction of error of collimation for the mean of the wires, taking, first, the case of a determination by employing a North mark and a South mark, and reversing the instrument (see Arts. 51, 52, 53). The selected instance was actually made use of in the reduction of some observations taken with the Cambridge transit instrument. It is to be understood that each of the subjoined micrometer-readings for bisection of a mark consists of an integral number of complete revolutions of the micrometer (signified by the letter r), together with decimal parts of one revolution (see Art. 31), and

4—2

52 PRACTICAL ASTRONOMY.

that it is the mean of the recorded readings for, at least, six bisections.

	Bisection of South Mark.	Bisection of North Mark.
	r.	r.
Micrometer-head eastward	25,784	17,948
Micrometer-head westward	22,626	30,495
Mean of each set of two readings	24,205	24,221
Mean of the four readings, or concluded reading for plane of collimation }		24,213
Micrometer-reading for coincidence with D		24,156
Difference		0,057

Hence as $1^r = 17'',06$, the error of collimation of $D = 0,057 \times 17'',06 = 0'',97$. As the micrometer-readings increase towards the micrometer-head, the true plane of collimation after the reversion was more *westward* than D, and the images of stars passed D after passing the plane of collimation. Hence, micrometer-head westward, the collimation-correction for D was $-0'',97$. Also it was known from previous calculation that at the time the mean of the wires was more westward than D by $0'',51$, or that the value of x (Art. 61) was $+0'',51$. Hence by the rules in Art. 61, the collimation-correction for the mean of the wires $= -0'',97 + 0'',51 = -0'',46$, for transits after the reversion, and $+0'',46$ for transits before the reversion.

63. It is usual to include in the correction of the error of collimation that of the small error due to *diurnal aberration*. The effect of this aberration is to throw the apparent places of celestial objects in the direction of the earth's rotation, and therefore eastward. The transits of stars above pole being thereby delayed, the sign of the correction for such positions is *minus*. The amount of the correction may be inferred from that for the annual aberration by multiplying the latter by the ratio of the velocity of the observer due to the earth's rotation to the mean velocity of the earth in its orbit. Let R be the mean radius of the earth's orbit, a the earth's equatorial radius, r the geocentric radius of the place of observation, λ the

geocentric latitude, and n the number of sidereal days in the sidereal year. Then that ratio is $\frac{nr}{R}\cos\lambda$, or $\frac{na}{R}\cdot\frac{r}{a}\cos\lambda$. Taking 8″,95 to be the Sun's Equatorial Horizontal Parallax, this being the value most recently obtained, it will be found that $\frac{na}{R}$ is the number whose logarithm is 8,20118, or, as often expressed for brevity, [8,20118]. Hence adopting 20″,445 for the constant of annual aberration, as determined by Struve, the general expression for the correction for diurnal aberration is — [9,51176] $\frac{r}{a}\cos\lambda$. For the Cambridge Observatory $\frac{r}{a}=$ [9,99909] and $\lambda = 52°.2'$. Hence the correction for that Observatory, or any place having the same latitude, is $- 0'',20$.

64. The method of determining collimation-error by two collimators, one to the North and the other to the South, without reversing the instrument, is virtually to make use of two marks at an infinite distance, and to secure that the axes of the beams by which they are seen in the transit telescope shall be strictly parallel, and approximately in the same straight line (see Arts. 54 and 55, and Fig. 13). In this case the micrometer-reading for plane of collimation is simply the mean between the two readings for bisection of the marks, whether by a micrometer-wire or the mid-wire, and the amount and sign of the collimation-correction for the mid-wire are found by the processes explained in Arts. 55—58. Then the additional correction for the mean of the wires might be determined precisely in the manner stated in Arts. 60 and 61. According, however, to present practice with the Greenwich and Cambridge Transit-circles, this last operation is not performed, the transit-times at the several wires being in all cases reduced to times of transit across the middle wire.

65. The foregoing methods of determining collimation-error are such as could hardly be put in practice except at well-appointed fixed observatories. There are, besides, at such observatories means of obtaining exact determinations of the error of collimation by making use of the collimating eye-piece, but as these are methods requiring independent determinations

of the Level error, they cannot be fully treated of till the correction of that error is under consideration. I propose, therefore, at present only to indicate a process which might be available for a portable, or temporarily mounted, transit instrument, or alt-azimuth instrument. In this method the instrument is, first, to be placed so that its plane of collimation is approximately coincident with the plane of the meridian. This might be done in the case of the transit instrument, possibly with sufficient accuracy, by using a spirit-level for horizontality of the axis of motion, and determining the direction of the meridian by a compass, or by the process described in Art. 10. Instead of this roughly approximate, but independent, method, if the Nautical Almanac and a timed chronometer were at hand, the time at which Polaris would be on the meridian might be ascertained, and by pointing the telescope to the star at that time, the axis being levelled, the instrument would be placed in the required position. For an alt-azimuth the process would be to adjust nearly the vertical and horizontal axes of motion by spirit-levels, and then determine the meridian direction by equal altitudes of the same star before and after its passage across the meridian.

66. The instrument having been thus put in position with sufficient exactness, transits of Polaris, or some other slow-moving star, are to be taken in the following manner. In Figure 15, $ADD'A'$ is the star's path *above* pole, as seen in the

Fig. 15.

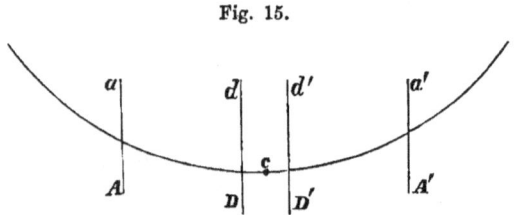

field of view, Aa is one of the transit wires, and Dd is the middle wire. Transits are taken at Aa and Dd, after which the instrument is reversed. After the reversion, whereby Aa

comes to $A'a'$ and Dd to $D'd'$, a transit is taken at $A'a'$. Let θ be the time-interval from A to D, and θ' that from D to A'. Then, according to the figure, $\theta' - \theta$ is the time of describing DD', and if δ be the star's polar distance,

$$\text{half the small arc } DD' = + \frac{15 (\theta' - \theta)}{2} \sin \delta,$$

which is the *correction* of collimation of Dd. If the interval θ be greater than θ', the positions of Dd and $D'd'$ would be reversed, and the sign of the collimation-correction, as also that of the above formula, would be changed. When the star is *below* pole, the apparent path is in the opposite direction, and the sign of the formula is changed on this account. Hence for the position the telescope had before the reversion, the general expression for the collimation-correction of Dd is

$$\pm \frac{15 (\theta' - \theta)}{2} \sin \delta,$$

the + or − sign applying according as the star was above or below pole. For the position after the reversion the same formula is to be used, with the signs ± changed.

Mechanical correction of Level Error by the Spirit-level.

67. Level Error, which may be defined to be deviation of an axis of motion from horizontality, may be corrected either by means of a plumb-line, or a Spirit-level. The former method will eventually come under consideration, but since, as respects the transit instrument, the other is most commonly used, it will suffice to give here only an account of the construction of the *Spirit-level*, and the mode of using it. It consists mainly of a cylindrical piece of glass in which is formed a uniform bore, nearly filled with spirit or ether, and closed at both ends. The axis of the bore is slightly curved, so as to be a small part of a circle of large diameter. There exist mechanical means of satisfying with great exactness the conditions required for trustworthy indications of the Level, namely, that the transverse section of the bore be uniform, and its axis be of uniform curvature in one plane. The sensibility of the Level is inversely proportional to the curvature of the bore, and it is, therefore, desirable to obtain a measure of the amount of this curvature. The following process gives the means of doing this. In Figure 16, O is the centre of curvature of the upper boundary

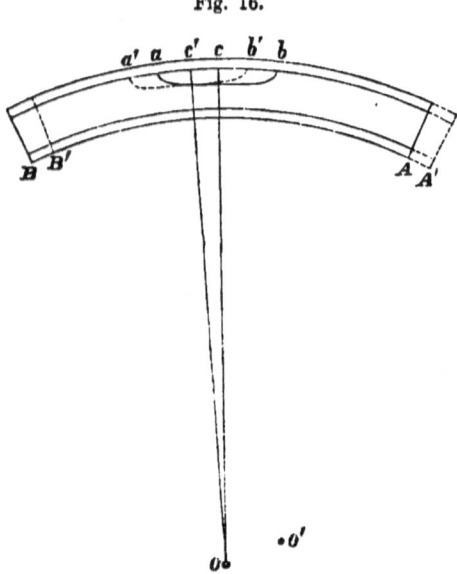

Fig. 16.

of a section of the bore by a vertical plane containing the axis, and Oc drawn vertically upwards passes through the central point of the bubble ab, which, as being the space not filled with the fluid, occupies the uppermost part of the bore. The tube being supposed to receive a small angular displacement in the direction from A towards B, whereby the bubble is made to pass from the position $a'b'$ relative to the bore to the position ab, let Oc' be drawn bisecting the original interval between the ends a' and b' of the bubble. The motion may be assumed to take place wholly about the centre of curvature O, because if that point were carried by the disturbance to some position O', it may be conceived to be brought back to O by a motion of translation, without change of the place of the bubble in the tube. From these explanations it will be seen that

$$a'a = b'b = c'c,$$

and that if $Oc = r$ and the $\angle cOc' = \alpha$, we have $r\alpha = cc'$. Hence r may be calculated if α and cc' be given. Taking, for example, the Spirit-level of the Cambridge transit instrument, let cc' be the interval between the scale divisions, which is one-tenth of an inch. Then, the corresponding value of α in arc having been ascertained to be $1'',3$ (by means that will be subsequently stated), we have for calculating the value of r,

$$r = \frac{0^{in.},1 \times 206265}{1,3},$$

which will be found to be 1322 feet. This length, the inverse of which measures the curvature of the axis of the bore, indicates a high degree of sensibility of the Level[1].

68. In order that a Spirit-level may be applied for measuring small inclinations to a horizontal plane, it is necessary that the position of the bubble which corresponds to horizontality should either be actually marked, or should be implicitly determined. This may be called the *adjustment of the Level*. Let us, at first, suppose the Level to have an adjusting screw and marks, but no scale; which is the case of most of the Spirit-levels pertaining to small or auxiliary instruments. In Figure 17 (1) and (2), such a Level is represented in reverse positions, with marks on the glass tube and an adjusting screw at S. In (1) the Level is applied to the plane db, m and n are the bubble-ends, after the bubble is brought by turning the screw (which alters only the inclination of the glass) to be, according to the judgment of the eye, in mid-position relatively to the marks, and ac is a straight line

---

[1] In the observations made by W. Struve for determining the constant of aberration with a transit instrument in the prime vertical, a Spirit-level was used, the sensibility of which was measured by a radius of 1527 feet, the scale-interval being one French ligne (= 0,0888 in.) and its value in arc not sensibly differing from $1''$. (See *Astronomische Nachrichten*, 1844, No. 468, col. 202.)

drawn parallel to the tangent to the bubble at its lowest point, and therefore horizontal. Conceiving the line ac to be fixed, relative to the brass case containing the glass, we may take ad

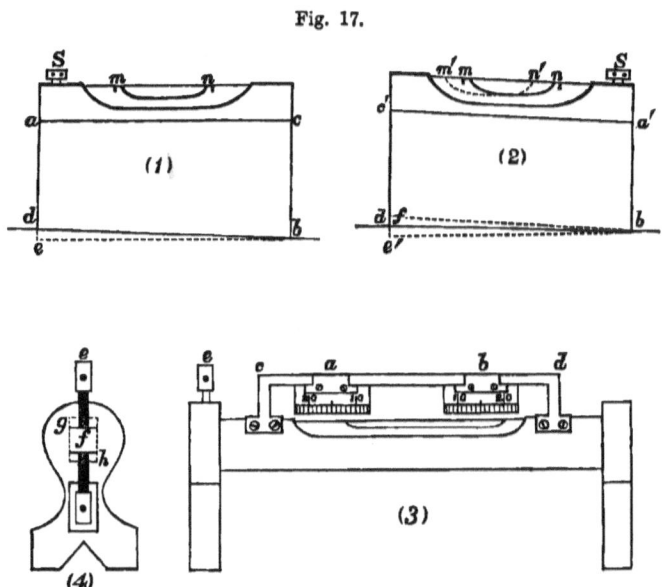

Fig. 17.

and cb to be legs supporting the Level; and drawing be horizontally, the angle dbe of inclination of the plane to the horizon, which the Level is required to indicate, will be subtended by de the excess of cb above ad. Now let the Level be reversed. Then, as shewn in (2), on applying it to the same plane db, ac comes to $c'a'$, ad to $a'b$, cb to $c'd$, and the bubble-ends move up to the positions m' and n'. This movement of the bubble, being due to the angular change of position of ac, measures the inclination of $a'c'$ to the horizon. Draw be' horizontal, and bf parallel, to $a'c'$. Then $fd = c'd - a'b = cb - ad = de = de'$. Consequently the angle fbe', which is equal to the inclination of $c'a'$ to the horizon, is *double* the ∠ dbe' of inclination of the plane. Hence if the bubble be brought *half-way* back by turning the screw, its deviation from mid-position will measure an inclination to the horizon equal to that of the plane, and will, therefore, indicate truly. If after this adjustment the Level be applied to a plane in reverse positions, the places of mark and bubble-end, both on the right-hand and on the left-hand, should be the same after as before the reversion. If this condition be not satisfactorily fulfilled, the operation for adjusting the Level must be repeated. It is evident that after completing the adjustment, a plane is shewn to be horizontal if, when the Level is applied to it, the bubble stands in a central position relatively to the marks.

69. We have now to take the case of a Level which, instead of having marks, has scales of equal parts so placed that the positions of the bubble-ends may be read off in scale-intervals. Figure 17 (3) and (4) will answer the purpose of describing the construction and mode of using such a Level, although the drawings were not made to represent any particular instance. There are two ways of arranging the scales for recording positions of the bubble-ends: (1) the graduation is along the edge of a single ivory plate fastened to the brass case of the Level in such manner that the edge is parallel and close to the axis of symmetry of the part of the surface of the glass tube which is left uncovered for seeing the bubble. In this arrangement the graduation-readings increase alike in both directions from a fixed zero which is situated at, or near, the centre of the aperture. This is the case of most Levels not of large size. (2) Two scales a and b are graduated, as represented in Figure 17 (3), along their edges, and can be shifted by hand on the bar cd, which is fastened to the brass covering of the Level. According to this arrangement, which allows of putting the scales on each occasion of levelling in positions suitable for reading off, a small number of graduation-intervals suffices. On each scale the graduation-numbers increase in the direction from the middle of the bubble towards the end. In the cases of both arrangements the graduation on one side of the ivory is the counterpart of that on the other, for the sake of reading off

alike in reverse positions of the Level without change of position of the observer.

70. Taking first the case of a *fixed* graduation-scale, let x and y be the scale-readings for the bubble-ends before inverting the Level, and let x' and y' be the readings in the same directions as x and y respectively after the inversion. Then if h be the scale-measure of the angle made by the plane to which the Level is applied with a horizontal plane, and the index errors for the arbitrarily selected positions of the scales be a and b, we shall have from the readings before and after the inversion respectively,

$$2h = (a + x) - (b + y) \text{ and } 2h = (b + x') - (a + y'),$$

the reading for one bubble-end being as much increased or diminished by a change of level as that for the other is diminished or increased, and the index errors being the same for the same halves of the scale in both positions of the Level. By adding the two equations a and b are eliminated, and the result is

$$h = \frac{1}{2}\left(\frac{x + x'}{2} - \frac{y + y'}{2}\right).$$

Hence, as the bubble runs towards more elevated parts, according as h is positive or negative, the plane is inclined upwards or downwards in the direction in which the x's are measured. If by turning the adjusting screws (in the manner indicated below in Art. 71) the reading for the bubble-end in the x-direction be made $\frac{x + x'}{2}$, the reading for that in the y-direction becomes $\frac{y + y'}{2}$; and accordingly the Level is adjusted so that on applying it to any plane, half the excess of one reading above the other measures the angle of elevation of the plane on the side of the greater reading, and the plane would be shewn to be horizontal by equal readings.

71. Figure 17 (4), which represents a vertical section of that support of the Level in which the screw e works, shews the

apparatus by which a change of inclination of the glass tube is effected. A piece of brass f, in close contact with the vertical sides of the concealed aperture gh, is attached to one end of the glass tube, and is held in position by the screw e, and an antagonist screw, the capstan-head of which is situated within a rectangular aperture cut through the brass support. It is evident that by this contrivance, when one screw is turned to take off its bearing, by the action of the other the piece f, together with the connected end of the tube, may be either elevated or depressed. After the bubble has been brought by these operations into the position proper for the adjustment of the Level (see Art. 70), both screws should be made to bear on f.—The same figure exhibits also the form of the Y at the foot of each support of a Level of this kind, which is called a *striding* Level.

72. When there are two *movable* scales, after taking, as before, bubble-end readings in reverse positions of the Level, we can, *by shifting the scales*, make the reading in the x-direction equal to $\frac{x + x'}{2}$, and that in the y-direction equal to $\frac{y + y'}{2}$. Then so long as the scales retain the positions thus given to them, the level error might be inferred from scale-readings for the bubble-ends just as in the case of the adjusted fixed scale considered in Art. 70. Suppose now the Y's at the feet of the Level (Art. 71) to be applied to the cylindrical pivots of a transit instrument, and from bubble-end readings thus obtained, let the positions of the scales be altered in the manner just indicated. Then by turning the screws provided for correcting the Level error of the instrument (see Art. 46), the readings for the bubble-ends may be made equal; which being done, the Level error would be, at least approximately, corrected. It might, however, happen that after these operations the bubble would not be in mid-position relatively to the aperture in the brass; and since it is desirable, as regards the process of levelling, that this condition should be nearly fulfilled, the adjusting screws of the *Level* have to be turned for this purpose.

After these *initial* approximate adjustments both of the Level and of the axis of motion of the transit instrument, there is no need, so long as the Level error continues of small amount, to take into consideration index errors of the scales, provided always the Level error be calculated, according to the formula for h given in Art. 70, from readings taken in reverse positions of the Level. After any pair of such readings, the places of the scales might be altered *ad libitum* for the sake of convenience in reading off.

73. On the supposition that the pivots of a transit instrument are exactly cylindrical and equal, the process for determining the Level error by a Spirit-level is unaffected by any small inequality either of the angles of the Y's of the piers, or of the angles of the Y's of the Level. For if the former angles are unequal, the axis of motion will still have a fixed position, the same after as before a reversion, and its inclination to a horizontal plane is determined by the usual mode of applying the Level; and if the angles of the Y's of the Level are unequal, the displacement of the bubble due to that circumstance in either the x or the y direction before the inversion of the Level will be just equal, and, as respects the graduation, opposite to the displacement after inversion. Hence the sums $x + x'$ and $y + y'$, and consequently the calculation of Level error, will be the same as if the inequality did not exist. This reasoning shews the necessity of always determining Level error by readings taken in reverse positions of the Level, as the angles of its Y's can only be supposed to be approximately equal.

74. But it is possible that the determination may be affected by *inequality in the size of the cylindrical pivots*, and to make the logic of the method complete, it is required to indicate means of correcting any error that may be due to this cause. The purpose of the following reasoning is to obtain a formula for calculating such correction. In Figure 18, $ABB'A'$ is a transverse section of either cylindrical pivot, cutting its axis in O; the $\angle BHA$ $(= 2\alpha)$ is the angle of the Y of the Level, and the $\angle B'H'A'$ $(= 2\alpha')$ is the angle of the Y of the pier (both supposed to be known by actual measurement), and a common vertical

line HOH' bisects both angles. Hence, HA being a tangent to the circle at A, $OH = OA \operatorname{cosec} \alpha$. So also $OH' = OA \operatorname{cosec} \alpha'$.

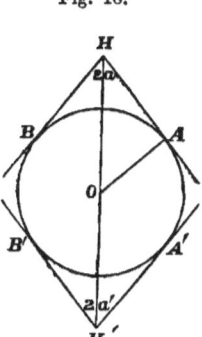

Fig. 18.

Let $\delta.OA$ be the difference of the radii of the pivots, $\delta.OH'$ the consequent difference of elevation of the axes of the pivots, and $\delta.OH$ the consequent difference of elevation of the corresponding ends of the Level above the common axis of the pivots. Then,

$$\frac{\delta.OH}{\delta.OH'} = \frac{\delta.AO \operatorname{cosec} \alpha}{\delta.AO \operatorname{cosec} \alpha'} = \frac{\operatorname{cosec} \alpha}{\operatorname{cosec} \alpha'} = k,$$

k being a known quantity because α and α' are known. Next let $x + y$ be the true Level error of the axis of the transit instrument in scale-measure before reversing it, y being the part due to inequality of the pivots. Then as only this part changes sign by the reversion (Art. 73), the true Level error after the reversion is $x - y$. Now $\delta.OH'$, as expressed in scale-measure, is the portion of Level error due to inequality of the pivots, and is therefore equal to y. Hence $\delta.OH$, expressed in the same measure, is ky. By this quantity the effect of the inequality of the pivots on the inclination of the axis is *apparently increased*, because of the simultaneous effect in the same direction on the inclination of the Level. Let, therefore, m and n be apparent Level errors obtained, by calculating according to the formula for

h in Art. 70, from indications of the Spirit-level in inverse positions recorded before and after the reversion of the instrument. Then, since $x+y$ and $x-y$ are the true Level errors before and after the reversion, we shall have

$$m = x + y + ky, \qquad n = x - y - ky.$$

These equations give for the value of ky, $\dfrac{k(m-n)}{2(1+k)}$, which is the formula for calculating the correction for inequality of the pivots, inasmuch as ky subtracted from m gives the true level error before the reversion, and added to n gives the same after the reversion. As the pivots are made by mechanical construction as nearly equal as possible, this correction will at first be of very small amount, and will afterwards only change slowly by unequal wear of the pivots. It should, therefore, be determined by repeated trials, levellings being taken for that purpose before and after each reversion, and a value of the correction (apart from its sign) should be calculated from each trial. The adopted value may then be the mean of several determinations. The proper sign of the correction, as respects the transit observations, will presently come under consideration.

75. As the formula for h in Art. 70 gives the Level error in scale-intervals, it is necessary, in order to express this quantity in celestial arc, to find the value in arc of the unit of the scale. In an observatory furnished with a Mural Circle, the readiest method of doing this is to attach the spirit-level firmly to the top of the circle, and after noting the scale-readings of the bubble-ends in two positions of the circle separated by an arc ascertained by microscopic reading of the graduation (as hereafter described) in each position, to compare this arc with the difference of the Level readings. The value of the scale-interval may then be calculated by a simple proportion. (In this manner Mr Airy obtained on October 10, 1834, the value $1''{,}3$ mentioned in Art. 67, having previously adopted $1''{,}0$ on the authority of the maker of the Level.) It is evident that with the same means the regularity of the indications of a Level might be tested by comparisons of its readings with a series of corresponding circle readings differing by a constant arc.

76. If these means should not be available, the required information may be obtained, although less perfectly, by an instrument called a *Bubble-trier*, constructed in the following manner. A brass bar, resting at one end on two legs at the extremities of a short cross-piece forming with it a T-shaped frame, is tapped at the other end for a screw to work in, which serves for a third leg. The intermediate part is furnished with two Y-supports on which the glass tube with its attached scale is placed. When the instrument rests with its three feet on a horizontal plane, the screw by being turned alters the elevation of the end in which it works, and consequently the inclination of the tube to the plane and the place of the bubble. Also to the stem of the screw is attached a graduated circular disk, the angular movement of which is read off from the graduation by the vertical edge of a brass index contiguous to the circumference of the disk and fixed to the bar. The purpose of the instrument is to compare arbitrary equal angular movements of the disk as indicated by the graduation, and by consequence *equal* differences of inclination of the Level, with the corresponding differences of scale-readings for the bubble-ends. If the latter differences be equal as well as the other, the performance of the Level is satisfactory. The maker of the instrument also arranges the graduation of the disk, with reference to the ratio of the interval between the turns of the screw to the length of the bar, in such manner that differences of readings from the disk give the actual differences in arc of the inclination of the Level. It is evident that by this apparatus the value of the scale-interval in arc may also be determined.

77. The *sign* of the correction for inequality of the pivots is determined by the following considerations. In the expression $\frac{k(m-n)}{2(1+k)}$ for ky obtained in Art. 74, m and n are the values of the level-correction h before and after the reversion of the instrument (as obtained by the process exhibited by the example given in the next article), and the difference between m and n depends only on the inequality of the pivots. According as m

is greater or less than n, the plane of collimation is more *eastward* before than after the reversion, and the pivot that was westward before the reversion is *greater* or *less* than the other, inasmuch as the eastward deviation of that plane is greater the greater the westward pivot. This rule determines which is the greater pivot. For instance, as respects the Cambridge instrument, it was found by many trials made before and after the date of the subjoined example, that the smaller pivot is that at the end perforated for the illumination apparatus. When therefore this end is *eastward*, the inequality *increases* the apparent amount of the level-correction (h) partly by increasing the inclination of the instrument and partly by increasing the inclination of the Level. The latter part only requires to be corrected, which is done by calculating its amount according to the expression for ky above, from data obtained at several reversions of the instrument, and applying an adopted mean value with a *negative* sign (see Art. 74). When the smaller pivot is westward, the correction is the same with contrary sign. Thus for the Cambridge instrument the correction for inequality of the pivots is *negative* or *positive* according as the illumination-end of the axis is *eastward* or *westward*.

78. I propose to give an example of the calculation of level-error from data furnished by the spirit-level, after making the following preliminary statements. First, the Level has a small cross-level attached to it near one end, with marks on the glass for putting its bubble in mid-position. This is done before every scale-reading of the large Level, in order that the plane of the axis of its bore, if not exactly vertical, may deviate from verticality always to the same amount, that thus the readings may be all consistent with each other. [The same use should be made of the cross-level in determining the value of the scale-interval by the method described in Art. 75.] Also this cross-level gives the means of indicating the position of the large Level as to East and West. Again, it is usual, for greater certainty, to infer the level error from several sets of double levellings, and accordingly, as may be seen from the expression

for h in Art. 70, if there be n such sets, the mean value h is given by the equation

$$h = \frac{\Sigma.x - \Sigma.y}{4n}.$$

If we now assume, regard being had (as in Art. 57) to the effect of the inclination of the axis on the time of a transit above pole, that this angle is *positive* or *negative* according as the *West* or *East* end of the axis is highest, the x-readings will pertain to the *west* end of the bubble, because the bubble rises towards higher parts, and the sign of h will be that proper for *correcting* the effect of the error. After these explanations the subjoined instance of calculating the correction of level error from data extracted from the records of the Cambridge Observatory will be easily understood.

1850, Oct. 21, 2^h. Illumination-end of the axis Eastward.

		East reading. $d.$	West reading. $d.$
Cross Level	E.	9,9	11,2
,,	W.	8,3	13,0
,,	E.	10,5	11,3
,,	W.	8,1	13,0
,,	E.	10,1	11,3
,,	W.	7,9	13,0
		$\Sigma.y = 54,8$	$\Sigma.x = 72,8$

Hence $h = \dfrac{\Sigma.x - \Sigma.y}{12} = 1^d,50 = 1'',95$, because $1^d = 1'',3$.

The correction is thus shewn to be *positive*, the sum of the West readings being greater than the sum of the East readings. By a reversion of the instrument on July 16, 1850, the amount of correction for inequality of the pivots was found to be $0'',408$, and by a reversion on May 6, 1851, $0'',491$. Adopting the mean of these for application to the levelling on Oct. 21, 1850, applying it with a negative sign because the illumination-end was eastward, and including the correction $- 0'',20$ for diurnal aberration, the concluded level-correction is

$$+ 1'',95 - 0'',45 - 0'',20 = + 1'',30.$$

79. In order further to ascertain the character of this correction for inequality of the pivots, I subjoin here the values obtained for it in the use of the Cambridge transit instrument from 1828 to 1854. In every instance the correction is given for the case of illumination-end eastward.

Year.	Correction. "	Year.	Correction. "	Year.	Correction. "
1828	+ 0,18	1843	− 0,33 ?	1849	− 0,48
1829	+ 0,13	1844	− 0,28	1850	− 0,53
1837	− 0,04	1845	− 0,26	1851	− 0,49
1839	− 0,10	1846	− 0,22	1852	− 0,68
1842	− 0,19	1847	− 0,22	1854	− 0,62

The first two values were obtained by my predecessor Mr Airy, the others by myself. That opposite the year 1843 was considered doubtful, as being the mean of three discordant values. The continuous change of value is most probably to be ascribed to dissimilar wear of the pivots due partly to turning the instrument for observing, and partly to the application of the Y's of the level to the pivots in the operation of levelling, which was most commonly performed with the telescope horizontal and pointing southward. By levellings taken in 1838 and 1839 with the telescope inclined from the Zenith 45° northward or southward I found that the additional correction + 0",30 was required. These circumstances all point to the desirability of being able to calculate the level error by some method that shall be independent of the form of the pivots, and not require the use of the spirit-level. Such a method I put in practice at the beginning of 1850, making use also from that time of *the collimating eye-piece* for determinations both of collimation error and level error. The spirit-level was used as an auxiliary to the end of 1854, but the adopted level errors virtually rested on data furnished by the collimating eye-piece. After 1854 the use of the spirit-level was wholly discontinued. The above-mentioned method of making the transit-observations independent of the forms of the pivots cannot be fully described till the meridian or azimuth error has been under consideration.

It will also be previously necessary to describe the collimating eye-piece, and the method of using it for determinations of collimation and level errors, which I now proceed to do.

The Collimating Eye-piece.

80. This important auxiliary instrument, which enables the observer to obtain instrumental corrections exclusively by optical means, was the invention of Bohnenberger of Tübingen, who has given a description of it in the *Astronomische Nachrichten* (Band IV., 1826, col. 327—336.) My attention was first called to it by Henderson, late Astronomer Royal at the Cape of Good Hope, who brought me a specimen (made apparently according to the above-mentioned description), having a metallic reflector with a hole at the centre, through which the wires and their reflected images were looked at with a Ramsden eye-piece. On trial I found this construction to be extremely inconvenient on account of the limited field of view, and the small interval between the eye-glass and the wires, rendering it difficult to hold a lamp for throwing light upon the reflector. On mentioning these circumstances to the late William Simms, he constructed for me the instrument represented by Figures 19 and 20, in which a three-glass eye-piece is substituted for the Ramsden eye-piece, and for the metallic reflector a piece of plate glass, the reflection from which, as will presently be explained, gives the means of seeing the wires together with their reflected images with quite sufficient distinctness. By these changes the above-stated inconveniences were entirely removed. As far as I am aware the collimating eye-piece has since been uniformly made according to this pattern. I brought it into use in the Cambridge Observatory in the year 1850; the next year it was adopted at Greenwich, when the new Transit-circle was first made use of. It had already attracted the attention of Bessel, Gauss, and Lamont, but had not, I believe, been definitively employed for exact determinations relating to meridian observations with the transit instrument and mural circle, before I made such application of it at the Cambridge Observatory.

81. Figure 19 represents the collimating eye-piece detached from the telescope; d is the plate-glass reflector, seen through an aperture made in the tube to allow the light from an external lamp to fall on the reflector; m and n are two small milled heads, one for turning the reflector to the required angle of inclination, and the other for acting by a screw to fix it when put in position. Figure 20 exhibits a vertical section of the telescope and attached collimating eye-piece, made by a plane containing the common axis, and cutting at right angles the plane of the plate-glass reflector D. H is a lamp, the light of which, being incident horizontally on the reflector adjusted to an inclination of $45°$ to the axis of the telescope, is sent down as a divergent beam to the object-glass FOF, after transmission through which it is reflected at the surface of mercury in the trough GG. At a is a portion of a micrometer-wire situated at the geometrical focus of the object-glass, and movable by the micrometer-head M and opposing spring at N. Regarding, for the moment, the telescope as a collimator furnished with a wire at a, if there were another collimator pointing oppositely to this in a direction parallel to its axis, a *dark* image of the wire at a would be formed at the focus of the second collimator precisely in the manner explained in Art. 54. But by the reflection at the mercury the rays that go to form this image are turned into the contrary direction, so that the telescope becomes, in fact, its own collimator, and the image is formed at the position a'. O being the optical centre of the object-glass, Oa is the axis of the incident centrical pencil, Oa' that of the reflected centrical pencil, and by reason of the reflection Oa and Oa' make equal angles with the vertical direction through O. The wire at a and the image at a' are both seen distinctly with the collimating eye-piece, provided the wire be accurately at the geometrical focus of the object-glass (see next Art.). Figure 20 represent the form of a pencil by which a point of the *image* is seen, its course being, first, through the plate-glass D, then through the third lens C, the second lens B, and the first lens A, to the eye at E. This may be called a *negative* pencil, as originating in a *stoppage* of light from the lamp by

THE TRANSIT INSTRUMENT.

Fig. 19.

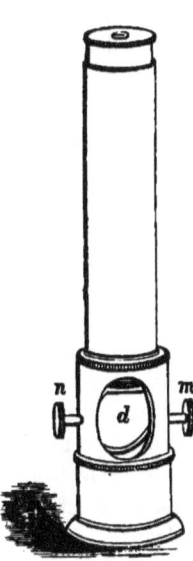

Fig. 20.

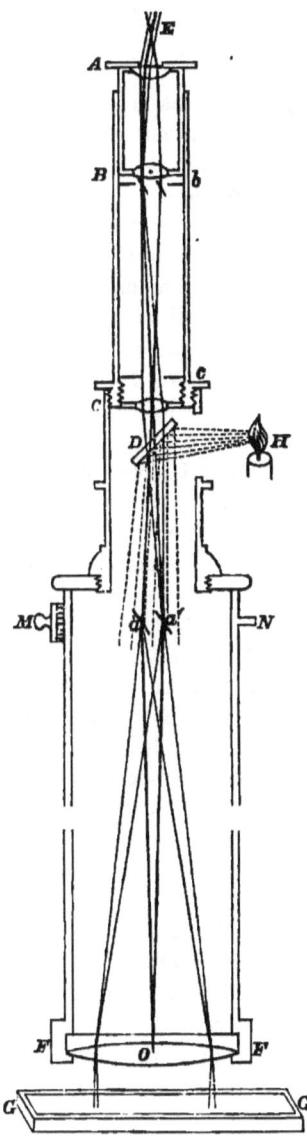

the wire at a. At the same time this wire, by reason of the incidence of the light upon it, might be expected to be seen by the eye at E by *positive* reflected pencils. This is found to be the case when the telescope is directed to the dark sky. (I have, in fact, taken advantage of this circumstance, by making use of the collimating eye-piece to illuminate wires in a dark field for taking transits of faint objects (see Art. 37). This twofold manner of viewing the wires has given occasion to misconception in some published accounts of the use of this instrument.) But when the telescope is directed to the trough of mercury, the moment the light reflected from the surface of the mercury illumines the field, the wires appear dark on a bright ground, because the very small amount of light that can be reflected from fine cylindrical wires is extinguished by the flood of the reflected light. Thus the wire at a and the image at a' are both viewed by *negative* pencils, and under favourable circumstances I have seen them so nearly alike, as to blackness and definition, that it was hard to distinguish one from the other. This, however, occurred with the telescope of the mural circle, but in no instance with that of the transit instrument, owing probably to some fault in the optical conditions of the lenses of the object-glass.

82. The optical effect produced by the collimating eye-piece may now be distinctly stated. In order that the Figure might not be too complicated, the negative pencil from a is represented only by its axis, which is exactly analogous to that of the negative pencil from a'. Both the wire at a and the image at a' are to be regarded as origins of pencils which by means of the lens at C, and the limiting diaphragm at c, form distinct images at b, where also there is a diaphragm. These images are viewed by the eye at E by means of the two lenses A and B of an adjustable Ramsden eye-piece. It hence appears that any movement of the wire at a by the micrometer-head M, and the consequent movement of the image at a', are seen by the eye at E, so that by turning the micrometer the wire and its image may be brought into coincidence. To effect this coincidence is the express purpose of the whole arrangement.

For ensuring accuracy the image should be brought into contact with the wire by just excluding the line of light between them an equal number of times on each side, and recording the micrometer-reading for each contact. Then the mean of all the readings is the reading for coincidence of the image with the wire. When the micrometer-wire is set to this reading, the plane passing through it and the optical centre of the object-glass is as nearly vertical as by the optical means it can be made to be. These explanations may suffice to shew how an essential astronomical element, namely, the direction of the vertical at the place of observation, formerly found by a plumb-line or a spirit-level, is determined by means of the collimating eye-piece.

83. As it is important that the wire and its image should be seen with equal distinctness, special means are provided for satisfying this condition. This is done, as already intimated by the reference in Art. 36 to Figure 6, by means of apparatus capable of moving, in the direction of the telescope-axis, the tube in which the frames of the fixed wires and the micrometer-wire are held, and fixing it in the position for which, according to the judgment of the eye, a wire and its image appear equally well defined. When this is the case the wires will have been placed at the geometrical focus of the object-glass by a method (alluded to in Art. 54) which is particularly adapted to this purpose, inasmuch as the deviations of the wire and its image from focus are equal in *opposite* directions, as is evident from optical considerations, and on that account any difference of definition is the more easily recognised by the eye-piece. From Figure 20 it is also to be seen that this is an *inverting* eye-piece, resembling in that respect the micrometer-microscope for reading off circle graduations.

84. The errors of collimation and level may be simultaneously obtained by measures taken with the collimating eye-piece in reverse positions of the instrument, as the following argument will shew. In Figure 21 O is the optical centre of the object-glass, the telescope pointing to the Nadir, and OZ is the vertical direction ascertained by making the micrometer-

wire and its image coincide by the process explained in Art. 82. Oc represents the plane of collimation, and OD is drawn from O to the position of the middle wire D before the reversion.

Fig. 21.

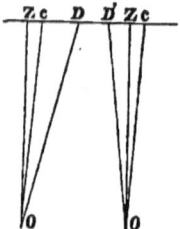

Hence the $\angle ZOc\ (=L)$ is the deviation of the plane of collimation from the vertical, and therefore the level error, and the $\angle cOD\ (=C)$ is the error of collimation of D. The $\angle ZOD\ (=\alpha)$ is measured by the difference of micrometer readings for the vertical direction and for coincidence with D. After the reversion, by which D is brought to D', the $\angle ZOD'\ (=\alpha')$ is similarly measured. Then since the angle cOD is, by the definition of collimation error, equal to the angle cOD', we have

$$\alpha = C + L \text{ and } \alpha' = C - L.$$

Hence $\qquad C = \dfrac{\alpha + \alpha'}{2}$ and $L = \dfrac{\alpha - \alpha'}{2}.$

Thus C and L are found by calculating according to these formulæ, the values of α and α' in micrometer revolutions being first converted into arc. For stars above pole, the sign of the level-correction is *positive* when the plane of collimation deviates from the vertical *eastward*, or the west end of the axis is highest, and, as already stated, the sign of the collimation-correction is positive when the pointing of the telescope is eastward from the plane of collimation. The position of the micrometer-head being given, the micrometer readings indicate the relative places of Z, c, and D.

85. If L be ascertained by the spirit-level, we shall have $C = a - L$, or $C = a' + L$; so that the collimation error might be determined without reversing the instrument. It is, however, to be remarked that these values, whether obtained by the spirit-level or the collimating eye-piece, would be affected by inequality of the pivots or irregularities of their forms. On this account, with prospective reference to a process (hereafter explained) for obviating this source of error in the Cambridge transit instrument, the level error was determined at each reversion by *both* methods, the telescope in every case of levelling being horizontal and pointing southward. As it was found that the two methods did not give the same results, from the differences accruing from comparisons made in a certain limited number of years, two mean differences were calculated from time to time, one applying to the case of illumination east, and the other to that of illumination west. By applying these differences to level errors found by the spirit-level, it was considered that the level error was virtually determined by the collimating eye-piece and reversing the instrument. This course was adopted because it would have been inconvenient to reverse the instrument on every occasion of finding the level error; and again, relatively to the proposed correction of the effects of irregularities of the pivots, it was essential that the level error should be uniformly determined by the collimating eye-piece. Accordingly this mode of calculating the level error of the Cambridge instrument was begun in 1850, when the collimating eye-piece was first used, and was continued to February 20, 1855; from which date the calculation was made on a different principle in the following manner. It was considered that as the collimation-correction varies but slowly, the same mean value of C might be used during a considerable interval, and the values of level error be calculated during each such interval by the formula $L = a - C$, or $L = C - a'$, the variations of the value of L being taken account of by frequent measures of a, or of a', in the same interval. In consequence of the adoption of the above-mentioned processes, all the corrections of collimation and level errors which I have employed in

the reduction of the Cambridge observations from the beginning of 1850 to the end of 1860 are in effect determinations by the collimating eye-piece.

86. The method of finding the errors of collimation and level which, as being found in practice to be the most exact, is adopted in well-appointed fixed observatories, is to obtain the collimation error by means of a north and a south collimator, as explained in Art. 54, and then to measure a or a' by the collimating eye-piece. The values of C, as well as those of a or a', might in this way be frequently determined without the trouble of reversing the instrument, and in using the formula $L = a - C$, or $L = C - a'$, instead of adopting mean values of C, the individual values would be used in order to take account of any sudden or irregular variations which the collimation error might be liable to. This process, as not requiring the instrument to be reversed, is constantly employed in the use of the transit-circles of the Greenwich and Cambridge Observatories. A method of calculating corrections for the forms of the pivots, applying alike to this mode of determining collimation and level errors, and to that described in Art. 85, will come under consideration after the discussion of meridian error.

87. The foregoing methods of finding collimation and level errors have not required astronomical observation, being conducted solely by instrumental means (with the exception of the process given in Art. 66 for obtaining the collimation error of a portable transit instrument by transits of Polaris). There is, however, a method of ascertaining the level error by observing transits of Polaris at meridian passage both directly, and by reflection at the surface of mercury, whereby good results are procurable without previously knowing either collimation or azimuth error. It is evident that this method, combined with the use either of north and south collimators, or of the collimating eye-piece, would give the level and collimation errors without reversing the instrument. But as it involves processes of observing and calculating which properly belong to *Division III.*, the account of the details of the method is deferred till the subjects of that Division are under treatment. At present we may

THE TRANSIT INSTRUMENT. 77

proceed to consider by what means meridian or azimuth error is calculated.

Calculation of Azimuth Error.

88. The azimuthal error is the deviation of the plane of collimation, corrected for level error, from the plane of the meridian, which is the plane that passes both through the Zenith of the place of observation and the Pole of the heavens. Consequently this error is necessarily determined by astronomical observation. For the purpose, however, of obtaining formulæ for calculating azimuth error, it suffices to take for granted that the observer is able to note with precision the time, according to an astronomical clock, of the transit of a star, whether near or distant from the Pole, across a wire of the telescope. In Figure 22, P is the pole of the heavens, Z is the

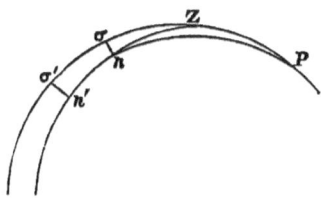

Fig. 22.

Zenith of the place of observation, Znn' the plane of collimation passing through the Zenith, error of level as well as that of collimation being supposed to be corrected, and σn is the path of a star between the plane of collimation and the plane of the meridian. Let $P\sigma = \delta$, $Z\sigma = z$, $\angle \sigma Zn = x =$ the azimuth error in seconds of space. Then

$$\sigma n = x \sin z, \ \angle \sigma Pn = \sigma n \operatorname{cosec} \delta = x \sin z \operatorname{cosec} \delta.$$

Hence $\angle \sigma Pn$ (in time) $= \dfrac{x}{15} \sin z \operatorname{cosec} \delta = hx,$

h being put for $\dfrac{1}{15} \sin z \operatorname{cosec} \delta$. Let t be the noted clock-time of transit of the star, corrected (according to formulæ that will be obtained under *Division III*.) for collimation and level errors, A the star's actual apparent Right Ascension, and T the excess of true sidereal time above the clock-time.

Then the true time of meridian transit $= t + T + hx = A$
The true time of transit of another star $= t' + T + \tau + h'x = A'$, τ being the clock's loss in the interval $t' - t$.

$$\text{From these two equations } x = \frac{A' - A + t - t' - \tau}{h' - h},$$

which is the general formula for the calculation of azimuth error. In order that it may give good results the stars should be selected so that the denominator $h' - h$ may have a large value. Hence one at least of the stars should be near the Pole; or, if both be near the Pole, h' and h must have different signs, which will be the case if one observation be taken above and the other below the Pole, because for a star below pole the polar distance (δ) is to be considered negative.

89. The general formula for azimuthal error may be safely applied either in its original form, or as modified by given conditions, in the four following cases.

(1) If one star be near the Pole and the other not far from the Equator, one of the factors h and h' will be large, and much larger than the other, so that $h' - h$ will not be small. In using the general formula for this case the Apparent Right Ascensions A and A' will have to be obtained by interpolation from catalogues of computed values, such as those for Upper Transits at Greenwich of a few stars very near the north or south pole, computed in the Nautical Almanac for every day of the year, and those for Upper Transits of a much larger number distant from the poles, computed for every tenth day. The Right Ascensions in the list of the Nautical Almanac severally depend on a very large number of transit observations taken at the Greenwich Observatory, and may safely be employed as data for calculating azimuthal error. For these reasons the apparent R. A. made use of for the calculation should be those of stars included in that list. The correction τ may be inferred from the clock's rate as determined by comparisons of the clock-

times of transit of the *same* star, or stars, on two or more consecutive days.

(2) If both stars be near the Pole, and their difference of Right Ascension be nearly 12^h, the transits may be taken, one above and the other below the Pole, nearly at the same time, and the correction τ would then be so small that it might be neglected. In this case $h' - h$ is large, because h and h' are both large and of contrary signs, and the general formula becomes

$$(h' - h) x = A' - A \pm 12^h + t - t',$$

the $+$ or $-$ sign applying according as A is greater or less than A'. The stars 51 (Hev.) Cephei and δ Ursæ Minoris are favourably situated for the application of this formula, and contemporaneous transits of these stars have not unfrequently been used in computing azimuth error for the reduction of Cambridge transit observations.

(3) If *two* transits of the *same* star near the Pole be taken, one above and the other below pole, $h' - h$ will be large, because for the transit below pole the polar distance is negative, so that h and h' will have different signs. Also $A' - A$ is the increment (ϵ) of the star's Right Ascension in the interval $t' - t$. Hence adding $\pm 12^h$ according as t' is greater or less than t, the formula for this case is

$$(h' - h) x = \pm 12^h + t - t' + \epsilon - \tau,$$

which is independent of the star's Right Ascension. The star being supposed to be in the list of the Nautical Almanac, the change (ϵ) of the R. A. in the interval $t' - t$ may be inferred from the computed apparent R. A. of the star at the times of transit.

(4) If *three* consecutive transits of a star near the Pole be taken, there will be another equation like the first, namely,

$$(h - h') x = \mp 12^h + t' - t'' + \epsilon - \tau,$$

it being assumed that the corrections ϵ and τ, for change of the star's R. A. and rate of the clock, are the same for the interval of 12^h between the second and third observations as those for the interval of 12^h between the first and second observations. Hence the two equations give

$$2\,(h'-h)\,x = \pm\,24^h + t + t'' - 2t',$$

the $+$ or $-$ sign to be used according as the sum of the first and third clock times is less or greater, by about 24^h, than twice the second. This result is independent both of the star's R. A., and of the change of its R. A., as well as of the clock's rate.

90. As these Lectures are intended to answer the twofold purpose of describing the best means at present available for taking exact astronomical observations, and at the same time indicating the fundamental principles on which the science of Practical Astronomy has been, or might be, raised, I take occasion to remark here, that if it were required to construct a Catalogue of the places of stars *ab initio*, the fourth of the above methods of calculating the azimuth error would be employed, inasmuch as it is independent of any results of antecedent observations.

91. Although, as we have seen, there is no necessity for *a meridian mark* in an Observatory furnished like that of Greenwich or Cambridge, still there are circumstances in the use of astronomical instruments in which a fixed distant mark is serviceable, if not necessary. I propose, therefore, to shew how to select a position for such a mark, and how, when the mark is set up, its distance in arc from the meridian may be ascertained. These determinations may be made without difficulty now that we are able by the foregoing methods to calculate the amount of the azimuth error. It is, in the first instance, requisite to correct this error *mechanically*. For this purpose the value in arc of one revolution of the screw by which the azimuth adjustment is effected must be found. (The screw is c or d in Figure 10, p. 38.) This may be done with sufficient accuracy by carefully measuring with compass and scale the interval, parallel to the axis of the screw, occupied by a certain integral number of its threads, and then measuring the length of the axis of motion intervening between the two middle transverse sections of the Y's. If h be the first measure em-

bracing n intervals, H the measured length of the axis, and R be put for 206265″, the value of one revolution in arc is $\frac{Rh}{nH}$. This quantity having been calculated, the correction of azimuth error is effected by turning the micrometer-heads (c and d in Figure 10) through the portion of a complete revolution corresponding to the amount of the error, and in the direction which, as indicated by its sign, is proper for correcting it. Let us now suppose that the micrometer-wire has already been put in the position of the mean of all the wires (see Art. 61), and that the micrometer reading for that position has been recorded; also that by adjusting the wire-frame the micrometer-wire in this position has been put into the plane of collimation (Arts. 48—50); lastly, that the level error has been mechanically corrected either by turning the screws a and b, fig. 9, p. 37, till the micrometer-wire and its image, as seen with the collimating eye-piece, coincide (see Art. 82), or by using the Spirit-level for the same purpose in the manner described in Art. 71. After these operations, if the telescope be pointed horizontally southward, the micrometer-wire will indicate the actual position of the plane of the meridian, and near this plane it is required to set up a firmly fixed mark, the image of which can be accurately bisected by the micrometer-wire. It may not be possible till after repeated trials to fix upon an appropriate mark, or place it in the most suitable position. When this has been effected, the mean of six or more micrometer-readings for bisection of the mark is to be recorded. Then the difference between this micrometer-reading and that for the position of the mean of all the wires measures the distance of the mark from the plane of the meridian. On account of uncertainties arising from atmospheric disturbances the adopted measure should be the mean result of several such determinations. This quantity being well ascertained, it is always possible afterwards, assuming the mark to be steady, to determine how far the plane of collimation is from the mark, and, by inference, how far it is from the meridian, and if the occasion required, to place it in the meridian. Also by bisections of the mark from time to time the variations of meridian error would be ascertainable, and any abrupt or excessive deviations from its ordinary value might be detected.

92. Having now shewn how to find the errors of collimation, level, and azimuth, we may proceed to determine the *time-correction* required to be applied to the time of transit across the mean of the wires to obtain that of transit across the meridian. Let a, b, c be respectively the three errors in seconds of space, inclusive of their proper signs, which, it should be remembered, have been given to them on the principle of taking each to be *positive* when it causes the observed time of transit of a star above pole to be *earlier* than meridian transit. Since by the law of coexistence of small errors, the value of each is the same as if the other two did not exist, the formulæ of calculation will be separately deduced by means of the diagrams (1), (2), (3) of Figure 23. In Fig. 23 (1), P is the pole of the heavens, Z the zenith, σ the point of transit of a star across the plane of collimation, which coincides with the plane of the meridian, because, by supposition, there is neither level nor azimuth error. The adopted line of collimation describes a small circle parallel to the meridian at the distance from it measured by the small equatorial arc Ee, which, therefore, is the collimation-error. The $\angle \sigma Pn$, subtended by the portion σn of the small circle described by the star, is the angular interval between the observed transit and meridian transit. Hence since $\sigma n = Ee = a$, if δ be the star's polar distance,

$$\angle \sigma Pn \text{ (in time)} = \frac{a}{15} \operatorname{cosec} \delta,$$

which, consequently, is the required correction for error of collimation.

In Fig. 23 (2), $HZ'h$ is the great circle described by the line of collimation supposing that there is no collimation or azimuth error, H and h being the points of its intersection with the meridian. The $\angle ZHZ'$, being the angular deviation of the plane of collimation from the plane of the meridian, is equal to

THE TRANSIT INSTRUMENT. 83

b the level-error. Hence if $P\sigma = \delta$ and $Z\sigma = z$, we have, since $\sigma n = b \cos z$ and $\angle \sigma P n = \sigma n \operatorname{cosec} \delta$,

$$\angle \sigma P n \text{ (in time)} = \frac{b}{15} \cos z \operatorname{cosec} \delta,$$

which is the correction for error of level.

Fig. 23.

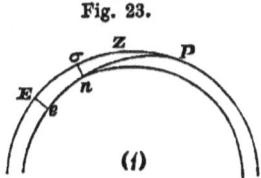

(1)

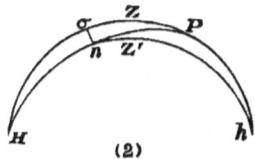

(2)

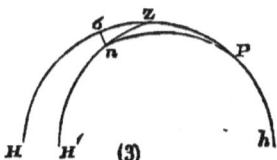

(3)

In Fig. 23 (3), HZH' is the azimuth error (c) supposing that there is no collimation or level error. Hence as $\sigma n = c \sin z$,

$$\angle \sigma P n \text{ (in time)} = \frac{c}{15} \sin z \operatorname{cosec} \delta.$$

6—2

Let τ be the whole reduction to the meridian. Then, since this is equal to the sum of the separate corrections, we obtain

$$\tau = (a + b \cos z + c \sin z)\,\frac{\operatorname{cosec} \delta}{15}.$$

93. This angle measures the inclination of the adopted line of collimation to the plane of the meridian for any position of the telescope, and might on that account be termed generally *the total error of collimation*, because when that line coincides with the plane of the meridian the instrumental error is zero. But the formula for calculating τ was obtained on the supposition that the pivots are equal and exactly cylindrical. If this be not the case, we may still call the deviation of the line of collimation from the plane of the meridian the total error of collimation; but the use of the transit instrument is not strictly logical unless we are able to calculate this error whatever be the forms of the pivots, or the changes of form they may undergo, assuming only that by reason of original construction they are always very approximately cylindrical and of the same size. This calculation I have succeeded in effecting by means of a practical solution of the following problem :—

A line being drawn to the optical centre of the object-glass from a selected point of the middle wire nearly mid-way between the two horizontal wires (called previously "the line of collimation"), it is required to find, whatever may happen to be the exact forms of the pivots, the small angle which this line makes with the plane of the meridian for any position of the telescope.

The solution rests on the principle of determining the position of the line of collimation by first finding by direct micrometer measures the position of another straight line fixed relatively to the instrument, and then after ascertaining the fixed geometrical relation between the two lines, to infer therefrom the position of the line of collimation. The subsidiary straight line may be conceived to be fixed relatively to the instrument by passing through two *dots* at the extremities of the pivots, and its position relatively to the piers to be ascertainable by means of micrometer-microscopes pointing towards the dots as exhibited in Figure 1 (page 21). This method was originally proposed by the Astronomer Royal, who applied it first in testing the forms of the pivots of the Alt-azimuth instrument of the Greenwich Observatory. I shall give here a description in detail of the mode in which I have employed it to obtain corrections for the forms of the pivots of the Cambridge transit instrument.

94. Two long micrometer-microscopes of considerable magnifying power, constructed expressly for the purpose by Mr William Simms, were fastened by screws in two rectangular cuttings at the tops of the piers, which were formerly occupied by the lever-counterpoises (see Art. 29). These microscopes were mounted so as to admit of a vertical adjustment, and of horizontal adjustments perpendicular to, and in the directions of, their axes, and thus could be made to point, according to the judgment of the eye, along the axis of motion of the instrument, and at the same time be placed at distances from the pivots proper for seeing the dots distinctly. In order to have the means of adjusting the positions of the dots, as well as fixing them relatively to the pivots, two *caps* were provided, which carried the dots, and fitted on to the ends of the pivots. Figure 24 represents one of the caps, and the mode of its

Fig. 24.

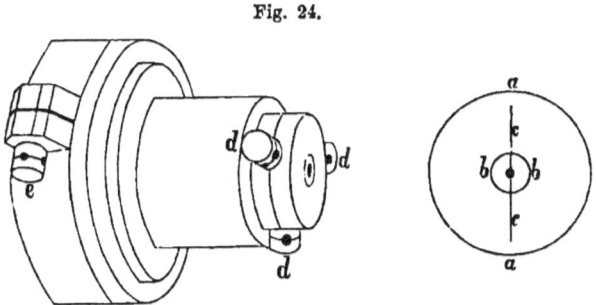

attachment to the pivot. By the capstan-headed screw at *e* the cap is firmly clamped to the pivot, and by the capstan-heads at d, d, d, pertaining to screws which work in a fixed cylinder covered by the movable cylindrical piece that carries the dot (their stems passing through elongated holes in the latter), this piece and the dot may be adjusted to any required position. The diagram accompanying the figure of the cap is a magnified representation of a circle aa described on a plate of white metal on the plane face of the movable cylinder, at the centre of which the dot, which is circular, is engraved, together with a smaller concentric circle bb of known radius, and a broken straight line cc passing through the centre of the dot. The uses of the small circle and straight line will be seen from the following account of the instrumental processes in which this apparatus is employed.

95. The first operation is to place the dots very nearly on the axis of motion of the instrument. This is to be done tentatively by the movements given to the dot by the screws d, d, d, till, as seen with the microscope, it describes a very small re-entering curve when the instrument is turned about its axis through a complete revolution. Next, the micrometer-wire is to be adjusted vertically. To do this a small plumb-line apparatus is applied in the following manner. The cap having been pushed close up to the end of the pivot and clamped, the microscope is adjusted in the direction of its axis so that the images of the dot and of the line cc are seen distinctly together with the micrometer-wire put roughly in a vertical position. The distance of the plane face of the cap from a shoulder of the pivot is then gauged, the cap is temporarily removed, and the plumb-line apparatus placed in such a position that the line hangs down at the distance measured by the gauge, and can be seen distinctly through the microscope. The micrometer-wire can now be placed with sufficient exactness parallel to the plumb-line by turning the microscope by hand about its axis. The cap being after this restored to its former position, the line cc will, by reason of the gauge, be seen distinctly, and may be placed parallel to the micrometer-wire by turning the *instrument* about its axis. When this is done the corresponding reading of the setting-circle should be observed and recorded. Then by turning the instrument from this position through 90°, the line

cc will be placed in a horizontal position, and will serve to adjust the micrometer-wire to horizontality. But as to do this requires the microscope to be turned about its axis, a new adjustment in the direction of its axis should be made in order to secure that the micrometer-wire and the line *cc* shall both be seen distinctly when the wire is finally adjusted horizontally. The readings of the setting-circle for the vertical and horizontal positions of *cc* being determined, the micrometer-wire can always be adjusted either vertically or horizontally by means of that line, care being taken that the wire and the image of the line be both well seen in focus. If these determinations of the vertical and horizontal directions be not quite accurate, no sensible error will arise provided the dot be always very near the axis of motion. These preliminary operations have to be performed in exactly the same manner for both dots.

96. We may now proceed to obtain a general formula for collimation-error independently of the particular forms of the pivots. In Figure 25, A and B are the positions of the dots,

Fig. 25.

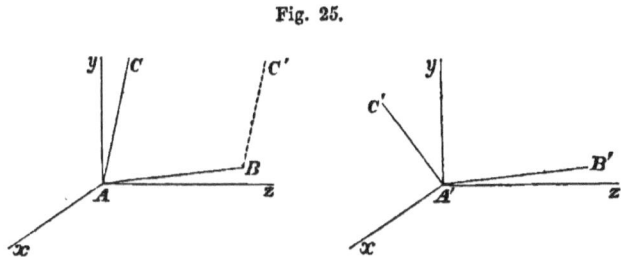

and Ax, Ay, Az are rectangular axes, Ax being drawn parallel to the plane of the meridian *southward*, Ay vertically *upward*, and Az in a horizontal direction *westward*, making a small angle with AB. The co-ordinates of A and B parallel to Ax and Ay are measured by the microscope-micrometers, adjusted, as described in the preceding Article, for either horizontal or vertical measures. Let the telescope be directed to the Nadir Point, and suppose AC in the plane yAz to be parallel to the line of collimation in that position. The angle yAC ($=\alpha$) is measured by the collimating eye-piece (see Art. 84), and is reckoned *positive* when AC deviates from the vertical *westward*. Let $F_1 Y_1$ and $F_2 Y_2 + k$ be the vertical ordinates of A and B, Y_1 and Y_2 being the recorded micrometer readings, F_1 and F_2 factors by which the micrometer readings are converted into seconds of arc, and k an unknown constant, the value of which is determined when the measures Y_1 and Y_2 are referred to the *same* horizontal plane. To find the value of F_1, let $D=$ the measured distance between the dots, $h=$ the diameter of the small circle bb (Fig. 24), as determined by the maker of the cap, and $M=$ the measure of that diameter in micrometer revolutions, which measure may be supposed to have been taken just after Y_1 was recorded. Then if $R=$ the number of seconds in an arc equal to radius, $\frac{hR}{D}$ is the measure, in seconds of space, of the angle subtended at B by the diameter bb. Hence the angle subtended by any smaller micrometer measure m is $\frac{hRm}{DM}$. Consequently the factor F_1 is equal to $\frac{hR}{DM}$. So for the opposite cap, $F_2 = \frac{h'R}{DM'}$, supposing the given diameters h and h' to be unequal. The values of F_1 and F_2 having been calculated by these formulæ, we shall have for the value in arc of the angle BAz, $F_2 Y_2 + k - F_1 Y_1$, in which k only is unknown. Hence, since $\angle yAC = \alpha$, if θ be the constant angle CAB, we shall have

$$\theta = 90° - (F_2 Y_2 + k - F_1 Y_1) - \alpha \ldots \ldots \ldots \ldots (1).$$

It will now be supposed that when the above measures were taken the Illumination-end was *westward*, so that, as B was by construction always at that end, $F_2 Y_2 + k$ and $F_1 Y_1$ are respectively the ordinates of the dots B and A for Illumination West. The instrument is next to be reversed and the same process repeated. Let Y_2' and Y_1' be respectively the micrometer readings for the bisections of the *west* and *east* dots (which are now A and B), and let $F_2' Y_2' + k$ and $F_1' Y_1'$ be their ordinates reckoned from the same horizontal plane as before, F_1' and F_2' being calculated by processes already indicated with respect to F_1 and F_2. Since instead of θ we have now its supplement, and instead of α an angle α' measuring, as before, the deviation of the line of collimation *westward* from the meridian, the analogous equation will be

$$180° - \theta = 90° - (F_2'Y_2' + k - F_1'Y_1') - \alpha' \ldots\ldots\ldots(2).$$

The equations (1) and (2) determine the unknown quantities k and θ. In fact, putting, for brevity, Y_x for $F_2Y_2 - F_1Y_1$ and Y_e for $F_2'Y_2' - F_1''Y_1'$, we obtain

$$k = -\frac{\alpha + \alpha'}{2} - \frac{Y_x + Y_e}{2} \ldots\ldots\ldots(3),$$

$$90° - \theta = \frac{\alpha - \alpha'}{2} + \frac{Y_x - Y_e}{2} \ldots\ldots\ldots(4).$$

97. Reverting to Figure 25, A' and B' are new positions of the dots (*before* the reversion) after turning the eye-end of the telescope through an angle (z) from the vertical towards the south; $A'x$, $A'y$, $A'z$ are rectangular axes, $A'y$ being vertical, $A'x$ horizontal southward, and $A'z$ westward. $A'C$ is parallel to the new direction of the line of collimation, so that the angle $C'A'B' = \theta$. Let $\xi =$ the angle by which $A'C$ deviates from the plane $yA'x$ towards $A'z$. Then y_1, y_2 being the micrometer readings in the vertical direction, and x_1, x_2 those in the horizontal direction, for bisections of A' and B' respectively, let f_1y_1 and $f_2y_2 + k$ be the measures of the vertical ordinates, and g_1x_1 and $g_2x_2 + k'$ the measures of the horizontal ordinates, the factors f_1, f_2, g_1, g_2, having to be determined, as already shewn, by means of micrometer measures of the diameter bb. Hence we have the following data for calculating ξ:—

∠ $C'A'z = 90° - \xi$,   ∠ $B'A'z = \epsilon$, a small angle,
∠ $C'A'y = z$, nearly,   ∠ $B'A'y = 90° - (f_2y_2 + k - f_1y_1)$,
∠ $C'A'x = 90° - z$, nearly,   ∠ $B'A'x = 90° - (g_2x_2 + k' - g_1x_1)$.

But by Analytical Geometry,

$$\cos C'A'B' = \cos C'A'z \cos B'A'z + \cos C'A'y \cos B'A'y$$
$$+ \cos C'A'x \cos B'A'x.$$

Hence, putting small arcs for their sines, and unity for cos ϵ in a small term, we obtain

$$\cos\theta = \xi + (f_2 y_2 + h - f_1 y_1)\cos z + (g_2 x_2 + k' - g_1 x_1)\sin z.$$

Now by the equation (4) we also have very nearly,

$$\cos\theta = \frac{a-a'}{2} + \frac{Y_w - Y_e}{2};$$

and if Z be the zenith distance of the pointing of the telescope reckoned southward, $Z - z = 180°$, $z = Z - 180°$. Hence by substituting in the above formula, and putting for k its value from the equation (3), it will be found that

$$\xi = \frac{a-a'}{2} + \frac{Y_w - Y_e}{2} + \left(f_2 y_2 - f_1 y_1 - \frac{a+a'}{2} - \frac{Y_w + Y_e}{2}\right)\cos Z$$
$$+ (g_2 x_2 + k' - g_1 x_1)\sin Z.$$

The deviation of the eye-end of the telescope from the plane of the meridian having been assumed (in Fig. 25) to be westward, that of the object-end will be eastward, and consequently ξ is *the total error of collimation with its proper sign.*

98. Supposing now the pivots to be cylindrical and equal, and the extremities of the axis of motion to be at A and B, let ξ_1 be the value of ξ in that case. Then, having regard to the expressions for which Y_w and Y_e were substituted in Art. 96, it is evident that as the axis of motion has the same position in space after as before the reversion, we shall have $Y_w = Y_e$, and also $f_2 y_2 - f_1 y_1 = Y_w$, the position being fixed. Moreover, the x's being reckoned positive southward, $g_2 x_2 + k' - g_1 x_1$ will be equal to the angular deviation of the west end of the axis *southward*, and will consequently be the azimuth error with its proper sign. Putting therefore c for the azimuth error, we have for equal cylindrical pivots,

$$\xi_1 = \frac{a-a'}{2} - \frac{a+a'}{2}\cos Z + c\sin Z.$$

By referring to Art. 92 it will be seen to follow as a corollary from this equation that $\frac{a-a'}{2}$ is equal to the collimation error, and $-\frac{a+a'}{2}$ to the level error. These values may also be obtained independently as follows. The $\angle yAC$ (Fig. 25), measured from the vertical towards Az, is equal to a, and if it were wholly due to level error, would have a *negative* value as respects the transit observations, because it would signify that the west end of the axis is *depressed*; but so far as it is due to collimation error it has a *positive* value, because a westward deviation of the *line* of collimation from the *plane* of collimation at the eye-end implies an eastward deviation at the object-end, and consequently a *plus* correction of the transit-time. Hence if L be the level error and C the collimation error, since the value of a is made up of both, we shall have $a = -L + C$. After the reversion nothing is changed except the sign of C, so that $a' = -L - C$. From these two equations it follows that

$$C = \frac{a-a'}{2}, \text{ and } L = -\frac{a+a'}{2}.$$

By subtracting the above expression for ξ_1 from that for ξ, the result is

$$\xi - \xi_1 = \frac{Y_w - Y_e}{2} + \left(f_2 y_2 - f_1 y_1 - \frac{Y_w + Y_e}{2}\right) \cos Z$$
$$+ (g_2 x_2 - g_1 x_1 + k' - c) \sin Z.$$

Consequently in order to calculate the correction $\xi - \xi_1$ for any given zenith distance, we have to find the value of the constant quantity $k' - c$. Before shewing how this may be done, the above formula will be modified for the purpose of making it applicable both to Illumination West and to Illumination East. The equation in its present form is assumed to apply to the case of the Illumination (or the dot B) being westward. Since Y_w has been put for $F_2 Y_2 - F_1 Y_1$, if we assume that the suffix w indicates that the expression applies to Illumination West, we can, by putting Y_e for $F_2 Y_2 - F_1 Y_1$, indicate that the same expression applies to Illumination East. Consequently the above value of $\xi - \xi_1$ is adapted to the case of

Illumination East by simply changing Y_w into Y_e and Y_e into Y_w. Hence it will be seen that

$$\xi - \xi_1 = \pm \frac{Y_w - Y_e}{2} + \left(f_2 y_2 - f_1 y_1 - \frac{Y_e + Y_w}{2}\right) \cos Z$$
$$+ (g_2 x_2 - g_1 x_1 + k' - c) \sin Z,$$

according as the Illumination is West or East. The value of $k' - c$ is found in the following manner.

99. In the case of equal cylindrical pivots the corrected plane of collimation coincides with the plane of the meridian, and the corrected line of collimation consequently passes through the pole. Hence if we introduce the condition that $\xi = \xi_1$ at the pole, the corrected line of collimation in the case of unequal and non-cylindrical pivots will be made to pass through the pole, and the satisfying of that condition gives the means of determining $k' - c$. For, after putting Y_0 and X_0 to represent the values of $f_2 y_2 - f_1 y_1$ and $g_2 x_2 - g_1 x_1$ respectively for the position of the pole, and $-l$ for Z, l being the co-latitude, we obtain

$$0 = \pm \frac{Y_w - Y_e}{2} + \left(Y_0 - \frac{Y_w + Y_e}{2}\right) \cos l - (X_0 + k' - c) \sin l;$$

or, $k' - c = \pm \left(\frac{Y_w - Y_e}{2}\right) \operatorname{cosec} l + \left(Y_0 - \frac{Y_w + Y_e}{2}\right) \cot l - X_0.$

After calculating the value of $k' - c$ from this formula, and substituting it in the general expression for $\xi - \xi_1$, this quantity can be calculated for any given zenith distance for *both* positions of the instrument, and the time-correction for the forms of the pivots may then be obtained from the equation

$$\text{Correction} = \frac{\xi - \xi_1}{15} \operatorname{cosec} \text{N. P. D.}$$

100. I proceed to give an example of the calculation of the constants in the formula for $\xi - \xi_1$, for the purpose of afterwards shewing how to obtain the time-correction for forms of the pivots in any given instance of a transit observation. We have, as fixed data, the length D of the straight line joining the dots

= $49^{in},80$, the diameter of the circle bb at the Illumination-end
= $0^{in},0465$, and the diameter of that at the other end = $0^{in},0458$.
[These values of the diameters were carefully determined by
Mr Simms: the angles which the diameters subtend at the
opposite dots will be found from the above data to be $192'',60$
and $189'',70$.] The data for calculating the constants in the
general expression for $\xi - \xi_1$ will be extracted from a complete
series of measures contained in pages lxxviii. and lxxix. of
Volume XX. of the Cambridge Observations, which were taken
on September 29, 1858, for the purpose of obtaining the pivot-
correction for any given N. P. D. It is to be understood that
each adopted measure of the co-ordinate of a dot is the mean of
three recorded measures, two for contact of the micrometer-wire
with the circumference of the dot on opposite sides, and the third
for an estimated bisection of the centre of the dot by the axis of
the wire. So also the adopted measures of the diameters bb are
in each instance means of three trials. The measures for the Pole
and the Nadir Point, being fundamental quantities, are all means
of *six* trials. The measures of diameters are designated by the
letter M with the suffix 1 or 2 attached. According as this
letter, or any other letter used in the subjoined example, has
the suffix 1 or 2, it pertains to the *east* or *west* end of the axis.
In order that the micrometer measures may be expressed in low
numbers, the recorded eastern *vertical* measures are all diminished by $10^r,000$, and the western *vertical* measures by $9^r,000$. For
the same reason constants are subtracted from the horizontal
measures, as will be indicated in the following account of the
calculations.

Illumination West.

$Y_1 = 10,336 - 10,000 = 0,336, \quad M_1 = 12,537, \quad h_1 = 0,0458,$

$Y_2 = 9,072 - 9,000 = 0,072, \quad M_2 = 12,516, \quad h_2 = 0,0465,$

$$F_1 = \frac{Rh_1}{DM_1} = [1,17988], \quad F_2 = \frac{Rh_2}{DM_2} = [1,18718],$$

$$F_1 Y_1 = 5'',084, \quad F_2 Y_2 = 1'',108, \quad Y_w = \frac{F_2 Y_2 - F_1 Y_1}{2} = -1'',988.$$

Illumination East.

$$Y_1' = 10{,}390 - 10{,}000 = 0{,}390, \quad M_1 = 12{,}711, \quad h_1 = 0{,}0465,$$
$$Y_2' = 9{,}028 - 9{,}000 = 0{,}028, \quad M_2 = 12{,}340, \quad h_2 = 0{,}0458,$$

$$F_1' = \frac{Rh_1}{DM_1} = [1{,}18047], \quad F_2' = \frac{Rh_2}{DM_2} = [1{,}18675],$$

$$F_1'Y_1' = 5''{,}909, \quad F_2'Y_2' = 0''{,}430, \quad Y_e = \frac{F_2'Y_2' - F_1'Y_1'}{2} = -2''{,}740.$$

For obtaining the values of $f_2 y_2 - f_1 y_1$ and $g_2 x_2 - g_1 x_1$ corresponding to the position of the pole, the data and the calculation are as follows:—

Illumination West, N. P. D. 0°.

$$y_1 = 10{,}597 - 10{,}000 = 0{,}597, \quad M_1 = 12{,}532, \quad h_1 = 0{,}0458,$$
$$y_2 = 8{,}667 - 9{,}000 = -0{,}333, \quad M_2 = 12{,}531, \quad h_2 = 0{,}0465,$$

$$f_1 = \frac{Rh_1}{DM_1} = [1{,}18005], \quad f_2 = \frac{Rh_2}{DM_2} = [1{,}18666],$$

$$f_1 y_1 = 9''{,}037, \quad f_2 y_2 = -5''{,}118, \quad f_2 y_2 - f_1 y_1 = -14''{,}155 = Y_0,$$

$$x_1 = 10{,}993 - 10{,}900 = 0{,}093, \quad M_1 = 12{,}551, \quad h_1 = 0{,}0458,$$
$$x_2 = 9{,}416 - 9{,}200 = 0{,}216, \quad M_2 = 12{,}569, \quad h_2 = 0{,}0465,$$

$$g_1 = \frac{Rh_1}{DM_1} = [1{,}17939], \quad g_2 = \frac{Rh_2}{DM_2} = [1{,}18535],$$

$$g_1 x_1 = 1''{,}406, \quad g_2 x_2 = 3''{,}310, \quad g_2 x_2 - g_1 x_1 = 1''{,}904 = X_0.$$

Hence $\quad k' - c = \dfrac{Y_w - Y_e}{2} \operatorname{cosec} l + \left(Y_0 - \dfrac{Y_w + Y_e}{2} \right) \cot l - X_0$

$$= +0''{,}752 \operatorname{cosec} l - 9''{,}427 \cot l - 1''{,}904$$
$$= -12''{,}836, \; l \text{ being} = 37° . 47' . 8''.$$

Illumination East, N. P. D. 0°.

$$y_1' = 10{,}250 - 10{,}000 = 0{,}250, \quad M_1 = 12{,}640, \quad h_1 = 0{,}0465,$$
$$y_2' = 8{,}745 - 9{,}000 = -0{,}255, \quad M_2 = 12{,}388, \quad h_2 = 0{,}0458,$$

THE TRANSIT INSTRUMENT.

$$f_1' = \frac{Rh_1}{DM_1} = [1{,}18290], \quad f_2' = \frac{Rh_2}{DM_2} = [-1{,}18307],$$

$f_1'y_1' = +3''{,}809, \quad f_2'y_2' = -3''{,}905, \quad f_2'y_2' - f_1'y_1' = -7''{,}714 = Y_0'.$

$$\begin{array}{cccc} \text{r.} & \text{r.} & \text{r.} & \text{in.} \\ x_1' = 10{,}945 - 10{,}700 = 0{,}245, & M_1 = 12{,}658, & h_1 = 0{,}0465, \\ x_2' = 9{,}174 - 9{,}300 = -0{,}126, & M_2 = 12{,}449, & h_2 = 0{,}0458, \end{array}$$

$$g_1' = \frac{Rh_1}{DM_1} = [1{,}18228], \quad g_2' = \frac{Rh_2}{DM_2} = [-1{,}18294],$$

$g_1'x_1' = 3''{,}728, \quad g_2'x_2' = -1''{,}920, \quad g_2'x_2' - g_1'x_1' = -5''{,}648 = X_0'.$

Hence

$$k' - c = -\frac{Y_w - Y_e}{2} \operatorname{cosec} l + \left(Y_0' - \frac{Y_w + Y_e}{2}\right) \cot l - X_0'$$

$$= -0''{,}752 \operatorname{cosec} l - 2''{,}986 \cot l + 5''{,}648 = +0''{,}569.$$

Consequently the final expressions for $\xi - \xi_1$ are,

Illumination West,

$$\xi - \xi_1 = +0''{,}75 + (f_2y_2 - f_1y_1 + 4''{,}73)\cos(\delta - l)$$
$$+ (g_2x_2 - g_1x_1 - 12''{,}84)\sin(\delta - l),$$

Illumination East,

$$\xi - \xi_1 = -0''{,}75 + (f_2y_2 - f_1y_1 + 4''{,}73)\cos(\delta - l)$$
$$+ (g_2x_2 - g_1x_1 + 0''{,}57)\sin(\delta - l).$$

101. The measures on September 29, 1858, were taken for every tenth degree of N. P. D. above pole, and for the polar distances $-5°$, $0°$, $+5°$, these three being included for the purpose of obtaining the values of $\xi - \xi_1$ for Polaris and δ Ursæ Minoris by interpolation. The order of proceeding was as follows: the Illumination being eastward, the micrometer-wires were first adjusted vertically, and all the horizontal measures were taken, and after adjusting the wires horizontally and to distinct vision of the dots, all the vertical measures were taken, *ending* with those for Nadir Point. The instrument was then reversed, and without altering the horizontal adjustments of the wires, all the vertical measures were again taken, *beginning* with those for Nadir Point; and lastly, after adjusting

the wires vertically and to distinct vision, all the horizontal measures were again taken. In other instances, as in that of September 30, 1858 (given in the same pages of Volume XX.), the measures were taken for every fifth degree of N. P. D. above pole, and for every tenth degree below pole. As the interval between the Y's is somewhat greater than that between the shoulders of the transverse axis which are contiguous to them, the axis is liable to a small displacement whereby the images of the dots are thrown out of focus. This fault was corrected by turning the instrument backwards and forwards about its axis, and at the same time pushing it moderately in the direction proper for the correction, till both dots could be seen distinctly for the same position of the instrument. This precaution having been attended to, it was not thought necessary to measure the diameters of the circles on each occasion of measuring the ordinates of the dots. On September 29 the measures of the diameters were taken only for 0°, 30°, 60°, 90°, 120° of N. P. D., and for the Nadir Point. After giving these explanations I proceed to adduce an example of calculating the time-correction for forms of the pivots, both for Illumination West and Illumination East, for a given N. P. D., selecting that of 60°.

Illumination West, N. P. D. 60°.

$y_1 = 10{,}406 - 10{,}000 = 0{,}406,\quad M_1 = 12{,}512,\quad h_1 = 0{,}0458,$

$y_2 = 9{,}169 - 9{,}000 = 0{,}169,\quad M_2 = 12{,}514,\quad h_2 = 0{,}0465,$

$f_1 = \dfrac{Rh_1}{DM_1} = [1{,}18074],\quad f_2 = \dfrac{Rh_2}{DM_2} = [1{,}18725],$

$f_1 y_1 = 6''{,}156,\quad f_2 y_2 = 2''{,}601,\quad f_2 y_2 - f_1 y_1 = -3''{,}555,$

$x_1 = 10{,}969 - 10{,}900 = 0{,}069,\quad M_1 = 12{,}524,\quad h_1 = 0{,}0458,$

$x_2 = 9{,}143 - 9{,}200 = -0{,}057,\quad M_2 = 12{,}551,\quad h_2 = 0{,}0465,$

$g_1 = \dfrac{Rh_1}{DM_1} = [1{,}18033],\quad g_2 = \dfrac{Rh_2}{DM_2} = [1{,}18597],$

$g_1 x_1 = 1''{,}045,\quad g_2 x_2 = -0''{,}875,\quad g_2 x_2 - g_1 x_1 = -1''{,}920.$

THE TRANSIT INSTRUMENT.

Illumination East, N. P. D. 60°.

$$y_1 = 10{,}223 - 10{,}000 = 0{,}223, \quad M_1 = 12{,}679, \quad h_1 = 0{,}0465,$$
$$y_2 = 9{,}354 - 9{,}000 = 0{,}354, \quad M_2 = 12{,}382, \quad h_2 = 0{,}0458,$$

$$f_1 = \frac{Rh_1}{DM_1} = [1{,}18157], \quad f_2 = \frac{Rh_2}{DM_2} = [1{,}18528],$$

$$f_1 y_1 = 3''{,}387, \quad f_2 y_2 = 5''{,}432, \quad f_2 y_2 - f_1 y_1 = 2''{,}045,$$

$$x_1 = 10{,}758 - 10{,}700 = 0{,}058, \quad M_1 = 12{,}665, \quad h_1 = 0{,}0465,$$
$$x_2 = 9{,}059 - 9{,}300 = -0{,}241, \quad M_2 = 12{,}428, \quad h_2 = 0{,}0458,$$

$$g_1 = \frac{Rh_1}{DM_1} = [1{,}18204], \quad g_2 = \frac{Rh_2}{DM_2} = [1{,}18367],$$

$$g_1 x_1 = 0''{,}882, \quad g_2 x_2 = -3''{,}679, \quad g_2 x_2 - g_1 x_1 = -4''{,}561.$$

Consequently, Illumination West,

$$\xi - \xi_1 = +0''{,}75 + (-3''{,}56 + 4''{,}73) \cos 22°.12'{,}9$$
$$\qquad\qquad + (-1''{,}92 - 12''{,}84) \sin 22°.12'{,}9$$
$$\qquad = +0''{,}75 + 1''{,}09 - 5''{,}58 = -3''{,}74.$$

Illumination East,

$$\xi - \xi_1 = -0''{,}75 + (2''{,}04 + 4''{,}73) \cos 22°.12'{,}9$$
$$\qquad\qquad + (-4''{,}56 + 0''{,}57) \sin 22°.12'{,}9$$
$$\qquad = -0''{,}75 + 6''{,}27 - 1''{,}51 = +4''{,}01.$$

Hence the time-correction, $\frac{\xi - \xi_1}{15}$ cosec 60°, is $-0^s{,}288$ for Illumination West, and $+0^s{,}309$ for Illumination East. These values, obtained by independent calculation, agree with those on page lxxviii of Volume xx. of the Cambridge Observations.

102. The pivot-corrections for all the particular N. P. D. for which micrometer-measures were taken having been calculated according to the foregoing example, and arranged in tabular form with N. P. D. for argument, the correction for any assigned N. P. D. might be obtained by simple interpolation.

As, however, it was found that the corrections varied gradually in amount from year to year, and the measures were subject to incidental irregularities, the table of values for any year was combined with those of two other years, one shortly preceding and the other shortly following, and from the three a mean table was formed, from which the correction for any given N. P. D. for the intermediate year was derived by simple interpolation. This pivot-correction may be immediately applied to the clock-time of transit across the mean of the wires (obtained as shewn in *Division III.*), and then the corrections for the errors of collimation, level, and azimuth may be obtained and applied just as if the pivots were exactly equal and cylindrical. The reasoning I have now gone through, although long and somewhat complicated, was yet necessary for shewing how transit observations may be freed from errors due to irregularities and changes of the forms of the pivots, and how, consequently, to make a completely *logical* use of the transit instrument[1].

103. Before proceeding to treat of the methods of taking and reducing transit observations, some account will here be given of the *Astronomical Clock*, and of the conditions it is required to satisfy when used in an Observatory. For details respecting the parts of chief importance in the construction of

---

[1] The method of obtaining the pivot-corrections which I put in practice with the means I had at command, is adopted in principle in the use of the Cambridge Transit-circle, but with various simplifications and improvements as to details which I shall take this occasion to point out. (1) The dots were engraved on the pivots themselves, and as nearly as possible on the axis of motion, so that no subsequent adjustment of their position was required, such as that I applied to the dots on caps (Art. 95). (2) A spring acting on the end of one of the pivots keeps them always to the same bearing, and thus the inconvenience I experienced from the images of the dots being put out of focus by displacements of the axis (Art. 101), is obviated. (3) The micrometer-wire of the microscope is adjusted horizontally by a small spirit-level, movable with the micrometer-apparatus about the axis of the microscope, and so adjusted by the instrument-maker that when the bubble is in mid-position the micrometer-wire is sufficiently horizontal, considering the small movement of the dot, for measuring its vertical ordinate. This expedient supersedes the use of the small plumb-line which I had recourse to (Art. 95). (4) As the Transit-circle is not reversed, the pivot-corrections are required for only one position of the instrument.

clocks for astronomical purposes, the student will do well to read Chapter III. of Godfray's "Treatise on Astronomy." As far as the Lectures are concerned it will suffice to refer only to the clock with Graham's mercurial pendulum, this being the form of pendulum which is found by experience to be best adapted for obtaining accurate astronomical measures of time. The particular advantage of this construction is, that it allows the astronomer himself both to alter the clock's rate and to correct imperfect compensation. These adjustments are effected by means which it is the object of the following statements to describe.

104. First, for altering the clock's rate, the mercury is put in a glass cylinder attached to the end of the pendulum-rod, and by turning a screw the cylinder and contained mercury may be raised or lowered relatively to the rod. As the weight of the mercury constitutes the chief part of the weight of the pendulum, this operation changes the position of the centre of oscillation, and, by consequence, the time of an oscillation of the pendulum. A small graduated circular disk, attached to the stem of the screw and turning with it, is read off by an index fixed to the rod. It being understood that the clock's rate is its gain or loss upon 24 hours in the sidereal interval between consecutive meridian transits of the same star, let the index be set to a certain reading, and the clock be rated, by transits of stars taken *ad libitum*, in the manner stated at the end of Art. 89 (1). After altering the index-reading by a certain integral number of divisions, let the clock be rated in the same manner again. The difference of the index-readings and the corresponding difference of the rates being noted, the index-reading which would correspond to a certain desired rate might then be found by a simple proportion, and the length of the pendulum might be altered accordingly. For convenience as respects the reduction of transit observations it is usual to maintain a small *losing rate*, the minute-hand being put forward when the clock's loss exceeds one minute.

105. Compensation for changes of the length of the pendulum produced by changes of temperature is effected by the

mercurial pendulum in the following manner. The cylindrical column of mercury, being uncovered at the top, is free to expand upwards upon an increment of temperature, which, therefore, has the effect of raising the centre of gravity of the mercury, and, by consequence, the centre of oscillation of the pendulum. At the same time the pendulum-rod, by expanding downwards from the point of suspension, produces the opposite effect as respects the position of the centre of oscillation. When these two results of the change of temperature just counteract each other, the required compensation is effected, the distance of the centre of oscillation from the point of suspension, and consequently the time of an oscillation of the pendulum, being thereby made invariable. Analogous reasoning applies to the effects of a decrement of temperature. Now the astronomer is in a better position than the clock-maker for telling whether or not the compensation be good, because for doing this completely exact transit observations carried on through the hottest and coldest parts of at least one year are required. It is the special advantage of the mercurial pendulum that it gives the astronomer the means of correcting the compensation if found thus to be imperfect. Supposing, for instance, the observations to shew that the losing rate comes to a maximum in the hot months, this will indicate that the length of the pendulum is at that time too great, and consequently that the effect of the expansion of the rod is greater than that of the expansion of the mercury. The clock is in that case *under-compensated*, and requires an addition to be made to the quantity of mercury; or, if it be found that the losing rate diminishes in the cold months till it attains a minimum, the pendulum is then too short, and the contraction of the pendulum-rod by the cold is not compensated for by the contraction of the mercury; that is, as before, the clock is *under-compensated*. But if the clock goes faster in the summer months, or slower in the winter months, than in Spring or Autumn, it is *over-compensated*; and requires that some mercury should be taken out of the cylinder.

106. The simplest practical method of adjusting the *quantity* of the mercury in the cylinder so as to effect complete compen-

sation seems to be the following. Suppose that *after* giving to the clock a small losing rate by the process explained in Art. 104, the clock were rated by transit observations during a whole year (in the manner that will be explained under *Division IV.*). Then, from what has been said in Art. 105, according as the summer losing rates are found to be *greater* or *less* than those of winter, the compensation requires mercury to be *added to* or *subtracted from* the quantity (Q) that was in the cylinder when the rate was adjusted. To calculate how much, let R the maximum losing rate, and r the minimum losing rate, in the course of the twelve months, be extracted from the reduced transit observations, and suppose that the contemporaneous temperatures are known to be T and t. At the end of that period let the quantity of mercury be altered to $Q \pm q$, and the observations be continued through another twelve months, and let R', r', T', t' be the new values corresponding to R, r, T, t. Then to reduce the two sets of values to the same conditions as respects temperature, let $R'' - r''$ be the value that $R' - r'$ would have had if the difference of temperature $T' - t'$ had been equal to $T - t$. Hence, as small differences of rate may be assumed to be proportional to the differences of temperature, we shall have $R'' - r'' = (R' - r') \dfrac{T-t}{T'-t'}$. Hence the additional mercury $\pm q$ has produced the change $R - r - (R' - r') \dfrac{T-t}{T'-t'}$ in the difference between the maximum and minimum rates. But there is still to be corrected the difference $R' - r'$. Let q' be the unknown quantity of mercury required for this purpose. Then we may assume that $\pm q$ is to q' as $R - r - (R' - r') \dfrac{T-t}{T'-t'}$ is to $R' - r'$. From which it follows that

$$q' = \pm q \cdot \frac{(R'-r')(T'-t')}{(R-r)(T'-t') - (R'-r')(T-t)}.$$

This formula gives in terms of known quantities the amount of mercury to be put in or taken out of the cylinder to complete the compensation, at least, approximately. If after this has been done, further correction appears to be needed, it may be calculated by means of a like formula, obtained by combining the transit observations of a third twelve months with those of the second. Should there still be residual variations of rate, whether from defect of compensation or other causes, they will have no appreciable effect, so long as they are neither large nor abrupt, on the final determinations of Right Ascension, supposing transit observations of fundamental stars to be carried on in the Observatory night and day during the year.

III. METHODS OF TAKING TRANSIT OBSERVATIONS, AND CALCULATIONS OF CORRECTIONS.

107. It will be proper to begin this Division of our subject by stating in what manner a setting-circle is used in preparing to take a transit observation. The construction and uses of setting-circles have been described in Art. 34, but it was not shewn how, before employing the setting-circle to point the telescope to a celestial object intended for observation, its spirit-level has to be adjusted. This may readily be done, if a collimating eye-piece be adaptable to the telescope, and the telescope be thereby pointed so that the reflected image of each of the two horizontal wires is coincident with the other wire; in which case the path across the field midway between the horizontal wires, which the object observed is usually made to traverse, will be exactly in the nadir direction. Under these circumstances the index which will always be used for settings should be made to point to an angle below pole equal to the latitude of the Observatory, for instance to $-52°.13'$ for the Cambridge Observatory. After this adjustment has been made by the tangent-screw and clamp (Art. 35), the bubble of the level is to be put in mid-position by turning the screws x and y (Figure 7 in p. 28). Then if the index be set to any given circle-reading, the telescope, after being moved till the bubble is in mid-position, will point to the north polar distance signified by that reading. It should, however, be noticed that the adjustment of the level was made independently of any effect of *refraction*, on which account in setting for a star or other object,

the N. P. D. as *altered by refraction* should be set for. For doing this readily the observer should have at hand a small table of approximate refractions with N. P. D. for argument. If the use of a collimating eye-piece be not available, the adjustment of the level might be performed by first setting the index to the apparent N. P. D., as altered by refraction, of a bright star, such as Arcturus, and after catching it as it enters into the field of view (which it is easy to do), to turn the telescope till the path of the star's image is mid-way between the two horizontal wires. Then, the telescope retaining its position, the bubble is to be put in mid-position by turning the screws x and y provided for this adjustment.

108. Proceeding, in the next place, to give an account of the methods of taking transit observations of the different kinds of celestial bodies, I shall assume that in every instance the Right Ascension and Polar Distance of the object to be observed, as also the error of the clock on true sidereal time, are, at least approximately, known. (By what antecedent operations and observations the clock's error, and the places both of stars and of the moving bodies, may have been approximated to, or how by beginning *de novo* they might be sufficiently well determined for being used to obtain places of higher degrees of accuracy, will be unfolded in the sequel of the Lectures, more especially in the treatment of what relates to meridional transit and circle observations under *Division IV*.) The polar distance being approximately known, we can set for the object in the manner described in Art. 107, and the clock's error being sufficiently ascertained we shall know when to look for the entrance of the object into the field of view. When fairly in the field, the image has to be brought, by shifting the telescope, into a position intermediate, as nearly as may be, between the two horizontal wires. Suppose, first, the image to be that of a *star*, which may be considered to be a point of light (see Art. 19). The observer, as it approaches the first wire, takes the number of a seconds' beat from the clock-face, and then goes on counting the beats mentally, looking at the same time intently at the central point of the image in order to note its position just at

the beat which next precedes the transit across the wire, and again its position at the beat next after the transit. Also to the best of his judgment he estimates the ratio of the space between the first position and the wire to that between the two positions in tenths of the interval[1], and the counted second of the first position with this proportional part of a second added is recorded as the clock-time of the transit. This operation is repeated at each of the wires, the intervals between the wires allowing time for recording the several transits. After the transit across the last wire, the clock-face is looked at to see whether the counting has been correct, and the hour and minute of the last transit are set down. This mode of taking transits is called for distinction the *eye-and-ear* observation.

109. The same method of observing to the tenth of a second is employed when the object is either the *First or Second Limb of the Sun*, care being taken in the setting to ensure that the Sun's centre shall pass as nearly as may be mid-way between the horizontal wires. On account of the Sun's heat the eye of the observer generally requires to be protected by darkened glasses; but occasionally satisfactory observations can be taken when the Sun is seen through cloud or mist. The Cambridge transit-telescope had a small wheel attached at the eye-end, carrying near its periphery a set of glasses of different degrees of opacity, which, on turning the wheel by hand, could be placed, as the occasion required, opposite the aperture through which the field of view was seen. Also to prevent any disturbing effect being produced by the incidence of the Sun's rays on parts of the instrument, a movable dark screen has to be provided, by shifting which the solar rays may either be wholly cut off, or be allowed to pass through a circular opening in the screen so as to fall on the object-glass during the interval

---

[1] For a long time astronomers were satisfied with noting the time of transit to the nearest second. Bradley in some cases estimated to the half or the third of a second. Maskelyne at the beginning of his career divided the second into eight parts; but eventually (in September, 1772) he began the practice of recording the times of transit to tenths of a second. (See Grant's *History of Physical Astronomy*, p. 489.)

occupied by taking the observation. When both limbs have been observed, the mean of the transit-times for the two limbs is the time of transit of the Sun's centre, and their difference is a measure of the Sun's diameter, the adopted value of which, as being an important astronomical element, should be the mean result of a large number of observations.

110. The *full moon* would be observed in the same manner as the Sun, and when, also, it is not quite full, the transits of both limbs would still be taken, because, after applying a correction for a small defect of illumination (the mode of calculating which will be subsequently explained), the transit-time of the Moon's centre, and a measure of the diameter, may be inferred from the times of transit of the limbs. As the Moon's diameter is an important datum, and the opportunities for measuring it are comparatively rare, advantage should be taken of all that are available.

111. Transits of *Jupiter* and of *Mars* are observed as follows. The number of wires being supposed to be seven, transits of the first limb are taken at the first, third, fifth, and seventh wires, and transits of the second limb at the second, fourth, and sixth wires. The mean of all these transit-times is strictly the observed time of transit of the planet's centre across the mean of all the wires. The excess of the mean of the transit-times of the second limb above the mean of those of the first limb is considered to be a measure of the planet's diameter. (More generally and strictly the diameter is measured by the excess of the mean of the times of transit of the second limb across three, or four, of the wires, reduced, as hereafter shewn, to time of transit across the mean of all the wires, above the mean of the times of transit of the first limb across the remaining four, or three, wires, similarly reduced.) The ansæ of *Saturn's Ring* (not the body of the Planet, excepting when the Ring disappears) are observed in the same manner, for the purpose of obtaining the transit-time of the planet's centre and a measure of the external diameter of the Ring. The observed times of transit of *Mercury* and *Venus* apply only to their illumined limbs. In any case of the observation of a single limb of a

planet, the reduction to sidereal time of transit of centre is made by adding or subtracting the sidereal interval occupied by transit of semidiameter, as derivable from those calculated in the Nautical Almanac for every second day "at transit at Greenwich." The assumed values in arc of the semidiameters at the Earth's mean distance from the Sun from which the calculations are made, are given in page v. of the Preface. When both limbs are taken and the planet is sensibly gibbous, a correction for defect of illumination, calculated by a process that will be given afterwards, is applied. When planets are very small, transits of their estimated centres are taken.

112. In observations of Polaris and δ Ursæ Minoris, or other stars very near the pole, it is not possible to judge with any certainty of the instant of transit to the fraction of a second. As, however, the instant at which the star is exactly bisected, or is struggling, as it were, to cross the wire, may sometimes be noted with great precision, in every case an estimate is made of the transit-time to the tenth of a second. Transits of stars near the pole are required to be taken from time to time at all the wires, because, as stated in Art. 60, by means of these observations the equatorial intervals of the several wires from the mean of all are calculated, and these intervals it is necessary to know for a reason which will be stated in the next Article. But because of the slow motion of stars near the pole, each set of such observations takes a long time; on which account, in general, a shorter way of obtaining their transits across the mean of the wires is adopted, by employing the micrometer-wire in the following manner. First, just before or after the observation, the coincidence reading of the micrometer-wire with the middle wire D is found by the process indicated in Art. 51. Then the micrometer-wire is placed successively at the distances of three revolutions, two revolutions, and one revolution from D on the west side, and at the distances of one, two, and three revolutions on the east side, and transits of the star are taken at each of the six positions and at the middle wire, the intervals between the positions allowing of time for recording the seven transits. The mean of

the transit-times is the time of transit across the wire D, and the interval of this wire from the mean of all being known by the method explained in Art. 60, the transit-time across the mean of the wires may be inferred. As the Cambridge and Greenwich Transit-circles have no micrometer-wire, additional sets of fixed wires, the intervals between which are small, are inserted, one on each side of the middle wire, by means of which the abbreviated observation may be made in the manner above described; and in order that the mid-wire may be readily distinguished, the intervals between it and the two adjacent wires are larger than the others.

113. When transits of both limbs of the Moon, or a Planet, are taken at the same meridian passage, generally a correction is required on account of *defect of illumination* of one of the limbs, in order to obtain the exact time of transit of centre, and an exact measure of the diameter. The object of the following investigation is to give the means of calculating such corrections. Figure 26, which applies more expressly to cor-

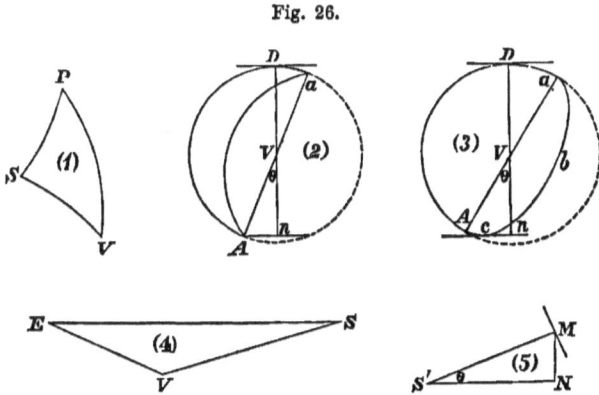

Fig. 26.

rections of measures of vertical diameters for defect of illumination, may be made use of for the present purpose. In (1) of Fig. 26, P is the pole of the heavens, S the Sun's place, and V

that of a planet at the time of taking the measures. The polar distance PS of the Sun and PV that of the planet, and the difference of their Right Ascensions SPV, will be supposed to be deduced from the Nautical Almanac, and from these data the $\angle PVS$ and the arc SV may be calculated. As transits of cusps are usually not taken on account of the large probable error of such observations, I shall suppose the Moon or planet to be gibbous, and to deviate little from a circular form. In that case the boundary of illumination opposite to the Sun, being the projection of a circle on a plane perpendicular to the line of vision, is an ellipse, as represented by the curve Aba in Fig. 26, (3). The angle of projection is equal to the inclination of the line of vision to the line drawn from the Sun's centre to the centre of the Moon or planet. In Fig. 26, (4) EVS is this angle, E, S, V being respectively the positions of Earth, Sun, and Planet in space, and according as the angle EVS is greater or less than $90°$, the illuminated body is horned or gibbous. Let Aa, the axis major of the ellipse, be equal to $2a$; then if the semi-axis minor $= b$, we shall have $b = a \cos EVS$, and $\sin EVS = \left(1 - \frac{b^2}{a^2}\right)^{\frac{1}{2}} = e$. Suppose $SE (= R)$ and $SV (= r)$ to be obtained from the Nautical Almanac. Then

$$e = \sin EVS = \frac{R}{r} \sin SEV = \frac{R}{r} \sin SV,$$

by which expression e can be calculated, SV being already known. Since DVn in (3) coincides with the circle of polar distance PV through V in (1), and AVa is at right angles to SV, it follows that the $\angle AVn (= \theta)$ is equal to $90° - \angle PVS$, and is therefore known. Hence the angle which a parallel of declination makes with VA is $\theta + 90°$. Now by Conic Sections, if θ' be calculated from the equality

$$\sin \theta' = e \sin (\theta + 90°) = e \cos \theta,$$

the length of the perpendicular from V on a tangent to the ellipse parallel to DVn is $a \cos \theta'$. Hence if I be the observed interval occupied by the transit of the diameter, and $2I_0$ be the true interval, $I = I_0 + I_0 \cos \theta' = 2I_0 \cos^2 \frac{\theta'}{2}$, assuming the observation to be exact. Hence $I_0 = \frac{I}{2} \sec^2 \frac{\theta'}{2}$. From this equation the interval of transit of semi-diameter, as given by the observation, may be calculated, and the value be compared with that in the Nautical Almanac. From a collection of such comparisons a correction of the tabular diameter in arc might be obtained, but not as accurately as from the micrometer measures of vertical diameters which will have to be considered in treating of the Mural Circle. When the transit-time of centre is inferred from the transit-times of both limbs, the correction required for defect of illumination is plainly $\mp \left(I_0 - \frac{I}{2}\right)$, according as the first or second limb is defective, and is therefore equal to $\mp \frac{I}{2} \tan^2 \frac{\theta'}{2}$.

114. The foregoing expressions may be applied in observations of Venus, Mars, and Jupiter, when the disks are approximately circular, and transits of both limbs are taken. For Jupiter the corrections will in any case be excessively small, and for the other planets they will often be far less than unavoidable errors of observation. It is, however, right in principle to correct errors however small, when they are due to known causes, and can be accurately calculated.

115. For the Moon the calculation of θ' may be simplified as follows. As the correction can be applied only when the Moon is nearly full, let in Fig. 26 (5) S' be the point opposite the Sun's centre, and M be the place of the Moon's centre; so that $S'M$ is a small arc. But in this case $\sin SV = \sin S'M = S'M$ nearly. Also since $\frac{R}{r} = 1$ nearly, and by Art. 113 $e = \frac{R}{r} \sin SV$, it follows that $\sin \theta'$ or $e \cos \theta = S'M \cos \theta = S'N$ nearly. Or θ' is nearly equal to $S'N$, which is the difference of R. A. of the Moon and the point opposite the Sun, multiplied by the cosine of the Sun's declination. After thus finding θ' with sufficient approximation, the corrections of the measure of diameter and

of transit-time of centre may be calculated by the formulæ obtained in Art. 113.

116. It will frequently happen that by the intervention of clouds, or from other causes, a transit observation is not taken at all the wires, in which case the mean of the observed transit-times is reduced (by formulæ which will presently be given) to time of transit across the mean of all the wires, and the weight of the observation is considered to have to that of a complete observation, the ratio of the number of the recorded times to the whole number of wires. According to Cambridge practice, the observation was inserted in the computing books, but not reduced, if it was taken at less than three wires. For the above-mentioned reduction of *broken* observations, it is necessary to be able to calculate the interval of any wire from the mean of all for any given polar distance. The method of obtaining these intervals for an equatorial star has already been indicated in Arts. 59—61. In the use of the formula,

$$\sin 15a = \sin \delta \sin 15t,$$

for this purpose (Art. 60), the polar distance δ was supposed to be taken from the Nautical Almanac (see Art. 89); but it should be noticed that this datum might be independently obtained by a contemporaneous observation with the Mural Circle, or one observer might determine δ and t simultaneously by means of a Transit-circle. We have next to ascertain how the intervals are to be calculated for any polar distance, and for any object, whether it be a star or a moving body.

117. For a star not very near the Pole, if $\delta =$ the star's polar distance, $h =$ a given equatorial interval, and $d =$ the required interval for the polar distance δ, we shall have, with sufficient approximation, $d = h \operatorname{cosec} \delta$.

118. For a star near the Pole, $\sin h = \sin \delta \sin d$. Suppose that in any year a considerable number of *complete* transit observations of Polaris, either above or below pole, are taken, and let Δ_0 represent a polar distance consisting only of degrees and minutes, chosen so as to be little different from the mean of the polar distances of Polaris in the period during which the

calculated intervals are to be used. It is first required to deduce from all the observations a mean value of the interval of any given wire from the mean of all for the polar distance Δ_0. Let d_1 be this interval, and let δ and d be the given polar distance and observed interval pertaining to any one of the observations. Then since $\sin h = \sin \delta \sin d = \sin \Delta_0 \sin d_1$, we have
$$\sin d_1 = \operatorname{cosec} \Delta_0 \sin \delta \sin d.$$

From the last formula, after converting d into arc, the value of d_1 in arc can be calculated. Each observation furnishes in this manner a value of d_1, and the mean of the several values, which mean will be called d_0, is the adopted interval corresponding to the polar distance Δ_0. The same process has to be gone through for each wire.

119. Now let $\Delta_0 + n$ represent *any* polar distance of Polaris, n being expressed in seconds, and let $d_0 + q$ be the corresponding interval. Then since $\sin(\Delta_0 + n) \sin(d_0 + q)$ is equal to $\sin \Delta_0 \sin d_0$, it will be found, if only first powers of the small quantities n and q be retained, that $q = -n \cot \Delta_0 \tan d_0$. The value of n may be taken from the pages of the Nautical Almanac containing the apparent places of Polaris; but as declinations are there given in place of polar distances, it will be convenient to substitute in the above expression $90° - D_0$ for Δ_0 and $-n$ for n. We shall then have for the interval $d_0 + q$ corresponding to the declination $D_0 + n$ the general expression
$$d_0 \text{ (in time)} + \frac{n}{15} \tan D_0 \tan d_0,$$

n being now taken immediately from the Nautical Almanac.

120. The same formula is applicable to any other circumpolar star in the Nautical Almanac, as δ Ursæ Minoris, or 51 (Hev.) Cephei; but in these cases we have the choice of determining D_0 by independent observations, or inferring its value from that adopted for Polaris. The latter method is perhaps to be preferred, the value for Polaris admitting of superior exactness of determination on account of the proximity of this star to the Pole. If d_0' and D_0' be corresponding values

of wire-interval and declination for the second star, those for Polaris being d_0 and D_0, we have

$$\sin h = \cos D_0 \sin d_0 = \cos D_0' \sin d_0';$$

or
$$\sin d_0' = \sec D_0' \cos D_0 \sin d_0.$$

Hence d_0', corresponding to a selected declination D_0', may be calculated, and then the interval corresponding to any other declination $D_0' + n$ may be obtained by means of a formula analogous to that for $d_0 + q$ given above.

121. The subjoined Table of the several intervals of the wires from the mean of all for the case of an equatorial star, and for given declinations of Polaris and δ Ursæ Minoris, was formed by calculating according to the rules indicated in Arts. 116, 118—120. The data for the calculation consisted of the transit-times of eleven complete observations of Polaris taken from October 3, 1853, to April 13, 1854, as contained in Volume XIX. of the Cambridge Observations.

Wire.	Interval for an Equatorial Star.	Interval for Polaris. Declination $= 88° 32' + n$.	Interval for δ Ursæ Minoris. Declination $= 86° 36' + n$.
	s.	m. s. s.	m. s. s.
A	− 40,352	− 26. 20,03 − 0,301n	− 11. 20,39 − 0,056n
B	− 26,867	− 17. 30,72 − 0,199n	− 7. 33,03 − 0,037n
C	− 13,577	− 8. 50,59 − 0,101n	− 3. 48,93 − 0,019n
D	+ 0,068	+ 2,68	+ 1,16
E	+ 13,521	+ 8. 48,37 + 0,100n	+ 3. 47,98 + 0,019n
F	+ 26,908	+ 17. 32,33 + 0,200n	+ 7. 33,72 + 0,037n
G	+ 40,310	+ 26. 18,35 + 0,300n	+ 11. 19,69 + 0,056n

The intervals (d_0) for Polaris corresponding to the declination $88° 32'$ are each the mean of the values derived by the process stated in Art. 118, from the eleven intervals given by the observations, and the intervals (d_0') for δ Ursæ Minoris corresponding to the declination $86° 36'$ were calculated from the values of d_0' by the formula in Art. 120. The above Table was used from September 21, 1853, to the end of 1854, on which account the selected values of D_0 and D_0' are 88° 32' and 86° 36'. The equatorial intervals were calculated from the same values of d_0 (see Art. 60). [The above Table was formed independently of that in page xxxiv of Volume XIX., from which it slightly differs, owing to differences of the data and details of the calculation.]

122. For observing transits of faint objects which admitted of very little illumination of the field, I found that *dark bars* of moderate breadth could be used with good effect. A set of such bars, two on each side of the system of wires, was inserted in the Cambridge transit-telescope between August 29 and September 21, 1853, and after that date transits of faint objects, such as the asteroids, were often taken at the bars, and generally at the *first* edges, the observations of disappearance being thought to be more certain than those of re-appearance at the *second* edges. These bar-transits are all reduced to transits across *the mean of the wires*, the intervals of the bar-edges from this mean being determined as follows. In all the instances, from September 26, 1853, to July 24, 1854, in which for this purpose transits of Polaris were taken at the wires and at the bars in the same meridian passage, the interval between each bar-transit and the mean of the wire-transits reduced in the usual way to the mean of all the wires, was determined. To each such interval for a given edge, and the corresponding declination of Polaris, a weight was attached proportional to the number of wire-transits, and from all the intervals and declinations, account being taken of the weights, a mean interval (t') corresponding to a mean declination (D'), was deduced. Then the equatorial interval (a') for the given edge was obtained by means of the formula, $\sin 15a' = \cos D' \sin 15t'$. The same computation was gone through for all the edges, and the eight intervals were arranged in the order of the transits of a star above pole for Illumination East. Then, for facility of calculation, a list was formed of the reductions to the mean of the wires for an equatorial star in all the cases that could occur of transits, broken and complete, at *first* edges, both for Illumination East and Illumination West, so that the reduction in

any instance of a bar-transit was obtainable by simply multiplying the proper equatorial reduction by the cosecant of polar distance. (See Cambridge Observations, Vol. XX., pages ii and iii).

123. In the case of the Sun, or any body of the Solar System, the diurnal motion is diminished by the apparent increment of Right Ascension due to the motion of the body in its orbit. If, therefore, I be the *horary increment* of its R. A., as given in the Nautical Almanac, or by other means obtainable, any interval for the moving body is greater than that for a star of the same polar distance in the ratio of 3600 to $3600 - I$. Hence for the Sun, or a Planet, or a Comet,

$$d = \frac{3600}{3600 - I} h \cosec \delta.$$

124. For the Moon another factor is required, in order to take account of parallax, the effect of which in this case is of sensible amount. In Figure 27, A is the place of the observer

Fig. 27.

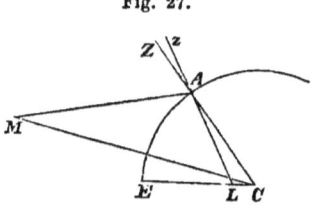

on the meridian AE, C the Earth's centre, CE the radius of the Equator, CAZ the direction of Geocentric Zenith, LAz that of Astronomical Zenith, and M the position of the Moon in its orbit, so that the angle AMC is the parallax (p). The effect of this parallax on the Moon's apparent motion may be determined as follows. If we suppose an angular motion about the Earth's axis, equal and opposite to the Earth's rotation, to be impressed in common on all bodies, the Earth will be, as it appears to be, at rest, and the motions of the other bodies will be just what they appear to be. But in that case if ω be the impressed angular motion and D the distance of the Moon from the Earth's axis, the motion impressed on the Moon will be $D\omega$; and if w be the Moon's orbital motion resolved parallel to the Equator, the whole motion in space about the Earth's axis is $D\omega + w$. This quantity, divided by AM the distance of the spectator from the Moon, gives the Moon's *local* angular motion, or that which is measured by a transit instrument at the place. The angular motion relative to C the Earth's centre is the same quantity divided by CM. Hence the above interval d is shortened by the parallax in the ratio of the distances, or the additional factor for the Moon is $\dfrac{AM}{CM}$. Now we have

$$\frac{AM}{CM} = \frac{\sin \angle ACM}{\sin \angle ZAM} = \frac{\sin(Z_1 - p)}{\sin Z_1},$$

also $\dfrac{AC}{CM} = \dfrac{\sin \angle AMC}{\sin \angle ZAM} = \dfrac{\sin p}{\sin Z_1}$; and $\therefore \sin p = \dfrac{AC}{CE} \cdot \dfrac{CE}{CM} \cdot \sin Z_1$.

Suppose that simultaneously with the transit observation the Moon's apparent zenith distance zAM was found by a Mural Circle or a Transit Circle to be Z, and that the local angle of the vertex $(= \epsilon)$, and ratio $\dfrac{AC}{CE} (= \rho)$, are known. Then $Z_1 = Z - \epsilon$; and if P be the Moon's Equatorial Horizontal Parallax, $\dfrac{CE}{CM} = \sin P$. Hence we have

$$\sin p = \rho \sin P \sin (Z - \epsilon),$$

by which formula, after deriving P by interpolation from the Nautical Almanac, p may be calculated. Then the required factor, being equal to $\mathrm{cosec}\,(Z - \epsilon) \sin (Z - \epsilon - p)$, is readily obtained. (In this calculation the arcs p and P should not be substituted for their sines.) For the latitude of the Cambridge Observatory $\rho = [9.9990916]$, and $\epsilon = 11' 12''$, the ratio of the Earth's axes being assumed to be that of 297 to 298.

If the Moon's apparent zenith distance be not determined by observation, the ratio of AM to CM may be calculated as follows. If $\lambda = \angle ALE$, the latitude of the observatory, $\angle ACE = \lambda - \epsilon$; and if $\Delta = \angle MCE$, the Moon's Geocentric Declination, to be obtained by interpolation from the Nautical Almanac, the $\angle ACM = \lambda - \epsilon - \Delta$.

Hence $\quad \dfrac{AM}{CM} = \dfrac{\sin \angle ACM}{\sin \angle ZAM} = \dfrac{\sin(\lambda - \epsilon - \Delta)}{\sin(\lambda - \epsilon - \Delta + p)}.$

Also $\quad \sin p = \rho \sin P \sin(\lambda - \epsilon - \Delta + p);$

or $\quad \tan p = \dfrac{\rho \sin P \sin(\lambda - \epsilon - \Delta)}{1 - \rho \sin P \cos(\lambda - \epsilon - \Delta)}.$

It will generally be sufficiently accurate to omit the small term in the denominator. After obtaining p from this equation, the ratio of AM to CM is obtainable from the formula above. (For the Cambridge Observatory the value of $\lambda - \epsilon$ to the nearest second is $52° 1' 40''$.) Putting F for this factor, we have for the Moon

$$d = \frac{3600}{3600 - I} Fh \operatorname{cosec} \delta,$$

I being taken immediately from the column of 'Variation of the Moon's R. A. in one hour of longitude' under the head of Moon-culminating Stars in the Nautical Almanac.

125. I take occasion to remark here that although for the Moon the interval of any wire from the mean of the wires is partly dependent on parallax, the measurement of the Moon's diameter by transits, spoken of in Art. 110, is not thereby sensibly affected. The reason of this is that the Moon's apparent diameter is greater to the spectator at A than to the spectator at C, in the same proportion that the apparent motion is greater, so that the duration of the transit of diameter is independent of the place of the observer. This is true also of transit measures of the Moon's diameter taken out of the meridian with an Equatorial Instrument.

126. Having shewn how to calculate the interval of the transit of any celestial body across any wire from the mean of the transit-times of a complete observation, we are now able to obtain a rule for applying a correction for reducing to that

mean the mean of the observed times of a broken observation. The rule may be inferred from a particular instance, such as the following. The number of wires being seven, suppose the times to be observed at four wires and to be omitted at three, and let m be the required reduced time of transit across the mean of all the wires. Then if $m+a$, $m+b$, $m+c$, $m+d$ be the observed times, and $m+e$, $m+f$, $m+g$ the omitted times, we shall have

$$m = \frac{(m+a)+(m+b)+(m+c)+(m+d)}{4} + \frac{e+f+g}{4},$$

supposing the observed times to be correct, because in that case $a+b+c+d+e+f+g=0$. The signs of a, b, c, &c. are positive or negative according as they designate intervals after or before mid-transit. (See Table in p. 112.) Hence the following rule:—Obtain the mean of the observed times, and add to it the sum of the calculated intervals of the lost wires from mid-transit divided by the number of the observed times.

127. The process of reducing an incomplete observation of any star to the time of mean transit having been indicated, we may now proceed to take into consideration the method of obtaining Level error by transits of Polaris which was mentioned in Art. 87. Suppose, first, that there is neither collimation error, nor azimuth error, and on that supposition let $HZ'hN'$ in

Fig. 28.

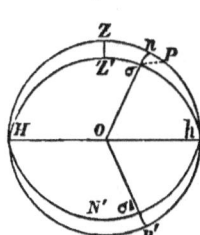

Figure 28 represent the great circle described by the line of collimation, cutting the great circle of the meridian at H and h

in the horizon. Then the inclination of the plane of collimation to the plane of the meridian is the ∠ ZHZ', or the small arc ZZ' drawn through the zenith at right angles to the two planes. A transit of Polaris is taken at four wires by reflection at the surface of mercury, and at the remaining three by direct observation, the slow motion of the star allowing time, after securing the reflection observation, to set for the other. The mean of the noted times is in each case reduced to time of transit across the mean of the wires by the method explained in Art. 126, a table of intervals being employed such as that for Polaris in page 112. Let t_1 and t_2 be the reduced times for the reflection and direct observations respectively. Then since by the principle of reflection ∠ $\sigma O h$ = ∠ $\sigma' O h$, the small arcs σn and $\sigma' n'$ are equal, and the time in which the star describes either is $\frac{1}{2}(t_1 \sim t_2)$. Let PZ the colatitude of the observatory = l, and Pn the star's polar distance = δ, so that $Zn = l - \delta$ for Polaris above Pole. Then σn, which is equal to $ZZ' \cos(l-\delta)$,

is also equal to $\qquad \frac{15}{2}(t_1 \sim t_2) \sin \delta.$

The correction for Level error is *positive* if the *West* end of the axis be highest (Art. 78), and in that case Polaris crosses the plane of collimation *before* passing the meridian in the *direct* observation, and *after* passing the same in the *reflection* observation. Hence t_1 is greater than t_2, and we have

$$ZZ' = \sigma n \sec(l - \delta) = \frac{15}{2}(t_1 - t_2) \sin \delta \sec(l - \delta).$$

If Polaris be below Pole, t_2 is greater than t_1 for a positive azimuth error, and the arc Zn is $l + \delta$. Hence the same formula applies by merely changing the sign of δ. Or, taking δ always positive, and the respective times for the reflection and direct observations to be always t_1 and t_2, the general formula is

$$\text{Level-correction} = \pm \frac{15}{2}(t_1 - t_2) \sin \delta \sec(l \mp \delta),$$

according as Polaris is above or below Pole.

The effect of collimation error, if it existed alone, would be to make the line of collimation describe a small circle of the sphere parallel to the plane of the meridian, and consequently, as the reduced transit-time would be the same for the direct, as for the reflection, observation, $t_1 - t_2$ would be zero. Also if there were only azimuth error, each reduced time would be the transit-time across the *same* vertical plane inclined at a small angle to the plane of the meridian, and hence as far as depends on azimuth error $t_1 - t_2$ will again be zero. Thus, since each small error is independent of the amounts of the other two, the determination of Level error by this method is not affected by actual small errors of collimation and azimuth.

128. When the Level error has been found either by the Spirit-level, or by the foregoing method, we have the choice between two ways of determining azimuth error. For we may either obtain the sum of the collimation and level errors by the collimating eye-piece (Art. 84), whence, the level error being known, the collimation error may be inferred; or we may find the collimation error directly by means of collimators (Arts. 53—56). The collimation and level errors being thus in either case both ascertained, the azimuth error may be calculated by the methods indicated in Arts. 88 and 89. It will be seen that these two ways of finding the three errors do not require the instrument to be reversed, and are, therefore, applicable when a Transit-Circle is used without reversion. It is, however, to be remarked, that as the determinations ought to be free from any effects of irregularities of the forms of the pivots, the correction of such effects should be provided for by special measures of the co-ordinates of the dots (Arts. 94—96) for all the positions of the telescope that are employed for calculating the three errors. Upon the whole, for a non-reversed Transit-Circle, the best method seems to be to determine the collimation and level errors by combining with the use of collimators that of the collimating eye-piece.

129. To complete what has to be said relative to the Transit Instrument under the *Third Division*, it remains to

give an account of *the Chronographic*[1] *method of registering transits by means of a galvanic circuit*. The apparatus employed at Greenwich for this purpose is described, with illustrations by plates, in the Appendix to the Volume of Greenwich Observations for 1856, to which "Description" recourse must be had by those who desire to be fully acquainted with all the details of this method. As far as regards these Lectures it will suffice to give a general account of the chief parts of the apparatus, such as that contained in the Introductions to Volumes of the Greenwich Observations (in the Volume, for instance, for 1874, pp. x and xi), which for the most part I have followed in the subjoined statements.

130. A clock being mounted in a suitable position, its movement is governed by the conical rotation of a sidereal seconds' pendulum, and to ensure regularity of the movement, which is an essential condition, apparatus is provided whereby, as the radius of the conical arc increases, a revolving plate is

---

[1] This is also called the American method, having been, both as to principle and working, invented solely by physicists and astronomers of the United States. I was first made acquainted with it by Professor Mitchel of Cincinnati, some time in 1848 or 1849, on the occasion of his visiting the Cambridge Observatory, when, if I remember rightly, he had come from America for the purpose of executing a commission to procure in Europe a large telescope for the Cincinnati Observatory. The method he described to me involved the principle of automatic registration of seconds by dots impressed on a revolving circular disk by a point acted upon by a galvanic current in connection with the seconds' pendulum of an astronomical clock; a radial shifting of the recording point after each revolution of the disk; and means of impressing dots by another point acted upon by the same galvanic current at the will of the observer. By this arrangement, which comprised all the more essential parts of the invention, the seconds' dots radiated from the centre of the disk, and the seconds' intervals were, therefore, measured by arcs increasing with the radius. Mr W. C. Bond, Director of the Harvard College Observatory, introduced a great improvement by making the records on a cylinder, covered with paper, which was caused by appropriate mechanism to revolve uniformly about its axis, and at the same time to have a uniform slow motion in the direction of the axis. By these means the seconds' dots are separated by equal intervals, and are arranged without confusion on the turns of a spiral. (For a history of the successive steps of this important invention, see *Annals of the Harvard Observatory*, Vol. I., Part I., pp. xxiv—xxviii.)

dipped deeper in an annular trough of glycerine, thus supplying increased resistance without affecting the centrifugal force upon the pendulum. A spindle of the clock gives a revolving motion to a brass barrel, about 20 inches long and 12 inches in diameter; and as the clock with conical pendulum moves without jerks, the barrel revolves with a motion sensibly uniform. A point on the cylindrical surface moves through 0,3 inch nearly in a second of time. The surface is completely covered with a sheet of paper, which, when filled by the registration, can be easily removed, to be replaced by another sheet. A different spindle of the clock turns two long screws, which are both parallel to the axis of the barrel, and cause a travelling frame to traverse its whole length. In one revolution of the barrel the frame moves through 0,1 inch. The travelling frame carries two levers, each armed at one end with a pricking point, and so mounted that when the opposite end is pulled away from the barrel, the pricking end is impressed on the barrel, and makes a permanent puncture on the paper. The prickers are mounted in such manner that when their points have entered the paper, they yield laterally to the motion of the revolving barrel, and do not scratch the paper. [In Bond's apparatus (see Note p. 120) a pen traces a continuous spiral line on the sheet of paper, and the records are indicated by small *offsets*, those for the observations being distinguished from the others by being somewhat longer.] Two galvanic magnets are fixed on the travelling frame, so as to attract the lever ends opposite the pricking points, and thus in order to produce punctures on the paper by these points, it is only required to send galvanic currents through the magnets.

One of the prickers registers the seconds of the Sidereal Standard Clock; for which purpose the wires of its magnet, after passing through a galvanic battery, are led to the Sidereal Standard Relay, which completes the circuit, and produces a puncture at each second, excepting the first second of each minute. The omission of the corresponding puncture serves to indicate the commencement of each minute. The circuit can be broken at pleasure, and the movement of the travelling-frame be suspended when not required for recording observations.

The other pricker is used for registering times of observation. The wires of its galvanic magnet, after passing through a battery, are led to the Transit-Circle-pier, and to insulated brass rings on the transverse axis of the telescope, whence conducting wires pass within the telescope-tube to a contact-piece near the eye-end, where the observer by a touch of the finger can complete the circuit, and thus make a puncture on the revolving barrel at the instant he judges the object observed to be on a transit-wire. The same arrangement is made available for transit observations with three other instruments, the Altazimuth, the Great Equatorial, and the Sheepshanks Equatorial. Also as the clock Hardy is still employed for eye-and-ear observations of stars near the pole, its contact springs are occasionally thrown into circuit with the magnet of the observations-pricker, so as to give the means of obtaining from the chronograph-register the difference between its time-indications and those of the Sidereal Standard Clock.

131. The chronographic method of recording transits has two special advantages; (1) it allows of taking the times of transit of an object across wires separated by small intervals, and, consequently, of observing several objects in quick succession, at the same time that each is taken at a sufficient number of wires; (2) the estimation of the fraction of a second can be made with greater certainty by this method than by the eye-and-ear observation, inasmuch as the position of the observations-dot relative to the antecedent seconds-dot may be estimated from the register at leisure by sight in tenths of the seconds-interval, or if great precision be required it may be read off by applying to the interval a scale of ten equal parts. Moreover, the discordance between transits taken by different observers, due to what is termed *personal equation*, is found to be less when the time of transit is recorded by means of the senses of sight and touch, than when the transits are observed by eye and ear. The mode of eliminating the influence of personal equation on the results of transit observations will come under con-

sideration in the next *Division*. I propose to add here, as an appropriate conclusion of *Division III.*, some remarks on the character of this peculiar error, and its amount under different circumstances.

132. In Volume XLIX. of the *Astronomische Nachrichten*, cols. 1—30, the editor, Professor C. A. F. Peters, has given an elaborate and interesting account of facts relating to personal equation, together with a discussion of its influence in chronographic registration of transits, as compared with that in the ordinary method of observing transits. The following particulars have been extracted from that article. Maskelyne, in his account of the Greenwich Observations of 1795, made the remark that although the transits of his assistant Kinnebrook agreed with his own at the beginning of that year, in August they were $0^s,5$ later, which difference increased in 1796 to $0^s,8$. Bessel, induced by that remark, subjected the facts at his command to a full enquiry, from the results of which (given in Part 8 of the Königsberg Observations) it appeared that Walbeck observed later than himself by $1^s,02$, and Argelander by $1^s,22$, and that, as shewn by direct comparisons, Struve's transits were later by $0^s,04$ in 1814, and by $0^s,77$ in 1834. In the same Article mention is made of other differences of personal equation, not quite so remarkable: Wolfers observed earlier than Nehus by $0^s,73$, Gerling later than Nicolai by $0^s,78$; Rogerson at Greenwich observed in 1854 earlier than Main by $0^s,70$, although in 1840 he observed later by $0^s,15$, whereas in the whole of the twelve years 1841—1853, between Main and Henry there was scarcely any difference. These facts shew that personal equation is very different in different individuals, and in the same individual is not a constant quantity. Professor Peters has also stated that although Bessel's observations of transits were earlier than Argelander's by more than a second, his observations of *sudden* phenomena, as occultations of stars by the Moon, were not earlier by more than $0^s,28$.

133. For judging of the comparative values of transits observed by the chronographic register and by the usual method Professor Peters has given the results of five sets of observations,

made by five combinations of two out of four observers, each set of observations consisting of transits of stars taken by eye and ear on one day, and on the day next before or after by galvanic registration. The stars observed by the two methods were for the most part the same in the same set, but different stars were observed in the different sets. Each observer took transits of each star at two or at three wires by the usual method, and at three or at four wires by the register-apparatus. The general result of 53 comparisons may be stated to be, that personal equation is less subject to variation, and the probable error of the transit-time of a star across any wire is considerably (erheblich) less, by the new method than by the old. It appears, in fact, from all the comparisons that the probable error of a single transit is $0^s,082$ by the usual method, and $0^s,060$ by the register-apparatus, according to one computation, and $0^s,087$ and $0^s,055$, according to another. It may also be mentioned, as farther indication of the superiority of the latter method, that it agrees better than the other with observations, such as those of Polaris, which are not affected by personal equation. From the foregoing considerations it is evident that differences of personal equation have to be determined, whether the transits be observed by the galvanic or the ordinary method, as is done, for instance, very fully from year to year for the galvanic method in the Introductions to the Greenwich Observations. (See the Introduction to the volume for 1873, pp. xliv—liii.)

134. From the experience I had as Director of the Cambridge Observatory I am able to add the following particulars respecting personal equation in eye-and-ear observations. From reduced transits taken in the year 1841, nearly contemporaneously, by myself (C) and one of my assistants, Mr John Glaisher (G), I was led to infer that our observations were "very nearly alike," and from like contemporaneous transits taken in the same year by G and my other assistant, Mr Baldrey (B), it appeared that B observed earlier than G by $0^s,28$. (See *Cambridge Observations*, Vol. XIII. for 1840 and 1841, p. xxxviii.) These observations were not made with any

reference to personal equation. In 1843, July 24, 12^h—13^h, transits were specially taken for determining the difference of the personal equations of C and G. Eight stars of various polar distances were selected, one observer took the transit of the star at the first four wires, and the other at the remaining three, and the successive observations of the transits were commenced by the two observers alternately. The mean result of the eight comparisons was, that C observed earlier than G by $0^s,004$, which is so small a quantity that the difference of the personal equations may be taken to be zero. The particular results, however, indicated a small difference depending on the polar distance of the star, C observing the slower moving stars later, and the quick moving stars earlier, than G. (See Vol. XV. for 1843, p. xii.) In the Introductions of the subsequent Volumes of Cambridge Observations, determinations of differences of personal equations are given in a large number of instances, in consequence of frequent changes of the observers having occurred. All these determinations were made by reducing independently the transits of each observer to the same epoch, 0^h of sidereal time, the clock's rate being in no case inferred from the transits of different observers. This method I consider to be preferable to that spoken of above as having been employed on July 24, 1843. The amounts of difference of personal equation range generally between $0^s,2$ and $0^s,4$, but in several instances are much smaller, not exceeding the twentieth of a second. I found, for instance, that the transits of Mr Charles Todd, Mr Carrington, my son J. L. Challis, and, as is stated above, those of Mr John Glaisher, agreed closely with my own. The last-named observer agreed with Mr James Glaisher of the Greenwich Observatory, who, as I have understood, agreed with the Astronomer Royal, as also did Mr Main, and it has already been mentioned that there was very little difference between Mr Main and Mr Henry. Here then are *nine* observers, under various circumstances, whose transits were found to be in accordance; from which it would seem allowable to infer that there exists a faculty, although not possessed by all observers, of taking transits nearly, or wholly, unaffected by personal equation; that, in short, there is a right way and a wrong way of observing transits by eye and ear.

135. Another peculiarity, which is of the nature of personal equation, has been detected in transit observations of the circular discs of the Sun, Moon, and Planets, some observers recording transit-times of the two limbs which are neither consistent with those of stars nor with each other. In the Introduction to Volume XIII. of Cambridge Observations, for 1840 and 1841, pp. xxii—xxiii, I have adduced a special investigation of corrections to be applied on this account to Mr Baldrey's observations of transits of the Sun's limbs, and to the inferred Right Ascension of the Sun's centre. Like corrections are also obtained, but with less certainty, for transits of the limbs of the Moon and of several Planets. The discrepancy appears to arise chiefly from discordant observations of second limbs. It may here also be mentioned that I detected a constant difference between my own microscopic bisections of a graduation-line of the Mural Circle and those of one of my assistants; which, probably, was to be attributed to certain physiological conditions of the organs of sight, differing in that respect from personal equation, which seems to have relation to mental conditions.

IV. CALCULATION OF APPARENT RIGHT ASCENSIONS FROM THE TRANSIT OBSERVATIONS, AND OF THE MEAN RIGHT ASCENSIONS OF STARS.

136. Before indicating the processes which in the existing state of Practical Astronomy are employed for inferring apparent Right Ascensions from reduced meridian transits, and obtaining concluded Mean Right Ascensions of stars, I propose, with the view of distinctly exhibiting the principles involved in those processes, to state historically how they have actually been discovered, and to shew also how they might be logically arrived at by beginning *de novo*. The requisites for freeing the transit-instrument from the effects of faulty mechanical conditions, and placing it exactly in the plane of the meridian,

were very imperfectly understood by astronomers who preceded Flamsteed, and even the methods which he employed for these purposes fell far short of attainable accuracy. It appears from what is said in *Division II.*, that it is possible to effect all the instrumental adjustments without knowing the Right Ascension of *any* star (see Art. 90). This having been done according to rule, an astronomer provided with a clock going pretty regularly, might, although his transit-instrument were imperfectly adjusted, ascertain *the clock's rate* with much accuracy by transits from day to day of the same unknown star, or stars, inasmuch as the transits of a given star might be supposed, for a few consecutive days, to be equally affected by defect of instrumental adjustment. The clock's rate, which is the difference between the clock-times of transit of the same star on consecutive days and 24 hours, being thus sufficiently determined, he might proceed to calculate the differences of *sidereal intervals* between the transits of different stars, and after fixing upon one of them as a point of reference, the sidereal intervals between the transit of that one and the transits of all the others on the same day might be inferred. These intervals, which would be affected by imperfect instrumental adjustment and other sources of error, might, however, be called provisionally *Right Ascensions* of the different stars, such determinations of Right Ascension having been actually made by the older astronomers. Flamsteed, having undertaken to observe Sun, Moon, and Planets, as well as stars, and perceiving the advantage of choosing a point of reference not defined by the place of a star, selected for his purpose the point of intersection of the Equator with the Ecliptic, which is named *the first point of Aries.* There is, besides, a particular reason why a star should not be chosen for point of reference, arising out of the well ascertained fact that stars have *proper motions*, in consequence of which the determinations of Right Ascension might be affected to an unknown amount, and in different degrees, by the differences of the proper motions of the selected star and those of the other stars. The adoption of the equinoctial point for the origin of arcs of Right Ascension, gets rid

128   PRACTICAL ASTRONOMY.

of this source of error, and at the same time allows of independent determinations of the proper motions of stars in R. A. Accordingly I proceed now to describe a method of determining the position of the first point of Aries, which, although differing in details from that employed by Flamsteed, is the same in principle.

137. It is not possible to determine the place of the Equinox except by combining transit observations of the Sun with simultaneous observations of its declination by a Mural Circle, or a Transit-Circle. Hence, although the mode of taking observations of declination has not yet been under consideration, I shall anticipate so far as to assume that transits of the Sun's centre and observations of its declination are taken at the same time at several meridian passages both preceding and following the *Vernal Equinox*. (As respects the *logic* of the present enquiry it may be remarked, that the supposed observations of the Sun's declination are simply obtained by instrumental means, and are in no way dependent on determinations of places of stars, or of their right ascensions.) Supposing, in accordance with what is stated in Art. 136, the errors of collimation, level, and azimuth, of the Transit Instrument to be ascertained prior to any determination of the right ascensions of stars, and the clock to be rated by transits of *unknown* stars, let t_1, t_2, t_3, &c. be the clock-times of the several transits of the Sun's centre observed near the Equinox, corrected both for instrumental errors and for clock-rate, so that $t_2 - t_1$, $t_3 - t_2$, &c. are true sidereal intervals between the transits. Then if d_1, d_2, d_3, &c. be the simultaneously observed declinations of the Sun's centre, and d the declination at any arbitrary time t, and if we take, for the sake of simplicity, only three transits, we shall have by a known formula of interpolation

$$d = \frac{(t-t_2)(t-t_3)}{(t_1-t_2)(t_1-t_3)} d_1 + \frac{(t-t_1)(t-t_3)}{(t_2-t_1)(t_2-t_3)} d_2 + \frac{(t-t_1)(t-t_2)}{(t_3-t_1)(t_3-t_2)} d_3.$$

Suppose now that $d = 0$, or that the Sun is at the Equinox. Then for determining the corresponding clock-time t_0, we have an equation of the form $0 = A + Bt_0 + Ct_0^2$, in which A, B, C

are given functions of t_1, t_2, t_3 and d_1, d_2, d_3. Hence, after obtaining a positive value of t_0 from the solution of this equation, if we subtract this value from each of the times t_1, t_2, t_3 we obtain sidereal intervals reckoned from the instant at which the Sun passed from the South to the North side of the Equator. (Similarly, sidereal intervals reckoned from the instant the Sun passed from the North to the South side of the Equator might be obtained by observations made near the Autumnal Equinox.) We have now to find the clock-error when the Equinox is on the meridian of the place of observation, which is equivalent to making the clock indicate *local* sidereal time.

138. The rate of change of the Sun's right ascension about the time of the Equinox may be inferred as follows from the above mentioned Transit observations. Let, for instance, a be the increment of the Sun's R. A. in the interval $t_n - t_{n-1}$ between two meridian transits, which, for the sake of simplicity, will be supposed to be the one next before, and the one next after, the passage of the Sun through the equinoctial point. Now this increment takes place in $24^h + a$ of sidereal time, so that $t_n - t_{n-1} = 24^h + a$; or $a = t_n - t_{n-1} - 24^h$. Hence the increment in the interval $t_n - t_0$ is $(t_n - t_{n-1} - 24^h) \dfrac{t_n - t_0}{t_n - t_{n-1}}$, which must be the Sun's actual right ascension at the epoch t_n, because at the beginning of that interval the R. A. was zero. Hence the *correction* of the clock-error at the time t_n is the excess of that R. A. above t_n, which is equal to $-t_0 - 24^h \dfrac{t_n - t_0}{t_n - t_{n-1}}$. So the correction of clock-error at the time t_{n-1} is $-t_0 + 24^h \dfrac{t_0 - t_{n-1}}{t_n - t_{n-1}}$. (The first or second of these expressions should be preferred according as $t_n - t_0$ is less or greater than $t_0 - t_{n-1}$.) When the clock has been actually, or virtually, corrected according to either expression, the seconds' hand is made to point to $0^h,0$ when the Equinox is on the meridian, and consequently the clock will indicate local sidereal time. The error of the clock at a given epoch being thus corrected, and its rate being supposed to be continuously determined, its error at the clock-time of the meridian transit of any

object can be deduced, and accordingly the clock and transit-instrument be employed to record the R. A. of any number of objects as referred to the Equinox of the given epoch. The rate of regression of the Equinox being supposed to be uniform and of known amount, the origin of these R. A. might be transferred to the position of the Equinox at any other given epoch; and it is evident that the R. A. of a given star, as determined by transits on different days, might also be referred to this origin, and the mean of several determinations be adopted as the value at the new epoch.

139. Flamsteed formed his catalogued right ascensions in the *Historia Cœlestis* for the most part in the above manner, making use of a very accurate value of the precession of the Equinoxes. But his results were affected by two sources of error with which he was unacquainted. Assuming the regression of the Equinoxes, as well as the earth's rotation about its axis, to be uniform, astronomers at that time inferred that the arc of the Equator between the Equinox and the Circle of Declination of a star might be exactly measured by a correctly regulated clock. Bradley discovered that for two reasons this is not the case. First, because the actual regression of the Equinox is not uniform, but oscillates about a mean regression, and again, because, by reason of the *aberration of light*, the apparent place of a star undergoes variations depending both on its position in the heavens and on the season of the year. Bradley, after allowing for regular precession, distinguished by his observations between these two kinds of irregularity, and not only determined the quantities and laws of both, but ascribed them to true causes, the variation of the rate of precession to nutation of the earth's axis, and aberration to a relation between the velocity of the earth in its orbit and the velocity of the propagation of light. The former cause has been confirmed by the theory of gravitation, Laplace having found that the attractions of the Sun and Moon on the protuberant equatorial parts of the earth produce a nutation of its axis agreeing in law and amount with the results of Bradley's observations. Respecting what I conceive to be the rationale of the observed effects of aberration

in telescopic observations I propose to introduce a Note in a subsequent part of this work. At present we may proceed to consider how transit observations are freed from the consequences of these two sources of error.

140. Admitting it to be proved by the theory of gravity that the uniformity of the earth's diurnal rotation is not affected by the nutation of its axis, it is evident that the difference of the meridian-transits of two stars, as shewn by a well regulated clock, does not exactly measure the difference of their R. A., inasmuch as the pointing of the telescope towards any star will be altered in a certain manner by such nutation. At the same time it is altered by the apparent displacement of the star by aberration. Bradley's discoveries gave the means of calculating precisely the amounts by which the visual direction of the star is changed in R. A. by each of these causes, and consequently of correcting the instrumental measurement of the apparent difference of R. A. of the stars so as to reduce it to the true difference, as measured by a clock regulated according to the rate of the earth's rotation. It should also here be noticed that the theory of gravity has detected another cause of error due to the circumstance that the attraction of the earth by the Planets changes the inclination of the plane of the earth's orbit to the plane of its Equator, thereby producing a small *progression* of the first point of Aries along the Equator, the law and amount of which have been determined by the same theory. This is *Planetary Precession*, as distinguished from *Lunisolar Precession*, the two together constituting *General Precession*. When the fluctuations of value arising from these several causes have been eliminated from the instrumental measures of the difference of R. A. of two stars, the resulting *mean* measures are comparable with each other, and from a large number of such determinations an exact measure of the difference of the mean R. A. of the stars at the epoch of the mean of the times of the measures may be inferred. The different measures, for greater certainty, may be extended over a considerable number of years. Maskelyne selected for such observations 36 stars suitably distributed over the heavens, and bright enough to be seen in day-

time, and compared each in R. A. a great number of times with one of them, α Aquilæ. The R. A. of this one was referred, by the intervention of observations of the Sun, to the mean place of the Equinox at a certain epoch, the process being mainly the same as that of Flamsteed, which has already been described in Arts. 137 and 138. After determining the absolute R. A. of the 36 stars in this manner, Maskelyne named them *Fundamental Stars*[1]. The Apparent and Mean Places of fundamental stars form an essential part of the contents of the Nautical Almanac[2]. At the present time this work contains the mean places of 149 standard stars for the beginning of the year, the apparent places of 144 for every tenth day of the year, and the apparent places of the remaining five, which are near the North or South pole, for every day. The mean places of 136 were derived from the Seven-year Catalogue forming Appendix I. of the volume of Greenwich Observations for 1862.

141. In the determination of the place of the Equinox by Flamsteed's method, only observations of the Sun taken near the times of the Equinoxes are made use of. In the Volumes of the Greenwich Observations, as also in those of the Cambridge Observations, the place of the first point of Aries and the obliquity of the Ecliptic, together with the mean error of the assumed R. A. of the fundamental stars, are calculated from *all* the meridian observations of the Sun taken in the course of any one year, beginning at January 1. The process by which this is done cannot be logically indicated till the reduction of Circle observations of the Sun's declination has been treated of; at present we may go on to enquire in what way the apparent R. A. of objects are deduced when their transits have been taken contemporaneously with those of fundamental stars.

142. The formulæ now generally employed for calculating apparent places of stars from mean places, or, reversely, mean places from apparent, depend upon astronomical researches of

[1] See Grant's *History of Physical Astronomy*, p. 489. Respecting the history of *Practical* Astronomy generally, much valuable and interesting information is given in Chapter xviii. of that work.

[2] The original publication of the Nautical Almanac, which was due to Maskelyne, took place in 1767.

Bessel contained in his *Fundamenta Astronomiæ* published in 1818, and his *Tabulæ Regiomontanæ* published in 1830. The former work consists principally of two catalogues of all the observations of stars made by Bradley at Greenwich in the years 1750—1762, reduced to the beginning of 1755, together with discussions relative to the constant of aberration and the Precession of the Equinoxes. In both works Bessel commences his investigations relative to Precession and the Nutation of the earth's axis with adopting Laplace's gravitational formulæ given in Tom. III. of the *Mécanique Céleste*, but without accepting the numerical values of the principal constants. These he undertook to correct, first, by adopting a more recent determination of the mass of Venus, and then by calculating *general precession* exclusively from comparisons of Bradley's places of stars 1755,0 with places in Piazzi's Catalogue of 1800,0, after applying a general correction to the latter deduced from *absolute* Right Ascensions of Maskelyne's 36 fundamental stars inferred from his own observations. By this means he finally obtained in the *Tabulæ Regiomontanæ*, p. v., for the indefinite epoch $1750 + t$, the subjoined formulæ, in which ψ is the *lunisolar precession*, or movement, in the interval t, of the point of intersection of the Equator with a *fixed* great circle with which the Ecliptic coincided at the beginning of 1750, ψ_1 is the *general precession* or movement in the same interval of the intersection of the Equator with the actual position of the Ecliptic, ω is the arc between the Pole of the Equator and the Pole of the fixed Ecliptic at the epoch $1750 + t$, and ω_1 that between the Poles of the Equator and the actual Ecliptic at the same epoch:

$$\psi = 50'',37572 t - 0'',0001217945 t^2,$$
$$\psi_1 = 50'',21129 t + 0'',0001221483 t^2,$$
$$\omega = 23° 28' 18'',0 + 0'',00000984233 t^2,$$
$$\omega_1 = 23° 28' 18'',0 - 0'',48368 t - 0'',00000272295 t^2.$$

According to these formulæ the constant of general precession at 1750,0 is 50'',21129, and the annual diminution of the obliquity of the Ecliptic is 0'',48368. For constant of annual aberration Bessel adopts Delambre's value, 20",255, as being nearly the same as that given by Bradley's observations of γ Draconis. Also taking the Sun's parallax to be 8",5776 as found by Encke, he obtains for the constant of diurnal aberration 0",30847 (*Tab. Reg.* p. XXII). The values of Solar Parallax and constant of annual aberration which I have employed in Art. 63 for calculating local diurnal aberration, give a value of that constant larger than 0",30847 by 0",01576.

143. Since the publication of the *Tabulæ Regiomontanæ*, astronomers have bestowed much labour on investigations for correcting as accurately as possible Bessel's values of the principal constants. This may be pre-eminently said with respect to three dissertations contained in the Memoirs of the Imperial Academy of Sciences of St Petersbourg, Tom. III. (1844), pp. 17—285, the first of which, by Otto Struve, is a discussion on the constant of Precession, the second, by Professor C. A. F. Peters, treats of the numerical constant of Nutation, and the third, by W. Struve, is on the constant coefficient of the Aberration of fixed stars. It will be proper to give here some account of the processes and results of each of these important contributions to Practical Astronomy.

(1) The value 50",21129 of general precession was obtained, as stated in Art. 142, by comparing Bradley's R.A. of 1755,0 with Piazzi's of 1800,0, the interval being 45 years. O. Struve compared Bradley's R.A. of the same date with R.A. observed at Dorpat reduced to the epoch 1825,0, and thereby found that in the interval of 70 years the movement of the Equinox was greater by 0",97 than that given by Bessel's constant. Accordingly this constant requires to be increased by one-seventieth part of the difference, or 0",01386. To calculate the value of Bessel's constant at the epoch 1790,0, which is the mean between 1755,0 and 1825,0, we have

$$\frac{d\psi_1}{dt} = 50",21129 + 0",0001221483 \times 80 = 50",22106.$$

The addition of 0",01386 to this result gives 50",23492 for the corrected value of annual precession at the epoch 1790,0. It should be farther stated that the correction $+0'',97$ was obtained by O. Struve after eliminating, with all the accuracy attainable, from the R. A. given by observation, errors due to the motion of the Solar System in space.

(2) Dr Peters in his investigations relative to Nutation, adopting formulæ derived from the theoretical researches of Laplace and Poisson, calculated the numerical values of the coefficients from 603 observations of transits of Polaris taken above and below Pole at the Dorpat Observatory from 1822 Nov. 4 to 1838 Oct. 15, of which number 249 were observed by W. Struve, and the remaining 354 by his assistant W. Preuss. After applying corrections for difference of the personal equations of the observers, and taking into account O. Struve's determination of the annual general precession, Peters found for the coefficient of the principal term of the obliquity of the Ecliptic a value which, combined with two other values deduced by independent computations of MM. Busch and Lundahl, gave as the mean result $9'',22305$ for the epoch 1800,0.

(3) For determining the constant of aberration W. Struve observed transits of seven stars with a transit-instrument set up in the Prime Vertical at the Imperial Observatory of Poulkova, the zenith distances of the stars at meridian passage in no instance exceeding $1°\,30'$. Having carefully got rid of instrumental errors, and taken into account Peters' value of the constant of nutation, he estimated at $0'',0111$ the total amount of probable error due to observation, to annual parallax and proper motion of the stars, and to nutation, and finally obtained for the constant of aberration $20'',4451$.

144. To complete this account of the fundamental quantities pertaining to observational astronomy, the following particulars are added relative to determinations of the Obliquity of the Ecliptic, and the amount of its annual variation. Bessel did not accept the values of the obliquity and of its annual diminution $(0'',48368)$ given by the theoretical expression for ω_1 in Art. 142, but availing himself of obliquities deduced from *observations* made by different astronomers in the interval from

1755 to 1815 (see *Fundamenta Astronomiæ*, p. 61), he obtained for the epoch $1800 + t$ the formula,

$$\text{Obliquity} = 23° 27' 54'',80 - 0'',4570t.$$

Peters, adopting the apparent obliquities deduced by Bessel from Bradley's observations, as given in page 58 of the *Fund. Astron.*, calculated therefrom by his own formulæ a mean obliquity $23° 18' 14'',06$ for 1757,3, by comparison of which with a value deduced by W. Struve from observations at Dorpat (Vol. VI. of Dorpat observations, p. LXIX.), namely, $23° 27' 42'',61$ for 1825,0 he found for 1750,0 the value $23° 28' 17'',44$, and for 1800,0, $23° 27' 54'',22$, giving a mean annual diminution of $0'',4645$ (see pp. 189 and 190 of the Memoirs cited in Art. 143). Leverrier, according to results exhibited in page 203 of Tom. IV. of the Annals of the Paris Observatory, gives for the epoch $1850 + t$, Obliquity $= 23° 27' 31'',83 - 0'',47594t$.

The details of observational means by which the constants treated of in the foregoing Articles may be determined or corrected, will come under consideration in the sequel of the Lectures.

145. I propose in the next place to give an account of the derivation of formulæ proper for calculating the *Annual Variations* of the mean R. A. and Decl. of any star at any epoch, these formulæ being auxiliary to those required for calculating the differences in R. A. and Decl. between the apparent and mean places of a star at the same epoch. Bessel deduced his formulæ in p. X. of the *Tab. Regiomont.* for the annual variations in R. A. and Decl. from the expressions for ψ, ψ_1, ω, ω_1 already adduced in Art. 142. By a like process Peters obtained formulæ differing from Bessel's only as to the numerical values of the coefficients, the new values having been deduced from O. Struve's corrected general precession, and, as respects terms containing t^2, by taking advantage of M. Leverrier's fuller development of the gravitational theory, and introducing corrected values of the masses of the planets. Putting, in accordance with Bessel's notation, ψ for lunisolar precession from 1800 to $1800 + t$ referred to the Ecliptic and Equinox of 1800, ω for the angle at

the epoch $1800+t$ between the earth's Equator and the Ecliptic of 1800, and λ for the angle subtended at the Pole of the Equator by the arc joining the Poles of the Ecliptics of 1800 and $1800+t$, we have the following results of Peters's researches:

$$\psi = 50'',3798t - 0'',0001084t^2,$$
$$\omega = 23° 27' 54'' + 0,00000735t^2,$$
$$\lambda = 0,15119t - 0,00024186t^2,$$

$$m = \frac{d\psi}{dt}\cos\omega - \frac{d\lambda}{dt} = 46'',0623 + 0,0002849t^2,$$

$$n = \frac{d\psi}{dt}\sin\omega \quad = 20'',0607 - 0,0000863t;$$

and the mean R. A. and Decl. of the star being α and δ,

Annual Precession in R. A. $= m + n \tan\delta \sin\alpha$,
Annual Precession in Decl. $= n \cos\alpha$.

Also putting ψ_1 for the *general* precession from 1800 to $1800 + t$, and ω_1 for the mean obliquity of the Ecliptic at $1800 + t$, the same researches gave

$$\psi_1 = 50'',2411t + 0,0001134t^2,$$
$$\omega_1 = 23° 27' 54'' - 0'',4738t - 0'',0000014t^2;$$

and consequently the annual diminution of the obliquity is by this process $0'',4738$. (See Memoirs, &c., before cited, pp. 193—195.)

146. After the foregoing discussion of the values of the principal constants involved in the *Formulæ of Reduction* which are required for the calculation of Apparent places of Stars from their Mean places, it remains to adduce these formulæ (given originally by Bessel in pages XXIX and XXX of the *Tab. Regiomont.*), with corrected coefficients, and to indicate suitable modes of applying them. The subjoined equations, which are adapted to these purposes, were taken from the Nautical Almanac of 1879, and it may be assumed that their numerical coefficients will be sufficiently accurate for at least ten years

from the present date (1877). For the sake of convenience in making use of the formulæ, the sets of equations for calculating Apparent Declination are exhibited together with those for calculating Apparent Right Ascensions, although the former have more immediate reference to observations with the Mural Circle. The values of the coefficients are all adopted from formulæ given by Professor Peters in pages 71—76 of his *Numerus Constans Nutationis* (= pages 195—200 of the Memoir before cited), in which his own constant of Nutation, and the constants of Precession and Aberration, determined, as already stated, by O. and W. Struve, are introduced.

The Nutation of Obliquity being represented by $\Delta\omega$, and that of Longitude by ΔL, the formulæ for computing these quantities, given in page v of the Nautical Almanac of 1879, are

$$\Delta\omega = 9'',2237 \cos \Omega - 0'',0895 \cos 2\Omega + 0'',5507 \cos 2\odot,$$
$$\Delta L = -17,2526 \sin \Omega + 0,2073 \sin 2\Omega - 1,2693 \sin 2\odot,$$

where Ω is the mean Longitude of the Moon's ascending Node, and $\odot$ the true Longitude of the Sun.

The Apparent Right Ascension of a star being a' *in arc*, and its Apparent Declination δ', at the time t reckoned each year in fractional parts of a tropical year from the instant at which the Sun's mean Longitude is 280° (which for 1879 is Jan. 0 + ^{d},227), the star's mean Right Ascension *in arc*, and its Mean Declination, being respectively a and δ, for the beginning of the year, and Δc, $\Delta c'$ its annual proper motions in R. A. and Decl., the equations for calculating $a' - a$ and $\delta' - \delta$ were reduced by Bessel to the following convenient forms:—

$$a' - a = Aa + Bb + Cc + Dd + t\Delta c,$$
$$\delta' - \delta = Aa' + Bb' + Cc' + Dd' + t\Delta c',$$

in which the factors A, B, C, D are the same at the same time for all stars, a, b, c, d and a', b', c', d' are quantities depending on the place of each star. The following equations, derived from the foregoing expressions for $\Delta\omega$ and ΔL, combined with the formulæ for the mean precessions in Right Ascension and

THE TRANSIT INSTRUMENT. 139

Declination (Art. 145), and the known law and amount of aberration, are employed for calculating the twelve quantities A, B, &c. above specified. (See Nautical Almanac for 1879, pp. 295 and 512) :—

$A = -20'',4451 \cos \omega \cos \odot,$ $\qquad B = -20''4551 \sin \odot,$

$C = t - 0,02519 \sin 2\odot - 0,34243 \sin \Omega + 0,00410 \sin \Omega,$

$D = -0'',5507 \cos 2\odot - 9'',2237 \cos \Omega + 0'',0895 \cos 2\Omega,$

$a = \cos \alpha \sec \delta,$ $\qquad a' = \tan \omega . \cos \delta - \sin \alpha \sin \delta,$

$b = \sin \alpha \sec \delta,$ $\qquad b' = \cos \alpha \sin \delta,$

$c = 46'',0848 + 20'',0539 \sin \alpha \tan \delta,$ $c' = 20'',0539 \cos \alpha,$

$d = \cos \alpha \tan \delta,$ $\qquad d' = -\sin \alpha,$

where $\odot$ is the Sun's true Longitude, Ω the mean Longitude of the Moon's node, and ω the obliquity of the Ecliptic, each for the time t. The values of $\log A$, $\log B$, $\log C$, $\log D$ are given in the Nautical Almanac for every day of the year.

147. The Astronomer Royal has obtained the following formulæ for calculating the excess of Apparent above Mean R. A. in time $\left\{ = \frac{1}{15}(\alpha' - \alpha) \right\}$, and the excess of Apparent above Mean North polar distance $\{= -(\delta' - \delta)\}$;

$$\frac{1}{15}(\alpha' - \alpha) + 300,00 = Ee + Ff + Gg + Hh + L + l$$

$$-(\delta' - \delta) + 300,00 = Ee' + Ff' + Gg' + Hh' + L + l'.$$

The particular advantage of these forms is, that all the terms on the right-hand sides of the equations have *positive* numerical values (excepting when the first equation is applied in the case of a star less than $3°.10'$ from the Pole), the relations between the new quantities E, e, F, f, &c. and the quantities A, a, B, b, &c. of Bessel's formulæ, being given by the following simple equations :—

$E = A + 25,$ $\quad F = B + 25,$ $\quad G = C + 1,2,$ $\quad H = D + 25,$

$L = 210 - 1,2 E - 1,2 F - 25 G - 1,2 H,$

$e = \frac{a}{15} + 1,2,$ $\quad f = \frac{b}{15} + 1,2,$ $\quad g = \frac{c}{15} + 25,$ $\quad h = \frac{d}{15} + 1,2,$

$$l = 210 - 25\,e - 25f - 1{,}2\,g - 25\,h,$$
$$e' = a' + 1{,}2, \quad f' = b' + 1{,}2, \quad g' = c' + 25, \quad h' = d' + 1{,}2,$$
$$l' = 210 - 25\,e' - 25f' - 1{,}2\,g' - 25\,h'.$$

To facilitate this method of calculation, the values of $\log E$, $\log F$, $\log G$, $\log H$ and L, which are independent of the particular star, are given in the Nautical Almanac for every day of the year. (See pp. (x) and (xi) of the Appendix to the Greenwich Observations of 1847, containing the Twelve-year Catalogue of 1836—1847.) Applications of both methods to a particular instance are exhibited in pages 512 and 513 of the Nautical Almanac of 1879.

148. The means of calculating the clock-time of transit of any celestial object across the meridian having been indicated in *Division III.*, it is now required to shew how the true sidereal time of transit is inferred from the clock-time by correcting for the *clock-error*. For this purpose, it is first necessary to form a catalogue of stars arranged according to their mean R. A. at a given epoch, and to determine their R. A. by a large number of observations, in order that they may be used as fiducial points of reference, or *fundamental* stars. These stars should be bright enough to be seen in day-time, at moderate distances from the Equator, and their distribution in Right Ascension should be pretty uniform. In the course of the foregoing discussions (Arts. 136—147), it was shewn how a provisional catalogue of R. A. might be formed *ab initio*, and how, after the effects of precession, nutation, and aberration on the apparent places of stars had been ascertained, the transit-observations of a given star extending over a large number of years might be all made to contribute to an accurate determination of its mean R. A. at a given epoch. By such means the R. A. in the catalogues of mean places of stars produced in successive Nautical Almanacs have been derived from numerous transit observations taken at the Greenwich Observatory during the superintendence of Maskelyne, Pond, and the present Astronomer Royal. In the more recent catalogues ascertained

changes of places of the stars due to proper motion[1] have been taken into account. For these reasons if the stars selected for use as fundamental stars be contained in one of the lists of the Nautical Almanac, their Mean and Apparent R. A., as there calculated, might be adopted without alteration. The process, however, I am about to describe, which has been employed at the Cambridge Observatory, although it takes advantage of the Mean and Apparent R. A. of the fundamental stars as calculated in the Nautical Almanac, really makes use of values that are independently determined by observations made at the Observatory.

149. The first list of mean R. A. of fundamental stars used at the Cambridge Observatory was adopted by Professor Airy, without alteration, from the Naut. Alm. of 1829, and the number of Stars was 34 inclusive of Polaris and δ Ursæ Minoris. For the next year the adopted fundamental R. A. were derived from the R. A. resulting from the observations of 1829 by adding the Annual Variations in R. A.; and so on through successive years to 1835 inclusive, at which date the number of the stars was 31. On taking the superintendence of the Observatory in 1836, I followed the same rule, deriving the fundamental R. A. for that year from the results of the transits of 1835 by adding the annual variations, and, with the exception of omitting Sirius, I adopted the same list of stars. The R. A. for 1837 were obtained in the same manner; but a new point of departure was established by the list of fundamental R. A. for 1838, which were calculated on the following principles. The difference between the mean R. A. of every two of the stars taken consecutively, was assumed to be the same as a mean difference derived from all the transit observations of the two stars in the eight years 1830—1837, by giving to the difference as determined by the observations of any one year a weight proportional to the product of the numbers of the observations of the two stars in that year. The epochs of the concluded mean differ-

---

[1] The proper motions now adopted in the Nautical Almanac are those determined by Main and Stone, contained in Vols. XIX., XXVIII. and XXXIII. of the *Memoirs of the Royal Astronomical Society*.

ences of R. A. thus obtained would be different for different stars, and would generally be nearer to 1834,0 than to 1838,0. As, however, the variations of the differences between the two epochs would in all cases be extremely small, it was thought sufficiently exact to refer all the concluded mean differences to the common epoch 1838,0, after applying corrections for *relative* proper motions in the instances in which the absolute proper motions obtained from the most reliable sources indicated that the difference of R. A. might thereby be sensibly affected. This being understood, let $a_1, a_2, a_3, \ldots\ldots a_n$ be the adopted mean R. A. 1838,0 in order of magnitude, and $\Delta_1, \Delta_2, \Delta_3 \ldots \Delta_{n-1}$ the successive differences determined as above said. Then

$$a_2 = a_1 + \Delta_1, \quad a_3 = a_2 + \Delta_2 = a_1 + \Delta_1 + \Delta_2, \text{ and so on.}$$

Consequently $\quad a_1 + a_2 + \ldots\ldots + a_n = na_1 + Q,$

Q being a given function of $\Delta_1, \Delta_2, \ldots \Delta_{n-1}$. If now we introduce the condition that the sum of the excesses of the adopted R. A. above the corresponding R. A. of the Naut. Alm. shall be zero, the left-hand side of the above equation will be equal to the sum of the Naut. Alm. R. A. Hence putting S for this sum, we have $S = na_1 + Q$, from which equation a_1 may be calculated, and then in succession $a_2, a_3 \ldots a_n$. It resulted from this calculation that the mean of the adopted R. A. was less by 0",083 than the mean of those which would have been obtained from the observations of 1837 by the former process. Consequently a change in the origin of the R. A. was made at the beginning of 1838. As the true position of the Ecliptic relàtively to the assumed origin is determined from year to year by the Circle observations of the Sun, this change is of no importance. The adopted R. A. of Polaris and δ Ursæ Minoris were those given by the observations of 1837 diminished by 0",08.

150. Subsequently to 1838 the assumed Mean R. A. of the fundamental Stars for any year were derived in general from the R. A. resulting from the observations of the preceding year by merely adding the annual variations of the Naut. Alm. When, however, the number of the observations was small, the assumed R. A. were deduced from the results of the observations

of two preceding years by applying the annual variations, the weight given to each year's result being proportional to the number of the observations by which it was obtained. Eventually the assumed R. A., beginning with those of 1844, were calculated by the following rule. If A be the Star's assumed R. A. for any year, and A' the R. A. resulting from a number (n) of observations in that year less than 20, the assumed R. A. for the next year is $A + (A' - A)\dfrac{n}{20}$ *plus* the annual variation. Having found it desirable to employ a larger number of clock-stars, I added nine in the fundamental Catalogue of 1844, the assumed R. A. of which were adopted from the Naut. Alm. with the correction $+ 0^s\!,02$ applied, excepting that for two stars observed in 1843 the corrections were $+ 0^s\!,03$ and $0^s\!,00$. As the mean excess of the R. A. of the fundamental stars above the R. A. of the Naut. Alm., resulting from the observations of 1843, was $+ 0^s\!,020$, it will be seen that the place of the Equinox would not be sensibly altered by employing the adopted R. A. of the additional stars.

151. We are now prepared to consider how the *clock's daily rate* and the *clock-error* at any given time may be calculated. It has been stated (Arts. 149 and 150) that after assuming (provisionally) approximate Mean R. A. of the clock-stars for the beginning of a certain year, the adopted Mean R. A. for the beginnings of successive years are inferred from the results of the yearly observations by the applications of *Annual Variations* of Mean R. A. Formulæ for calculating the Annual Variations for any epoch are given in Art. 145. The Apparent R. A. of any clock-star for any year, at the time of its meridian transit on any day of the year, is derived from its adopted Mean R. A. at the beginning of the year by applying a quantity termed the *Correction*, the formulæ for calculating which are those contained in Art. 146. As already mentioned (Art. 146) both sets of formulæ are the same as those used for the computations of the Naut. Alm. Consequently, after applying to the Apparent R. A. of the Naut. Alm. the excesses of the adopted Mean R. A. for the beginning of the year above the corresponding Mean

R. A. of that work, we may regard the results as Apparent R. A. independently determined by observations at the Observatory. A clock regulated to sidereal time ought to indicate these Apparent R. A. when the stars to which they pertain are on the meridian. The difference between the clock-time of the transit of a fundamental star and its calculated Apparent R. A. is presumed to be the error of the clock. It is a great part of the work of a well-appointed Observatory to determine such clock-errors by observations carried on night and day. If a clock-star has proper motion, it will plainly give an erroneous clock-error if this motion is not taken into account in calculating the Apparent R. A. For this reason the Annual Variations of the Naut. Alm., which include ascertainable proper motions, should be adopted.

152. For calculating the *correction for clock-error* to be applied to the clock-time of meridian transit, the observations are divided into groups severally containing observations of stars proper for giving clock-errors. These groups are frequently separated by intervals during which no observations were taken, but, as often as may be, they consist of the night and day observations made in consecutive intervals of 24 hours. The mean of the clock-errors in each group is considered to pertain to the mean of the times of transit of the stars which furnish them. From a comparison of this mean error with errors similarly derived from the next preceding and following groups, a preceding and a following rate are calculated, each expressed as proportionate loss of the clock in 24 hours, and applying to the mean between the epochs of the errors from which it is calculated. From the two rates a rate is inferred which is assumed to be uniform throughout the middle group. No definite rule can be given for deducing the adopted rate; attention should be paid to the probable accuracy of the rates on which it depends, and also to the proportion of the intervals separating the preceding and following mean clock-errors from the intermediate one. The rule which I have most frequently adopted was obtained in the following manner. Let e, E, e' be the three mean clock-errors, $t - \tau, t, t + \tau'$ the corresponding mean

clock-times, and r, r' the computed rates. Then $r = \dfrac{24}{\tau}(E-e)$ = the rate at $t - \dfrac{\tau}{2}$, and $r' = \dfrac{24}{\tau'}(e'-E)$ = the rate at $t + \dfrac{\tau'}{2}$. Hence, supposing R to be the rate at the time t, and the rate to *change uniformly* in the interval from $t - \dfrac{\tau}{2}$ to $t + \dfrac{\tau'}{2}$, we shall have

$$\frac{R-r}{r'-r} = \frac{\tau}{\tau+\tau'}, \text{ and } \therefore R = \frac{\tau r' + \tau' r}{\tau+\tau'}.$$

The error E and rate R at the clock-time t being thus known, the clock-error at the next preceding 0^h of sidereal time may be thereby calculated. Then the correction of clock-error for the observation of any object whose approximate R. A. is A, is the clock-error at 0^h, together with the proportional part of the rate for the interval A, which is *additive* supposing the clock, as is usually the case, to have a *losing* rate.

153. It should be noticed that if the observations from which any mean clock-error is calculated were made by two observers, the calculation takes account of the difference of their personal equations, and thus virtually gives the error as determined by the observations of one of the observers. Also, if in calculating the clock-rates r and r', the mean clock-errors e, E, e' are derived from groups of observations made by different observers, they are reduced to errors as obtained by one observer by applying corrections for differences of personal equations (see Art. 134).

154. From the Apparent R. A. of a *star* thus found, its mean R. A. for the beginning of the year may be obtained by applying with opposite sign the correction calculated by the formulæ given in Arts. 146 and 147. If the star be in the list of the Naut. Alm., the correction with the proper sign may be obtained by subtracting its mean R. A. given in the Table of "Mean Places of Stars," from its Apparent R. A. obtained by interpolation from the Table of "Apparent Places of Stars." The Apparent R. A. of stars are reduced to Mean R. A. for the purpose of forming a Catalogue of their Mean R. A., the results of all the observations of the same star in any year being thus

made available for deducing a mean result. Also by applying the annual variations the results obtained for the same star in several years may be brought to the same epoch, and thus the concluded R. A. of the Catalogue be made to depend on a large number of observations. The same process of reduction to Mean R. A. is applied to clock-stars for the purpose of farther correction of the assumed R. A.; but, in Cambridge practice, this is not done unless three of them at least are in the group. It is from the mean of all the Mean R. A. of a clock-star obtained by this process in the course of a year, that the adopted Mean R. A. for the next year is deduced by applying the annual variation (see Arts. 149 and 150). The Apparent R. A. of Polaris and δ Ursæ Minoris are not independent determinations except they are inferred from two or more alternate transits of the star above and below Pole, and if this be not the case the Mean R. A. are not calculated [1].

[1] Chapter XIX. of Grant's *History of Physical Astronomy* contains an historical account of Catalogues of the Fixed Stars, beginning with the earliest, and extending to those of recent date. I thought it might be useful to give the following particulars extracted from this account. The several catalogues of Ptolemy (A.D. 138), Ulugh Beigh (1437), Tycho Brahé (1600), Hevelius (1660), and Halley's Catalogue of Southern stars (1677), after being subjected to a careful revisal by Francis Baily, were reprinted, and now form Vol. XIII. of the Memoirs of the Royal Astronomical Society. Flamsteed's British Catalogue was published in 1725, in Vol. III. of the *Historia Cœlestis*, and republished, with additions, by Baily at the end of his "Account of the Life and Correspondence of Flamsteed." Bradley's observations were published at Oxford in 1798, and from those made between 1750 and 1762 Bessel formed his *Fundamenta Astronomiæ*, containing 3112 stars reduced to the epoch 1750,0. Then follow in order of time, Lacaille's three catalogues, the principal of which, consisting of southern stars, was reduced by Henderson to 1750,0, Mayer's Catalogue of 998 zodiacal stars, published in 1800, Maskelyne's fundamental catalogue of 36 stars, Wollaston's "Fasciculus" of circumpolar stars (1800), the *Histoire Céleste Française*, published, unreduced, in 1801, and in a reduced form by the British Association in 1837, Piazzi's catalogue of 7646 stars for the epoch 1800,0, Bessel's zone observations, reduced by Weisse to 1825,0, and like observations of Santini, reduced to 1840,0 and published in vol. XII. of *Mem. of the Astron. Soc.* Of subsequent catalogues, the principal are, Taylor's Madras Catalogue of 11,015 stars formed from observations in 1822-43, Pond's catalogue of 1112 stars reduced to 1830,0, the British Association catalogue of 8377 stars reduced to 1850,0 and the Greenwich twelve-year catalogue (1840,0 and 1845,0) and seven-year catalogues (I. 1860,0, II. 1864,0) formed by the present Astronomer Royal. Interesting details respecting the above, and various other catalogues too numerous to mention here, are given in Grant's History, as above cited.

THE TRANSIT INSTRUMENT.   147

155. The Apparent R.A. of *bodies of the solar system*, as usually calculated in the Tables, are immediately comparable with the R.A. determined observationally by the processes above indicated, excepting that the Tables and the observations may not reckon the R.A. from exactly the same position of the equinox. The consideration of this point must be reserved till observations with the Mural Circle have been under treatment, the Transit Instrument being incapable of itself of determining absolute Right Ascensions.

156. An example of the complete Reduction of the transit-observation of a star, extracted from one of the Calculation-books of the Cambridge Observatory, is here subjoined:—

Transit of α Aquilæ, 1851 January 27, 23^{h} mean time.

Illumination-end of axis East.

N.P.D. of star $= 81°.31'$. The line of collimation pointed $0'',7$ westward, and the west end of the axis was $4'',1$ high and $0'',9$ southward. Wires A and B lost.

	h.	m.	s.		s.
Wire I			,	Interval of wire I} from mean	$-40,344$
„ II			,	„ Wire II „	$-26,892$
„ III			57,1		
„ IV			10,8	Sum $= -67,236$	
„ V			24,7	$\delta = 81°.31'$ $z = 43°.44'$	
„ VI			38,0	$\tfrac{1}{15}\cosec \delta = 0{,}067$	
„ VII			19.43.51,7	$\tfrac{1}{15}\cos z \cosec \delta = 0{,}049$	
mean of the times		19.43.24,46		$\tfrac{1}{15}\sin z \cosec \delta = 0{,}047$	
Corr^{n} for lost wires			$-13{,}59 = -\dfrac{67{,}236}{5} \times 1{,}101$		
			10,87		
Corr^{n} for form of pivots			$+0{,}33$		
„ collimation error			$-0{,}05 = -0'',7 \times 0{,}067$		
„ level error			$+0{,}20 = +4'',1 \times 0{,}049$		
„ azimuth error			$+0{,}04 = +0'',9 \times 0{,}047$		
Clock-time of transit		19.43.11,39			
Clock slow at 0^{h}		16,65			
Corr^{n} for losing rate $1^{\text{s}},17$		$+0{,}96$			
Apparent R.A.		19.43.29,00			
Corr^{n} to Mean R.A.		$+1{,}84$			
Mean R.A. 1851,0		19.43.30,84			

148   PRACTICAL ASTRONOMY.

Plate I. is a representation in outline of the Cambridge Transit Instrument, copied from that attached to Woodhouse's Description in the Philosophical Transactions for 1825, referred to in page 5. *A* is the position of the apparatus for adjusting the axis of motion, a full description of which, illustrated by the Figures 9 and 10, is given in pages 37—39. *B* is the eye-end of the telescope, of which the Figure 6 in pages 27 and 31 is an enlarged representation, accompanied by a description in detail of the different parts. At *C* apparatus was provided for regulating the amount of surface of the object-glass exposed for observations of the Sun, but in practice it was not found to be required, the use of dark glasses of different degrees of opacity at the eye-end being sufficient for all purposes (see Art. 109).

THE MURAL CIRCLE.

157. A Mural Circle is constructed for the primary purpose of measuring the angular elevation of a celestial object above a horizontal plane at the instant of its passing the meridian of the place of observation. Transit-observations are required for obtaining one of the spherical co-ordinates by which the position of an object in the heavens is determined, and observations with a graduated Quadrant or Circle are proper for obtaining the other. The way in which this ultimate object is reached by the use of a Mural Circle expressly constructed for the purpose just mentioned, and the processes employed for ensuring accuracy in the determination of the co-ordinate, will form the principal subject matter of the portion of the Lectures now entered upon. As in the case of the Transit instrument, the Circle will be treated of under four *Divisions* (see Art. 25).[1]

---

[1] The Mural Circle is the simpler of the two instruments, inasmuch as it is only concerned with measurement of space, whereas the other measures space by the intervention of *time*. I have followed usage in treating of the Transit instrument first, and apart from the Circle; but as respects astronomical purposes neither instrument is complete by itself, on which account in the treatment of the Transit instrument, there was occasion several times to refer by anticipation to what is effected by the Mural Circle. The use of the Transit-Circle, which is a complete astronomical instrument, is capable of being explained in a perfectly logical manner.

The Cambridge Transit Instrument.

PLATE I.

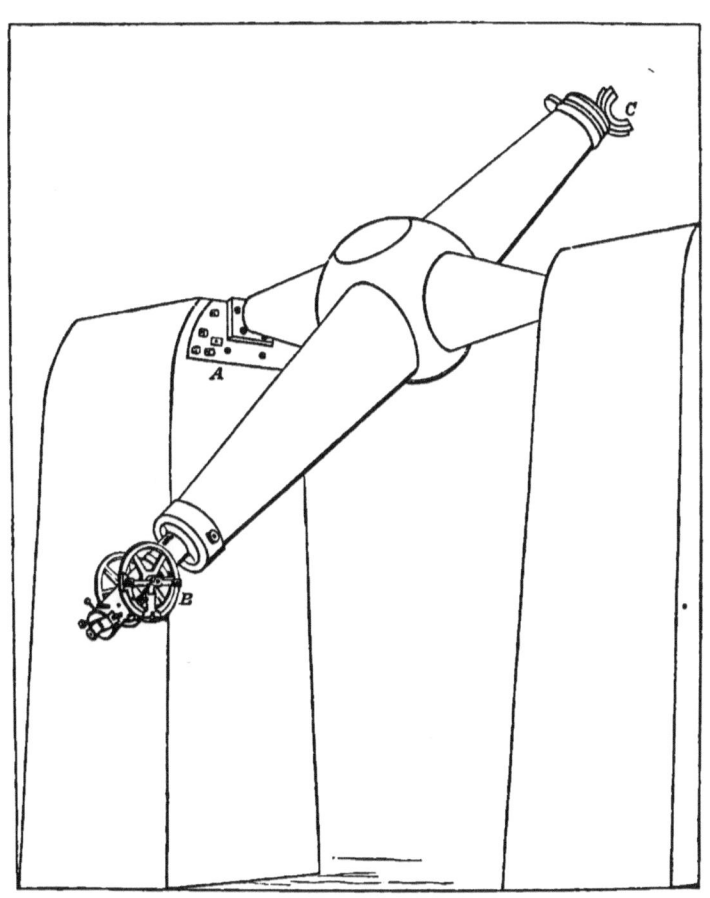

I. DESCRIPTION OF THE PARTS AND MOUNTING OF THE INSTRUMENT.

158. In describing the Mural Circle (so called because when mounted it is usually contiguous to the vertical face of a massive wall) I shall refer mainly to that of the Cambridge Observatory, respecting which, after long use of it, I am prepared to say that in respect to efficiency and accuracy it has probably not been surpassed by any other instrument of the same kind. It was constructed by Troughton and Simms, and mounted in October 1832. The divisions, having been approximately marked, were finally cut by Mr Simms in November of the same year, it being thought advisable, regard being had to the weight and size of the instrument, to perform this operation after it was mounted. It came into use at the beginning of 1833, and in the Introduction to the Volume of Cambridge Observations for that year Professor Airy gave a description of it, which for the most part I have followed in that which is produced here. The circle is eight feet in diameter, but in all other respects the Cambridge instrument is similar to the mural circles of the Greenwich Observatory. The limb is connected with the centre piece by sixteen spokes, or hollow cones, and these are joined together at about the middle of their lengths by consecutive bars forming a circle whose diameter is about half that of the exterior circle. The limb was cast in several pieces, which were afterwards united by a process technically called *burning together*, which, when skilfully performed, makes a connection as perfect as if the whole had been cast at once. The axis on which the circle turns is a hollow cone, $4\frac{1}{2}$ feet long, passing through a very massive stone pier, the circle being on its eastern face, and both bearings of the axis on the same side of the circle. At each bearing the axis is armed with a steel ring which turns within a steel cylinder. A considerable part of the weight at the bearing of the larger ring, that near the circle, is supported by lever counterpoises. The divisions of the limb are on the external edge, so that the microscopes by which they are read are in the

same plane as the circle. The microscopes, six in number and separated by equal intervals, are attached to the stone pier by brass supports, the height of each of which above the point of attachment is intended to supply a thermometrical expansion nearly equal to the upward expansion of the central support of the instrument. The microscopes are fitted with micrometers, by which the interval of 5' between the divisions is subdivided, and the circle-reading can be taken to integral seconds, and by estimation to tenths of a second. The telescope is carried by a steel rod passing through the hollow conical axis, and its position on the circle is secured by frames which are clamped to the limb at both ends of the telescope, one being as near as possible to the object-glass cell, and the other as near as possible to the wire-frame. The instrument is fixed, or moved slowly, by clamps which act immediately on the limb by tangent-screws, and are so disposed about the circle that they can be severally used according to the convenience of the observer. Originally five tangent-screws were attached to the wall, but after it was found that three were sufficient, one at the nadir point and the other two about $30°$ below the horizontal diameter, the two uppermost were taken off.

159. To prevent any injurious effect from the Sun's heat in mid-day observations, a wooden screen is provided, by which all the parts of the limb on which the Sun's rays might fall are shaded. That part of the screen which corresponds to the interval between the tropics, consists of a sliding board with a circular hole in it, a little larger than the aperture of the telescope, to admit of the incidence of rays from the Sun and planets on the object-glass when the screen is let down.

160. The principle of the *tangent-screw and clamp* may be gathered from what has been said in Art. 35. I propose to give here the description of one specially constructed for application to the Mural Circle. Figure 29 represents the form of the clamps of the Cambridge Circle. The screws *a, a* fasten to the wall a brass plate which carries the clamp; *e, e* are milled heads, which are fixed to the ends of a straight rigid stem for turning it; *b, b* are supports of the stem fastened to the

THE MURAL CIRCLE. 153

brass plate; the stem passes through two cylinders c, c, which are fixed by being rigidly connected with the supports b, b; the left-hand cylinder is tapped for the screw portion of the stem

Fig. 29.

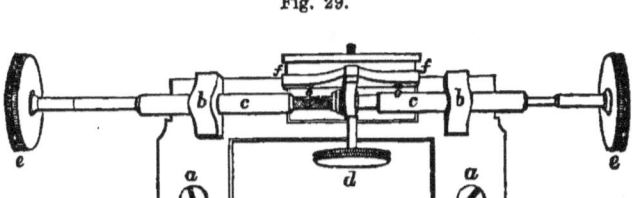

to work in; the right-hand cylinder contains a strong spiral spring, which surrounds the stem, and is compressed between a stop in the cylinder and the end of an interior movable cylinder. The latter projects from the other, and abuts on a transverse part of the clamp opposite the end of the screw, so that the screw and spring in action are antagonistic to each other. When by turning either milled head, the screw is drawn out, the clamping pieces and the transverse screw d are moved in the direction from left to right according to the figure, and on turning the milled head in the opposite direction they are moved by the force of the spring from right to left. The limb of the Circle is nipped in the space ff between two clamping-pieces by turning the screw d, which draws the farther clamping-piece close up to the other. On turning that screw back to release the hold, the Circle is unclamped, and the farther piece is pushed out by two small springs to allow of free passage of the limb when the Circle is moved, that piece being at the same time kept in place by two screw-heads seen at o and o. When the Circle is clamped it partakes of the motion given to the clamping-pieces by turning one or the other of the milled heads either way. But for the final motion in taking an observation, the milled head is always to be turned in the direction which draws out the screw from the cylinder c, and therefore produces motion of the clamp from left-hand to right-hand. The reason for this is, that the bearing on the threads of the screw may be presumed to be more complete and uniform when the motion is opposed to the action of the spring than in the contrary case[1]. That the observer may be able to decide at once which way he is to turn the milled head, half of its edge is cut (as shewn in the Figure) in such manner as to be felt by the finger passing over it rougher in one direction than in the other. The direction of the turning is required to be that in which the greater roughness is felt.

161. It is particularly to be noticed, as being the principal advantage of this form of clamp, that although the stem is rigid, and its supports b, b are rigidly connected with the brass plate, the clamping apparatus is so arranged that it is capable of receiving after the clamping, together with the motion from left to right, a small motion of rotation, in consequence of which no *strain* is produced by the circular movement of the Circle. For this purpose the apparatus is made to have a *sliding* motion on the brass plate, and the stem passes through an *elongated* hole in the transverse brass support of the screw d. [The clamping apparatus of the Greenwich mural circles, which differed from this, will be described subsequently.]

162. It is now required to take into consideration the construction and use of the *micrometer-microscope*, by means of which, as stated in Art. 158, the graduation intervals are subdivided, and the angular indications of the Circle are read off. Figure 30 represents a Troughton reading microscope, the construction of which in all essential points is the same as that of the microscopes of the Cambridge Circle, although it differs from them in form. The arrangement of parts in the interior

---

[1] In order that the Circle may be easily turned about its axis, the pivots and their collars are from time to time greased by a composition, in certain proportions, of oil, tallow, and bees' wax, to diminish the friction. It sometimes happened with the Cambridge Circle, when the composition required renewing, that the force of the spring was insufficient of itself to overcome the friction. This is another reason for not relying on the action of the spring in moving the Circle to make a bisection. Oil alone is a better lubricant.

of the box *a* of the micrometer, which is transverse to the axis of the microscope, may be understood by referring to Figure 4 in page 24, so far as relates to the movement of the micrometer-

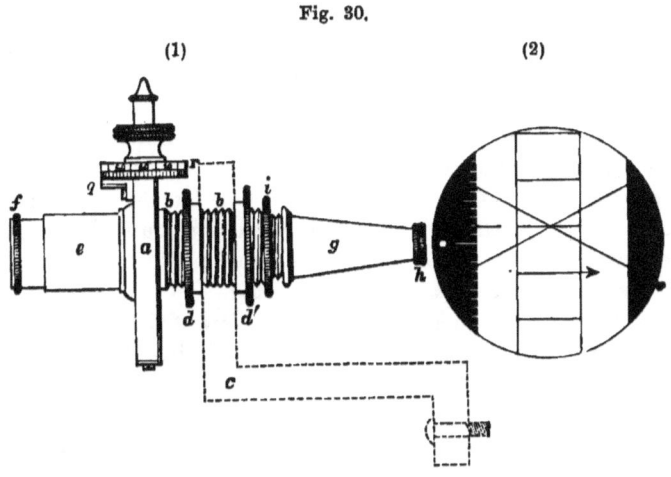

Fig. 30.

wire by turning the micrometer-head; but in the present case, instead of a single transverse wire, the turn of the micrometer-head moves two wires crossing each other, so as to be suitable for bisecting a line graduation in the manner shewn in Figure 30 (2). The graduated rim *r* is divided into 60 equal parts for reading off to seconds of space by means of the contiguous fixed index *q*. The tube of the microscopes consists of three parts: (1) the part *e* is a cylindrical tube, screwed to the box of the microscope, and carrying a Ramsden eye-piece *f*, which slides in it with friction, so as to be adjustable for seeing the cross wires distinctly; (2) the cylindrical tube *bb*, which is screwed fast to the plate of the micrometer-box, has on its outside a coarse screw for action with the tapped nuts *d*, *d'* with milled heads, and also an interior fine screw from end to end; a portion of the same tube has, besides, at the end a fine screw on the outside for action with a third circular nut *i* with milled head; (3) the part *g*, having the form of a frustum of a cone, carries at one end the cell *h* of the object-glass, and terminates at the other in a short cylindrical part, which is inserted in the tube *bb*, and has a fine screw on the outside for action with the fine interior screw of that tube, this action giving the means of adjusting the distance of the object-glass from the cross wires. When the adjustment has been made, the nut *i* is screwed up to a shoulder, and the position of the object-glass is thus fixed. The dotted figure *c* has reference to the mode of support of the microscope, and the connection of the support with the wall. The upright part of the support *c* has a hole for the cylindrical part *bb* to pass through, so that when the nuts *d*, *d'* are screwed up, the attachment of the microscope to the support, and by consequence to the wall, is effected.

Figure 30 (2) represents the field of view of the microscope, shewing the image of a portion of the graduation of the Circle, the bisection of one of the lines of graduation by the cross wires, and the toothed comb for reading off the integral number of micrometer-revolutions reckoned from the zero position marked by a circular hole in the comb. The rules for reading off will be given in *Division II*.

163. The field of view of the telescope, the apparatus at the eye-end for adjusting the wire-frame, and the mode of connecting the telescope with the limb of the Circle, are exhibited by Figure 31. In the field of view there are five vertical wires and a horizontal wire all fixed on the same frame, and an additional micrometer-wire movable by turning the micrometer-head M. The fixed horizontal wire is adjustable to horizontality by the apparatus at C consisting of two screws supported by a frame and acting oppositely on a projection *e* rigidly connected with the wire-frame, by which means a rotational motion about the axis can be given to the wire. The parallelism of the movable wire to the fixed wire depends wholly on mechanical construction. The micrometer-head M has on its rim a scale of 100 equal parts, pointing to which there is a fixed index for reading off; and by turning M the micrometer-wire may be turned either way by intervention of the antagonistic spring at

m. For a reason already mentioned (end of Art. 160) the turning of M is always in the direction that makes the micrometer-wire approach the micrometer-head, and the nearer it is

Fig. 31.

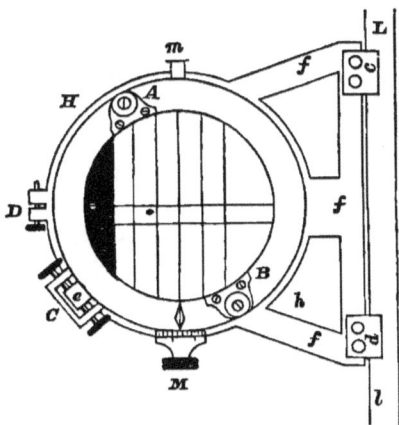

to the micrometer-head the greater is the reading of the micrometer-index. The comb on the left-hand side of the field, with its indentures and circle for zero-indication, serves for reading off the integral number of micrometer revolutions. Hh is a brass ring, rigidly joined to the frame *fff*, and capable of being clamped to the Telescope-tube by the capstan-headed screw at D. That frame is clamped to the limb of the Circle by the two clamps c and d, each of which consists of two rectangular clamping pieces drawn together by two capstan-headed screws, and nipping between them the frame f and the limb Ll of the Circle. The attachment of the Telescope to the limb at the object-end is exactly the same. After releasing the four clamps the Telescope may be shifted to a new position on the Circle. The screws for adjusting the wire-frame for correction of collimation-error (not seen in the Figure), have the same position as those of the Telescope of the Transit-instrument. At A and B are two screws by means of which the distance of the wire-frame from the object-glass may be adjusted and fixed. For this purpose they pass through two small plates, fastened by screws to the tube which carries the eye-end apparatus, and are worked either way in the frame Hh by capstan levers, the tube being finally held in position by two screws and circular plates, as shewn in the Figure. The reason for having the means of making this adjustment is the same as that already given in Art. 83 with reference to the apparatus applied to the Telescope of the Transit-instrument for effecting the same adjustment.

164. The *Telescope* of the Mural Circle is the usual astronomical telescope constructed to fulfil the optical conditions stated in Arts. 11—13, and is generally equal in length to the diameter of the Circle, that, for instance, of the Cambridge Circle having the focal length of eight feet. In case a stage be placed in front of the Circle for convenience in taking reflection-observations, the *illumination of the field* is best effected by putting a lamp on the stage opposite an aperture at the middle of the Telescope-tube for the incidence of rays from the flame on a plane reflector inclined at an angle of $45°$ to the axis of the Telescope. This is the arrangement in the case of the Cambridge Circle, and the degree of illumination is regulated by turning a milled head, which acts by rack-work on two rods connected with two angular plates forming a square opening, the size of which is by this means under command. A single *setting-circle*, constructed like the setting-circles of the Transit instrument, and attached to the east side of the Telescope-tube near the eye-end, serves for setting the Telescope according to the approximate N.P.D. of an object to be observed. In Cambridge practice, however, the setting-circle was seldom made use of, advantage being taken of the graduation of the Circle itself to perform the operation by means of a *pointer*, which is fixed to the wall, and points in the direction of the lines of graduation. These lines are inscribed on a *gold* band (white metal would have been better) at intervals of 5', and every alternate line is distinguished by circular dots to enable the eye readily to catch the divisions of a degree into intervals of 10'. Near the gold band there is a band of composite whitish metal, on

THE MURAL CIRCLE. 159

which the integral degrees are numbered correspondingly to the graduation-lines, so that by means of the two bands the indication of the pointer, to the accuracy of about a minute, may easily be seen, and the Circle, consequently, be quickly set to that degree of accuracy[1]. This process implies that the observer has approximately inferred the *setting-angle* from the polar distance of the object by previous calculation. The method of obtaining formulæ for this calculation will be explained in *Division III*.

165. It has already been stated (Art. 158) that the *pivots* of the Cambridge instrument turn in two steel collars, a larger one close to the Circle, and a smaller one at the end of the conical axis. The axis of motion may be considered to be the common axis of these two pivots. The weight of the instrument is principally sustained by the larger collar, on which account the pressure on this collar is taken off in great degree by lever *counterpoises* applied in the small interval between the collar and the Circle. The counterpoise action is effected by the upward pressure of two friction-wheels suspended by iron bars from the ends of two levers whose fulcrums are on the pier, while from their other ends heavy weights hang down in cylindrical holes made in the pier large enough for free suspension of the weights. The smaller collar is held in position within a fixed frame by being pressed in the vertical and horizontal directions by two screws, working in the frame, and by two counteracting springs. Turning the vertical and horizontal screws produces respectively vertical and horizontal angular movements of the axis of motion, and thus gives the means of correcting mechanically errors of level and azimuth.

The parts and appendages of the Mural Circle having now been sufficiently described, I proceed to treat of it under the *Second Division*.

---

[1] In the mode of setting practised with the Cambridge Transit-Circle, an additional microscope, directed for this purpose to the graduation, is made use of, and by this means the setting, as often as the occasion requires, is performed with a degree of certainty and exactness not attainable by the ordinary setting-circles. This method will be particularly described in treating of the use of the Transit-Circle.

II. THE ADJUSTMENTS OF A MURAL CIRCLE, AND CORRECTIONS OF INSTRUMENTAL ERRORS.

166. The Mural Circle being intended for measuring the altitudes of objects at their meridian-transits, the line of collimation of its Telescope is required to be adjusted to the plane of the meridian. But since the Circle is only used for measures of arcs of the meridian, and as these measures are not sensibly altered by small deviations of the line of collimation from the meridian-plane, the errors of collimation, level, and azimuth may generally be corrected with sufficient accuracy by mechanical means, without calculating their exact amounts. Also these corrections may be supposed to be made as if the pivots were perfectly cylindrical, and the axis of motion had a fixed position in space, because the effects of non-fulfilment of these conditions on the measures of arc are taken account of by the microscope-readings, as will afterwards be shewn. This being understood we may proceed to the consideration of methods of adjusting the line of collimation so that it shall move in the plane of the meridian.

167. In order to explain the requisite processes *ab initio* (as was done in treating of the transit-instrument), I shall begin with shewing how to select the place of the wall, or pier, which is to support the Circle, and how, after the wall is fixed in its position, it is prepared for the mounting of the instrument. First, a spot having been chosen near the transit-room, eastward or westward, a meridian line has to be drawn to mark the direction of the face of the wall. This may be done with sufficient accuracy either by means of a compass, the magnetic declination at the locality being known, or by the method described in Art. 10. Then the face of the wall may be made vertical, and the perforation for receiving the axis of the instrument be executed at the proper height and in the required horizontal direction, by ordinary mechanical means. Greater precision is required for fixing within the perforation, in sufficiently approximate positions, the supports of the rings, or collars, in

which the pivots are to turn (see Art. 166). Supposing this to have been done with as much accuracy as may be practicable by measurements taken with reference to the bearing of the face of the wall, and the known distance between the pivots, and the supports to have been temporarily fixed so as to admit of placing the instrument in position, the next operation is to *level* the axis of motion. The circumstances of the mounting not admitting of the use of a spirit-level for this purpose, the levelling is effected by a *plumb-line* employed with *Ramsden's Ghost Apparatus*. As this method depends essentially on an ingenious application of optical principles, and is found in practice to be accurate enough for a final horizontal adjustment of the axis of a Mural Circle, I propose to give here an account in detail of the construction of the apparatus, and the mode of using it.

168. The principle of the *Ghost* apparatus is to form, in rigid connection with the axis to be levelled, an *optical image*, with which the plumb-line may be brought into coincidence without interference with the freedom of suspension. The contrivance invented by Ramsden for this purpose is represented by Figure 32. MN is a brass cylindrical tube, which is

Fig. 32.

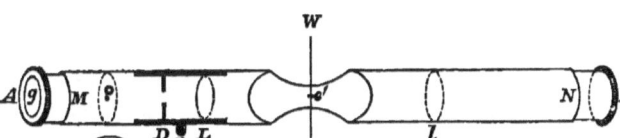

fastened to the telescope-tube so as to have its axis transverse to the direction of the axis of the telescope, and, as shewn in the Figure, is partly cut away to allow the plumb-line Ww to pass through or very near its axis. A and B are two movable parts, of cylindrical form with milled heads, inserted into the two ends M and N of the fixed tube. The inserted end of A, represented by the adjacent figure C, consists of a circular pearl disk, on which is fixed, *eccentrically*, a small brass ring, the interior diameter of which is extremely small. At the other end of A is a lens g by which the pearl is illuminated, so that the interior of the ring appears as a small bright circle at e. An image of this circle is formed at e' by means of the lens at L, A being first pushed close up to the end M of the fixed tube. By means of the projection n, movable in a slit, the lens L (with which is connected a diaphragm D for contracting the pencils) may be placed in a position proper for forming the image e' where the plumb-line can conveniently be brought into coincidence with it. The other movable part B carries at its inserted end a single lens l, with which the image e' is to be viewed by the eye at the other end, when the position of the lens for distinctly seeing e' has been tentatively ascertained. After the foregoing account it is, farther, to be mentioned that for the purpose of levelling the axis of rotation of the Circle, *two* ghost apparatus's, exactly alike, are fixed to the telescope-tube, one at a small distance from the eye-end, and the other at the same distance from the object-end. Before shewing how they are both made use of for effecting the levelling, it is necessary to explain the construction of apparatus specially designed for adjusting the position of the plumb-line.

169. The parts of this apparatus and the mode of using it may be indicated by means of Figure 33. At one end is a transverse brass plate AA, which, being let into a vertical socket fastened to the top of the wall, determines the horizontal direction of the intermediate part connecting this plate with another at the other end, to which the plumb-line is attached by the hook e. By turning the screw B, the tube to which the latter plate is fastened, is made, by rack-work within an exterior tube, to move through large spaces, for the purpose of placing the plumb-line at the distance from the wall proper for being viewed with the ghost apparatus, or removing it farther off to any required distance. By this means, and by the tangent-screw movement of the Circle, the plumb-line is brought *nearly* into coincidence with the ghost, and the position is fixed by the

clamping-screw C; after which an exact adjustment is made by the following means. The plumb-line *efg* passes at f between threads of a fine horizontal screw, which is attached by brackets to a *second* plate fastened to the other near the top, but capable

Fig. 33.

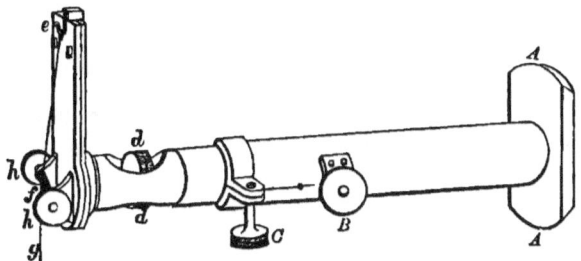

of a small separation from it at the lower part. This separation is effected by turning the milled head *dd*, which, by the intervention of a fine screw, gives slow motion to the plate and plumb-line in the horizontal direction perpendicular to the face of the wall. Then, since the line passes between threads of the screw, by turning either of the milled heads *hh* pertaining to the screw, the line is moved horizontally in a direction parallel to the wall. By the two rectangular movements the plumb-line may be brought into exact coincidence with the image. Supposing this to have been done with the *upper* apparatus, since by the previous arrangements the axis of the instrument is approximately horizontal, and, when the axis of the telescope is vertical, one ghost apparatus is exactly below the other, the line may be expected to be seen in the lower apparatus not far from the image. If, however, it be found to be inconveniently distant, the line is to be brought *half-way* towards the image by turning the vertical adjusting screw at the back of the wall, and the operation of bisecting the upper image must be repeated. The line, as seen with the lower apparatus, may always by such means be brought sufficiently near the lower image to admit of a coincidence being effected, in consequence of the eccentric position of the point e, by simply turning the cylindrical piece A about its axis (see Fig. 32). By this process the plumb-line is made to *pass through two points rigidly connected with the Circle.* We have next to consider how by having the means of satisfying this condition, the axis of motion of the Circle (as also that of any mass to which the same apparatus could be applied) may be horizontally adjusted.

170. The plummet is usually a brass weight of cylindrical form, immersed in water, and largely indented at its vertical surface, for the purpose of putting a stop to oscillatory and rotatory motions. In Figure 34, let a and b be the two bisected points joined by the vertical line aCb, which may be supposed to pass through a point C on the axis of rotation. Let CA be the direction of this axis, and draw CH in the horizontal direction, so that ACH is the angle to be corrected. After effecting the first bisections of the two points, the screw C (Fig. 33) is to be turned for unclamping, and then by turning the screw B the plumb-line can be moved far enough from the wall to be free from the ghost apparatus's, so as to allow of turning the telescope through 180°. By this motion let a come to a' and b to b', so that aa' and bb' are perpendicular to AC, and the angle Caa' is equal to the angle ACH. Next the plumb-line is to be restored to its place by turning the screw B, and the observer may then proceed to bisect b' by precisely the same process as that by which a was bisected in the first instance. This having been done, let the plumb-line hang down in the direction $b'p$. If the axis had been horizontal p would have coincided with a', instead of being seen with the lens l (Fig. 32) at the distance pa'. Now let the dotted line Ch bisect the angle bCa'; then will $\angle hCa' = \angle Caa' = \angle ACH$, the angle to be corrected. But $pa' = $ twice $ba' = $ four times ha'. Hence that angle will be corrected by turning the adjusting screw at the back of the pier till the point a' is moved in the direction from a' towards p through *one-fourth* the interval pa', because by

Fig. 34.

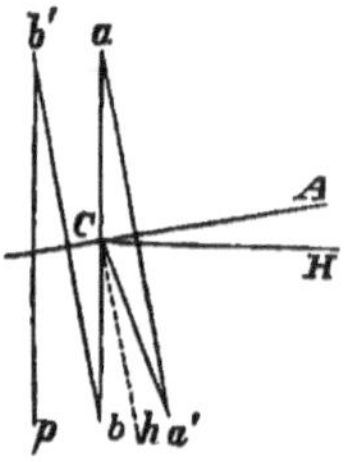

this motion the straight line Ca' will be moved through the angle $a'Ch$, and therefore CA through the angle ACH. The level-error of the axis will thus be corrected; but because no micrometer is used, and the judgment by the eye of one-fourth that interval is uncertain, the correction can only be considered as roughly approximate.

171. The use of the ghost apparatus is perhaps the best method of effecting a *first* levelling of the axis of a Mural Circle, and by repeating the process a sufficiently accurate final adjustment might be made. But it would be preferable, after the first approximation, to correct the result by means of reflection and direct observations of Polaris obtained in the manner stated in Art. 127. Since, however, this method gives the level-error in angular measure, in order to correct it mechanically (as proposed in Art. 166), the following process might be adopted. The angular error in seconds being ϵ, the length of the axis in inches l, and R being the number of seconds in an arc equal to radius, the length of the small line which at the position of the screw for level adjustment subtends the angle ϵ is $\frac{l\epsilon}{R}$. Suppose a pair of compasses to be applied to the screw so as to include in the direction parallel to its axis a number n of threads, and let the opening of the compasses measured on a scale be h in inches. Then will $\frac{h}{n}$ be the common interval between the threads, or the motion in the vertical direction produced by one revolution of the screw. The required correction will consequently be made by turning the screw, in the proper direction, through the fractional part $\frac{nl\epsilon}{hR}$ of a revolution, which might be done by estimation, or, for greater certainty, by appropriate scale measure.

172. After this correction of level-error, the error of collimation may be mechanically corrected by using the collimating eye-piece, first, for placing the telescope in a vertical position by bringing the fixed wire and its reflected image into coincidence by the tangent-screw motion (see Art. 84), and then for making

the mid-wire and its reflected image coincide by means of the screws for adjusting the wire-frame horizontally. Again, by the use of two collimators[1] in the manner described in Arts. 53—55, we might *begin* with correcting mechanically the collimation-error, by adjusting the wire-frame till the mid-wire is made to collimate with each of two wires, one in the North collimator, and the other in the South collimator, these wires having been previously made to collimate with each other through an aperture formed at the middle of the telescope-tube for this purpose. The level-error can then be corrected by means of the collimating eye-piece, by turning the screw for level adjustment till the mid-wire coincides with its image. After the mechanical corrections of the collimation and level-errors have been made by either of these methods, the azimuth error may be calculated from transit observations taken under any of the conditions mentioned in Arts. 88 and 89, an astronomical clock being placed in the Circle-room for taking such observations. The error being thus obtained in angular measure, the movement of the azimuth-adjustment screw for correcting it may be ascertained just as in the case of the mechanical correction of level-error, the account of which is given in detail in Art. 171.

173. Although the line of collimation may by the mechanical means above indicated be made to move in the plane of the the meridian with sufficient exactness for observations of zenith distances, it is not possible under the usual arrangements pertaining to the Mural Circle to obtain by calculation the *amount* of the collimation-error, for the two reasons that the Circle does not admit of azimuthal reversion, and is not provided with a vertical micrometer[2]. However, during a portion of June, 1858,

---

[1] The Cambridge Observatory was furnished, at my instance, with two collimators which, not being permanently fixed, could be temporarily set up for the purpose here mentioned. They were originally procured to be employed for measuring the effect of flexure on the indications of the Circle for any position of the telescope, by a process which there will be occasion to speak of in Division III.

[2] The large mural quadrant of eight feet radius, constructed by Bird, which for a long time was used at the Greenwich Observatory, was too ponderous for azimuthal reversion, and, consequently, could not furnish data for calculating

while repairs were going on in the Transit-room, being desirous of keeping up the determinations of clock-rate and clock-error, I made use of the Circle and the clock in the Circle-room for taking transits, and to ensure accuracy calculated the values of the three corrections by the following means. The intervals of the five wires from the mean of all were first obtained by transits of Polaris; then the level-error was found by reflection and direct transits of the same star; after which the telescope was pointed to the nadir, and the distance of the micrometer-wire from the fixed horizontal wire was adjusted so that, by means of the collimating eye-piece, the two wires, with the mid-wire and its reflected image, could be seen to form, according to the judgment of the eye, *a small square*. The interval in arc between the fixed and micrometer-wires being ascertained (by means that will be subsequently indicated), that between the mid-wire and its image became known, the half of which measured the angular distance of the mid-wire from the vertical plane through the optical centre of the object-glass. As this distance was made up of the level and collimation-errors, and the level-error was already calculated, the amount of collimation-error was inferred. The azimuth error was then calculated in the usual way. It is plain that such an operation would not be required in the case of an irreversible Transit-circle, because by means of its vertical micrometer, and the use of the collimating eye-piece, or of the two collimators, the error of collimation of the mid-wire, and, consequently, that of the mean of the wires,

its own error of collimation; on which account a reversible *Zenith Sector* was employed as a subsidiary means of effecting the calculation. This was, in fact, an altitude and azimuth instrument, limited as to range of altitude to small distances from the zenith, and having a radius considerably larger than that of the quadrant in order to diminish the probable error of comparisons of observations of the same star with the two instruments. In Woodhouse's *Astronomy* (1821), Vol. I., Part I., p. 68, an example is given of the calculation of collimation-error from a comparison of the means of the zenith distances of γ Draconis derived from observations made in the year 1812 with the quadrant and zenith sector of the Greenwich Observatory. In the erection of the Cambridge Observatory a room was prepared for the reception of a zenith sector of 25 feet radius, but the instrument was not procured, an exact calculation of the amount of the collimation-error of the Mural Circle not being required.

might always be calculated. (See Cambridge Observations, Vol. XX. for 1855—1860, p. lxxxv.)

174. In order to test the position of the Circle relatively to the plane of the meridian, it was usual at Cambridge to take transits of known stars of various polar distances from time to time with the Circle-telescope and clock (Molyneux) in the Circle-room, and at the same time to compare the clock (Hardy) in the Transit-room with Molyneux. The error of Hardy being known, and that of Molyneux inferred, the sidereal times of the transits were computed, and by comparing them with the known apparent R.A. of the stars, the errors in time due to deviation of the pointing of the telescope from the plane of the meridian were obtained. By these it would be seen whether the adjustment of the plane of collimation to the plane of the meridian was sufficiently exact.

The Microscope-reading.

175. The Circle being supposed to be put in position, we have next to consider the *adjustments* required to prepare the *micrometer-microscopes* for reading off the graduation, their parts and mounting being already described in Art. 162. It is, first, to be stated that the graduation proceeds in the direction from the South side of the Circle over the upper part to the North side, in order that the Circle readings, as shewn by a fixed index, may increase as the telescope is moved from the Pole southward, that is, as the polar distance increases. The micrometer-heads are all turned in the direction of the graduation, so that Figure 30 (1), in page 155, represents the position of the micrometer-head of that microscope which is near the *south* end of the horizontal diameter of the Circle. In the field of view of the microscope, represented by Figure 30 (2), is seen a portion of the gold band on the rim of the Circle, with lines of graduation across at intervals of 5', the graduation proceeding, since the microscope inverts, in the *downward* direction. One of the lines, that which is nearest the hole of the comb and towards the micrometer-head, is bisected by the cross-wires of the microscope. There are screws (not seen in the Figure) by

which the axis of the microscope is adjusted so that the intersection of the cross-wires is brought to the middle of the gold band. Also when the tongue pointing to the intersection of the cross-wires is placed, by turning the micrometer-head, across the centre of the circular hole in the comb, the index q by which the graduation on the rim r is read off, should point very nearly to zero. If this be not the case, the micrometer-head is to be set to the zero reading, and then the comb is to be shifted, by turning a screw provided for that purpose, till the tongue crosses the centre of the circular hole. This adjustment of the *comb* is to be performed for each of the microscopes, the Circle being in the mean time *clamped*.

176. After adjusting the combs, and while the micrometer-indices still point to the zero readings, a line of graduation is to be brought by the tangent-screw to be bisected by the cross-wires of *one* of the microscopes. This having been done, and the Circle remaining clamped, the cross-wires of the other microscopes are to be made to bisect, at the middle of the gold band, the lines of graduation to which the tongues point, by means of screws provided for adjustments of the microscope-axes in the vertical plane, in addition to those for the horizontal adjustment mentioned in Art. 175. The original mounting of the microscopes is intended to secure that the lines of graduation thus bisected shall be separated by equal arcs, for instance, of 60° in the case of six microscopes. If an inspection of the two graduation bands should shew that the proper line has not in every case been selected, new adjustments of the axes of the discordant microscopes will have to be made, every thing else remaining as before. What is done by these operations is called *the setting of the microscopes*.

177. The rim of the micrometer-head being, as stated in Art. 162, divided into 60 equal parts, and the common interval between the graduation lines of the Circle being 5', suppose the cross-wires of any microscope to be moved from the above-mentioned position for bisecting the line opposite the hole of the comb, through five complete revolutions, as indicated by the teeth of the comb and the micrometer-index, and let the

170   PRACTICAL ASTRONOMY.

direction of the motion be that of *increasing* index-readings, and therefore towards the micrometer-head. Then if the next line of graduation be found to be exactly bisected, the interval between the divisions of the micrometer-head will correspond to one second of the arc of the Circle. This condition is approximately satisfied at the first mounting of the microscope; but as it is liable to gradual change, or sudden disarrangement, the astronomer is provided with means of making the adjustment himself, or, at least, of securing that the deviation from its exact fulfilment shall always be small. The way in which this is effected by means of the apparatus described in Art. 162, may be shewn as follows by reference to that description, and to Figure 30 in pages 155 and 170. The first step is to ascer-

Fig. 30.

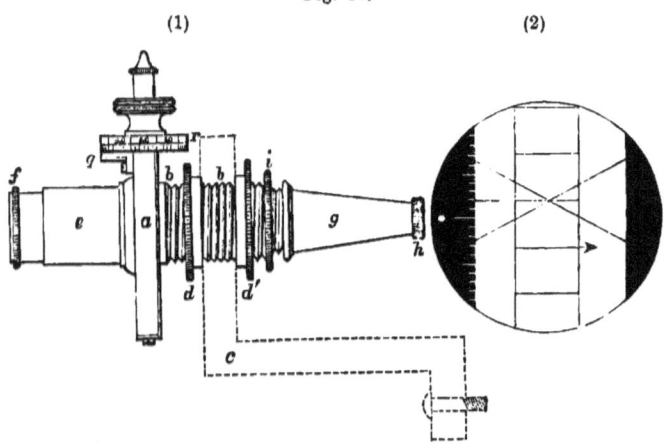

tain whether the image of the graduation in the field of the microscope, and the cross-wires, can be seen at the same time distinctly with the eye-piece f. If not, the distance of the object-glass at h from the graduation has to be altered, after releasing the clamping nut i, by turning the tube g (see Art. 162) till the required condition is fulfilled, and then clamping again. Next, the number of the division-intervals of the micrometer-head corresponding to the graduation-interval of 5' is to be found by bisecting two consecutive lines in the manner just stated, and noting the difference of the index-readings in the two cases. Supposing that in this way the difference is found to be $5 \times 60 + x$, and that x is a considerable number whether positive or negative, the distance of the microscope from the graduation will in that case have to be changed. The quantity x is called the *Run* of the microscope. According as x is *positive* or *negative* the image of the graduation-interval is *too large* or *too small*, and the microscope requires to be moved bodily *from* or *towards* the circle. Suppose that by means of the tapped nuts d, d' the microscope is moved, in the direction required for correcting the Run, through a given space k, which might be the interval between consecutive threads of the external screw. Then after adjusting the object-glass for seeing the graduation-lines distinctly together with the micrometer-wire, let the interval of 5′ be again measured as before, and suppose the number of micrometer-intervals now to be $300 + x'$. Then the additional space through which the microscope requires to be moved to correct the Run x' has to x' the same ratio that k has to $x - x'$, or is equal to $\dfrac{kx'}{x - x'}$. According as this quantity is *positive* or *negative*, the second movement is to be in the *same* direction as the first, or in the contrary direction. By such means the Run of a microscope can always be approximately corrected. For reasons that will be given in the next two paragraphs, the correction is not required to be made with extreme accuracy. After correcting the Runs of all the microscopes, the settings of the microscopes should be repeated.

178. It is now required to define precisely a *microscope-reading* of the Circle-graduation. The reference position of the cross-wires of the microscope is that for which the micrometer-index is at zero when the intersection of the wires is opposite to the hole of the comb. As shewn in Figure 30 (2), the line selected for bisection is that which is nearest the hole of the comb on the *same* side as the micrometer-head, the reasons for the selection being that the order of the graduation, as seen in the microscope, is *from* the micrometer-head, and the purpose

of the microscope-reading is to measure the interval from the bisected line to the reference position just defined. The degrees and minutes reckoned from the zero of the Circle-graduation to the bisected line are given at once by the two graduated bands described in Art. 164, and the addition of the microscope-reading, consisting of the minutes inferred from the number of indents of the comb between the hole and the projecting tongue, together with the seconds read from the micrometer-graduation, makes up the whole of the arc from the zero of the Circle-graduation to the reference pointing of the microscope. Let, for example, the arc of the circle from zero to the bisected line be $130°. 25'$, and the microscope-reading (which as represented in the Figure is nearly $2'$) be exactly $1'. 54'',7$; then the Circle-reading given by the particular microscope is $130°.26'. 54'',7$. It is to be noticed that the micrometer-graduation is always read off to estimated tenths of a second.

179. Here it is to be observed that the above microscope-reading is not accurate unless the value of the micrometer-intervals be exactly $1''$, or the *Run* of the microscope be zero. If it be found that the arc of $5'$ is measured by $300 + x$ micrometer-intervals, the value of one interval will be $\frac{300}{300 + x} \times 1''$. Hence if the recorded microscope-reading be h, the required correction is $\frac{300h}{300 + x} - h$, or $\frac{-hx}{300}$ nearly, x being by supposition small compared to 300. If h be expressed approximately in minutes and tenths of a minute, this amount is $-\frac{hx}{5}$, which is the formula for the *correction of the Run* of any one of the microscopes. In the example given in the preceding article, $h = 1',9$, so that if, for instance, x were found to be $+3'',5$, the correction would be $-\frac{1,9}{5} \times 3'',5 = -1'',3$ to tenths of a second. The rule is the same if h should somewhat exceed $5'$; but if, as is sometimes the case, a line on the *negative* side of zero be bisected, the micrometer-reading is set down as if the reading at zero were $5'$. Then if h be the recorded reading, and if h' be the

distance in micrometer-measure of the *positive* line from zero, we have $h' = x + h$, and the correction for Runs $= -\dfrac{h'x}{5}$ $= -(h+x)\dfrac{x}{5}$. Consequently the corrected micrometer-reading for the positive line is $h + x - (h+x)\dfrac{x}{5} = h + x\left(1 - \dfrac{h}{5}\right)$ nearly, $x \times \dfrac{x}{300}$ being omitted, because x does not exceed a small number of seconds. Thus the correction for Runs in this case is $+ x\left(1 - \dfrac{h}{5}\right)$, and is called a "Negative Correction."

The Circle-reading.

180. We are now prepared to obtain rules for calculating accurately the mean of all the microscope-readings, or, as it is called, a *complete Circle-reading*. Supposing the number of the microscopes to be six, let the recorded readings be $m_1, m_2, ...m_6$, and the Runs of the respective microscopes be $x_1, x_2,...x_6$. Then since by the setting of the microscopes h has nearly the same value for all, the corrected microscope-readings are $m_1 - \dfrac{h}{5}x_1$, $m_2 - \dfrac{h}{5}x_2$, &c., and their sum may be written $\Sigma . m - \dfrac{h}{5}\Sigma . x$. The Runs of all the microscopes, as being subject to continual changes, probably owing chiefly to the effect of changes of temperature on the radius of the graduation-band, are taken once or twice a week, or even nightly, and the values of $\Sigma . x$ thereby obtained are used for observations made both before and after the several times of taking the Runs. The quantity $-\Sigma . x$ is called the *Correction for Runs*; a proportional part of which in the ratio to the whole of h to 5, or, in the case of a *negative* line being bisected, in the ratio of $h - 5$ to 5, is added up with the six microscope-readings, and the sum, divided by 6, gives the complete Circle-reading.

181. In consequence of the setting of the microscopes, it suffices to record the Circle-reading for the bisected line in the case of only *one* of the microscopes, and the one most convenient

174   PRACTICAL ASTRONOMY.

for this purpose is that at the north end of the Circle's horizontal diameter. As a Circle-reading has of itself no astronomical signification, it may be reckoned from the zero of the graduation to an index in an arbitrary position; and hence, for the sake of convenience in recording, we may take, instead of the Circle-reading for the bisected line, that for a line distant from the bisected line by a constant arc. In order that this may be readily done, the *pointer* spoken of in Art. 164, as being used for setting for an observation, is so placed that it points to a graduation-line when there is one coincident with the zero position of the selected microscope. Then the *Pointer-reading* to be recorded is the circle-reading for the line which has the same position relative to the pointer, as the bisected line of graduation has to the zero of the microscope, and is, therefore, the circle-reading for the line which in the order of graduation comes *next before* the pointer (see Art. 178). In the case of the Cambridge Circle the pointer is placed *below* the north microscope, or farther on in the order of graduation, and the arc between the zero of the microscope and the pointer is $10°.40'$. Accordingly in the instance in Art. 178, the pointer-reading would be $130°.25' + 10°.40'$, or $141°.5'$, and the concluded Circle-reading $141°.6'.54'',7$ for the particular microscope. It is to be understood that the graduation proceeds from $0°$ to $360°$ round the Circle.

182. Besides the error of Runs, the microscope-readings may be affected by two other kinds of error due to the following causes. First, the centre of the graduation, defined to be the centre of a circle coinciding with the middle of the gold band, may not be on the axis of motion, in which case the Circle is said to be *eccentric;* and again, the axis of motion may not have a fixed position in space in consequence of a non-cylindrical form of the pivots. If the pivots be exactly cylindrical the path of the centre of graduation is a circle; otherwise it is a re-entering curve the deviations of which from the circular form depend wholly on the shape of the pivots. In Figure 35 let $OPQR$ represent the path described by the centre of graduation when the Circle is turned on its axis completely round; and supposing the centre

of the graduation to be at any point O of the path, let COC' be drawn through this point in the common direction of the axes of the opposite microscopes A and B; and at the same time, the

Fig. 35.

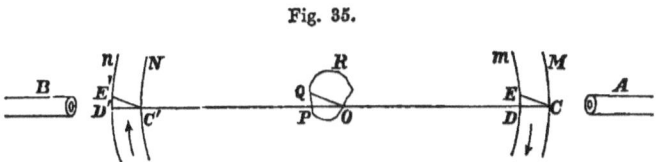

opposite limbs being at MC and NC', let the Runs of the two microscopes be taken. Suppose also that by turning the Circle about its axis the position of the centre of graduation is changed from O to Q, and that the limbs are brought into the new positions at D and D'. The motion of the Circle by which these changes are produced may be conceived to consist of a motion of rotation about the centre of graduation at O, and a motion of translation of every point of the Circle equal and parallel to the translation of the centre of graduation from O to Q. The former motion produces no change of the position of the periphery of the Circle; and the change produced by the other is such that if CE and $C'E'$ be drawn parallel and equal to OQ, the points which by the rotation alone would be brought to C and C' at the end of the interval during which the centre of graduation passes along the curve $OPQR$ from O to Q, are brought, at the end of the same interval, by the motions of rotation and translation combined, to the positions E and E'. It is evident since by the construction $CD = C'D'$, that the Run of A is as much diminished, as that of B is increased, by the change of the distances of the graduation from the microscopes, and their sum is consequently unchanged. Also, the graduation proceeding in the direction indicated by the arrow-heads, since $DE = D'E'$, the pointing of A is as much in advance of the position E as the pointing of B is behind that of E', so that the sum of the opposite readings remains unaltered by the change of position of the Circle. Since A and B may be *any* two opposite microscopes, the same inferences apply to the complete Circle-reading. Hence we may conclude, that after obtaining the correction of the error of Runs for a given position of the Circle, the remaining errors, due to eccentricity of the graduation and irregularity of the forms of the pivots, are corrected in the means of the readings of opposite microscopes[1]. It is advisable, for greater certainty, to determine the correction for error of Runs for more than one position of the Circle at the same time.

183. The position of reference in the field of view for observations with the Mural Circle is that of a fixed horizontal wire, which in Figure 31 (p. 157) is represented by the straight line passing through the hole of the comb, and the object observed is bisected by this wire by means of the tangent-screw movement, generally when it is crossing, or very near, the middle of the five vertical wires. As, however, the bisections cannot always be made at this part of the field, the fixed wire requires an exact *equatorial adjustment*, in order that the reference position which it indicates may be the same for bisections made at different distances from the middle wire. This adjustment is effected by means of the apparatus C (see Art. 163), which serves to give to the fixed wire a motion of rotation about the axis of the telescope. By joining this motion with the movement of the Circle by the tangent-screw, a small star, on or near the Equator, may be made to traverse the wire during its passage through the field. If an equatorial star should be judged to be well bisected by the wire throughout the course, the required adjustment may be considered to be

---

[1] According to this theory the errors of the graduation-readings of a Mural Circle might be corrected by the use of only two opposite microscopes. In reliance on this principle, Pond made trial for a time of a single pair of microscopes: but having found, after comparing the results with those obtained by six microscopes, that the latter were the more consistent and trustworthy, he finally adopted the use of six. Considering the changes of form the Circle may be liable to from the mechanical effect of its own weight, and from differences of temperature at different parts, it might be expected that the larger number of microscopes would have the effect of eliminating, at least in part, residual errors due to those causes.

made. That, however, is a condition it is not easy to satisfy, requiring generally a succession of trials by different stars, which, for greater certainty in bisecting, should all be of about the ninth magnitude. When the required condition has been by degrees fulfilled, the wire is to be fixed in position by making the opposite screws of C bear fully on the projecting piece e (Fig. 31). As it is important that this adjustment should be exact, it would be well, after completing the operation, to test the result by additional bisections of small equatorial stars.

184. The use of the fixed wire might be dispensed with by making all bisections with *the micrometer-wire* (Art. 161), supposing the reference position to be indicated by this wire when it crosses the hole of the comb and the micrometer-index is at zero. For convenience the micrometer-reading for this position may be taken to be exactly 10^r. This method requires the micrometer-wire to be equatorially adjusted by the process described in Art. 183 relatively to the fixed wire. This process may in the present case be facilitated by bisecting the star soon after its entrance into the field and shortly before its departure, and judging from the difference of the micrometer-readings for the two bisections, as to the amount of angular deviation of the wire that remains to be corrected. Although simplicity is gained by using only the micrometer-wire, it is still to be said that a fixed wire may be advantageously employed in certain observations, such as measures of the diameters of planets, and generally for measuring small differences of N. P. D. of adjacent objects. In any case, however, it will be required to reduce an observation made by bisecting the object with the micrometer-wire in any position, to what it would have been if the bisection had been made in the reference position. This is done when there is only a micrometer-wire by subtracting 10^r from the micrometer-reading for the bisection, and multiplying the excess (positive or negative) by the value in arc of one micrometer-revolution, as ascertained by processes that will be given in the next paragraph. The result with its sign changed is to be added to the mean of the microscope-readings to obtain the concluded Circle-reading for the observation. In the case of a fixed wire

178 PRACTICAL ASTRONOMY.

equatorially adjusted, the micrometer-wire (Fig. 31) is not assumed to be exactly parallel to it, because, as stated in Art. 161, the apparent parallelism depends wholly on mechanical construction. Hence, on the principle of testing and correcting mechanical adjustments by optical means, the readings for *coincidence of the micrometer-wire with the fixed wire* (see Art. 51) are taken at the five vertical wires and at the comb, and those for intermediate positions are inferred by interpolation. Then the distance of an object bisected by the micrometer-wire from the fixed wire is measured by the excess of the reading for bisection above the coincidence-reading corresponding to the place of bisection, and this excess, converted into arc, is to be applied negatively to the mean of the microscope-readings.

185. For finding the value in arc of one revolution of the micrometer, a meridian mark, such as that described in page 40, note 1, might be made use of in the following manner. The micrometer-wire, being set at a certain number of integral revolutions from zero, as -10^r, is brought by clamp and tangent-screw to bisect the mark, and the circle-reading is taken. Then after another setting, as at $+10^r$, the wire is again brought to bisect the mark, and another circle-reading is taken. The difference between the circle-readings is the value in arc of the difference between the micrometer-readings, and being divided by the number of the micrometer-revolutions (20, in the supposed instance), gives the value in arc of 1^r. The adopted value should be the mean of several such measures. In making use of this method I found that the measures were vitiated by variation of terrestrial refraction in the interval between the bisections, the apparent altitude of the mark being thereby altered, especially if the operation was performed in the evening. This source of error may in great measure be got rid of by employing the following process. Supposing the micrometer-wire to be set at $+10^r$, -10^r, and $+10^r$ in succession, and the mark to be bisected by the wire in each position, let the respective circle-readings be a_1, a_2, a_3. Then if by the increment of refraction in the interval between the first and second bisections the angular elevation of the mark be increased by ϵ, the reading

a_2, as compared with a_1, is *too small* by ϵ, or, corrected for change of refraction, becomes $a_2 + \epsilon$. So if the interval between the second and third bisections be the same as that between the first and second, and the variation of refraction during the two intervals be supposed to be uniform, the corrected third reading, or that which would have been obtained if the refraction had not varied, is $a_3 + 2\epsilon$. We have, therefore, from the three Circle-readings for bisection of the mark, when the micrometer-readings and the corrections for variation of refraction are taken into account, the following equations:

$$a_1 + 10^r = a_2 + \epsilon - 10^r = a_3 + 2\epsilon + 10^r.$$

Hence $20^r = a_2 - a_1 + \epsilon$, and $20^r = a_2 - a_3 - \epsilon$.

$$\therefore 20^r = a_2 - \frac{a_1 + a_3}{2}.$$

If the successive settings of the micrometer be $-10^r, +10^r$, -10^r, the value of 20^r is $\frac{a_1 + a_3}{2} - a_2$. Supposing the mark to be bisected any number of times n exceeding 3, and the micrometer settings to be alternately $+10^r$ and -10^r beginning with either, we might obtain, in the manner above shewn, a measure of 20^r from the Circle-readings a_1, a_2, a_3; another from a_2, a_3, a_4; and so on. *Each* of the $n-2$ values thus calculated would be in great measure free from the effects of variation of refraction. I have found that the several values thus obtained are consistent with each other, and that their sum divided by $n-2$ gives a very probable mean result[1].

---

[1] This method of *alternate* measures for extracting the true measure of a quantity from observations affected with small continuous changes, admits of various applications. We have had an instance in the determination of the azimuth-error of a transit-instrument by observations of the clock-times of transit of Polaris alternately above and below Pole. (See Art. 89 (4).) The method may be generalized as follows. Let $a_1, a_2, a_3...a_n$ be quantities recorded in order of time, as given by the alternate observations, and let $d_1, d_2, d_3...d_{n-1}$ be the differences (without respect to sign) between a_1 and a_2, a_2 and a_3, &c., that is, the several measures uncorrected for the above-mentioned changes. Then if $\Delta_1 = d_1 + d_2$, $\Delta_2 = d_2 + d_3$, ...$\Delta_{n-2} = d_{n-2} + d_{n-1}$, the corrected measure is $\frac{1}{2(n-2)}(\Delta_1 + \Delta_2 + ... + \Delta_{n-2})$.

180   PRACTICAL ASTRONOMY.

186. On account of the above-mentioned effect of the variation of terrestrial refraction, and the general unsteadiness of the image of a distant mark, the use of a *Collimator* for determining the value of the micrometer-revolution, whether it be temporarily or permanently fixed, is much to be preferred. The collimator may be at a short distance from the Circle-telescope, and even in the same room, and at its geometrical focus there should be two connected parallel wires, like those which in Figure 13 (p. 44) are represented as being in the field of view *bb'* of the collimator *B*. In fact, this collimator is well adapted for the present purpose, the parallel wires being first brought into position across the middle of the field by turning the Collimator's micrometer-head. These wires have to be adjusted to geometrical focus, and placed parallel to the micrometer-wire of the Circle-telescope, by the processes indicated in Art. 54. Then the micrometer-wire, after being set either at -10^r or $+10^r$, is to be brought by clamp and tangent-screw to the middle of the fixed space between the images of the two parallel wires (a condition which by the judgment of the eye can be satisfied with great precision), and the Circle-reading is to be taken. The same operations having been performed with the micrometer-wire set at $+10^r$, or -10^r, the difference of the two readings is a measure of 20^r, whence the value in arc of 1^r may be at once inferred. The adopted value should be the mean of several such determinations; and it is advisable, in this method also, to read off the Circle for observations made in alternate positions of the wire, as probably tending to diminish the effect of incidental errors.

187. It has been shewn (Art. 182) that corrections for the error of Runs, obtained in an arbitrary position of the Circle, are applicable to any other position, notwithstanding eccentricity of the graduation and irregularity of forms of the pivots, if the graduation-band be always circular and its radius remain unchanged. But if these conditions be not fulfilled, the sum of the Runs might be different for every different position of the Circle, or be different for the same position at different times. Such changes might be produced by the effects of change of tem-

perature both on the Circle and the microscopes, or by the deformation of the Circle due to flexure caused by its weight, or by the two causes combined. As far as regards the bodily movements of the Circle and the microscopes by the effects of temperature, it is evident that if the heights of the supports be arranged so that there is no motion of the microscopes relative to the Circle, the amount of Runs will not be changed. (See Art. 158, p. 152). Also supposing the change of height of the supports by temperature to be such as to make the Circle move relatively to the microscopes, it appears from the argument in Art. 182, that no change of the amount of Runs will thereby be produced, if only the motions of opposite microscopes be equal and parallel. But so far as by the effect of temperature the displacements of opposite microscopes are not the same in amount and direction, or the radius of the graduation-band is changed, or its curvature is caused to vary from point to point, the amount of Runs undergoes change, because the simultaneous changes of the Runs of two opposite microscopes do not in these cases neutralize each other. The errors with which the Circle-readings will in consequence be affected, can only be obviated by frequently taking the Runs, and in various positions of the Circle, with special reference to changes of temperature. The effect of the deformation of the Circle by its weight, and of the flexure of the telescope-tube thence resulting, will come under consideration in *Division* III.

The Zenith Point.

188. Before proceeding to treat of the mode of using the Mural Circle for observing celestial objects, it is required to find the value of an arc which may be called the *index error* of a Circle-reading. It has been stated (Art. 181) that a Circle-reading obtained, according to the foregoing rules, for any arbitrary position of the telescope, has no signification of itself; but a reading obtained (by means which will be presently indicated) for the particular position in which the telescope points to the zenith of the place of observation, gives the means of inferring from other Circle-readings *Zenith Distances*. It is

conceivable that by shifting the telescope on the Circle through an arc exactly equal to the Circle-reading for zenith-direction, that reading might be reduced to zero, and the index error thus be corrected. But there is no need to make this *mechanical* correction, because the index error is virtually corrected, as will afterwards appear, by the process of *calculation* applied to each observation. The uncorrected index error, as having reference to the zenith-direction, is called the *zenith point*.

189. There are various ways of obtaining the Circle-reading for the zenith-direction. The use of *floating collimators* for this purpose, proposed by Kater, is simple in principle, and gives trustworthy results, but does not admit of the degree of accuracy attainable by employing the collimating eye-piece in a manner which will presently be indicated. Floating collimators are of two kinds, vertical and horizontal, the principle of floating being applied alike in both. The collimator, consisting simply of an achromatic object-glass, with cross-wires at the geometrical focus, is firmly attached to a heavy iron plate which floats on mercury contained in a shallow iron vessel. In the case of the *vertical collimator*, the axis of which is nearly vertical, the object-glass is directed downwards through an aperture at the central part of the bottom of a vessel of circular form, the shapes given to the interior containing surface of the vessel, and to the iron plate, being such as to allow of collimating with the telescope through the aperture. A contrivance is added whereby the vessel and its contents, together with the collimator, may be readily turned about a vertical axis through an arc which by a check is limited to 180°. For use with a Mural Circle this instrument requires to be supported by an appropriate frame just above the highest part of the Circle, in order to perform the collimation with the telescope directed nearly to the zenith. The collimation is effected when, by moving the Circle and adjusting the position of the cross-wires, their intersection, as seen in the field of the telescope, is brought to be coincident with a defined point near the centre of the field. The Circle-reading is then taken, and after turning the collimator through 180°, the collimation is performed as before, the *same* two points being brought into

coincidence, and the Circle is again read off. Since by the principle of floating the axis of the collimator in the two positions is equally inclined to the vertical in opposite azimuths, the same will be the case with respect to the axis of pointing of the telescope, and the mean of the two Circle-readings will consequently give the reading for zenith-direction, or the zenith point. In Figure 36, O, o, c are respectively the positions of

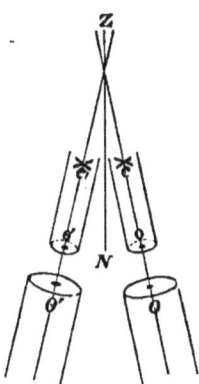

Fig. 36.

the optical centres of the telescope's and collimator's object-glasses, and the intersection of the cross-wires, before reversing the floating apparatus, and O', o', c' the positions of the same points after. The three points may in each case be assumed to be on the same straight line representing the common direction of the axes of the telescope and collimator, and the straight line NZ bisecting the angle which the two directions make with each other gives the direction of the zenith.

190. In the case of the *horizontal floating collimator*, the iron vessel containing the mercury is of rectangular form, and the float is a rectangular plate of iron, to which the vertical supports of the two ends of the collimator are attached. When floating the collimator has its axis very nearly horizontal, and by means of pins fastened to opposite edges of the plates and

184   PRACTICAL ASTRONOMY.

fitting into fixed slits, horizontal displacements of the apparatus are sufficiently prevented. After the collimator has been suitably placed on a level with the centre of the Circle to the North or South, the collimation with the telescope is effected just as in the case of the vertical collimator, and the Circle is read off. The collimator is then transferred to the opposite side of the Circle and similarly placed, and after a second collimation with the telescope, the Circle is again read off. For the same reason as in the case of the vertical collimator the axis of the telescope in the two positions is equally inclined in opposite azimuths to the vertical, and the mean of the Circle-readings is therefore the zenith point. The way in which the geometrical conditions of

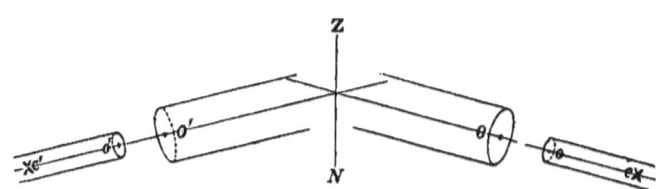

the operation are illustrated by means of Figure 37 may be stated in the very same terms as those applied above to the vertical collimator by reference to Figure 36.

(The method of obtaining the zenith point by direct and reflection observations of the same star belongs to the next *Division.*)

191. It only remains to shew how to determine the zenith point by means of the collimating eye-piece described in Art. 81, which, as used at the Cambridge Observatory, could be adapted both to the telescope of the Transit-instrument and to that of the Mural Circle. There are several ways of employing the collimating eye-piece for this purpose. Care should first be taken in each instance to perform the adjustment of the distance of the wire-frame from the object-glass by which a wire and its image are made to appear equally well defined (see end of Art. 161). If the telescope has only a micrometer-wire, for deter-

mining the zenith point it is only required to obtain the micrometer-reading for coincidence of the wire with its reflected image, the Circle being previously set so that the coincidence may take place near the middle of the field, or the coincidence reading not differ much from 10". The coincidence of a wire with its image may be judged of with considerable certainty by noticing the definition when they are being brought into conjunction, which becomes sensibly sharper as the coincidence is more exact. But in general it is best to place the image either in contact with the wire, or at small distances from it judged by the eye to be equal, an equal number of times on one side and on the other, and to take the mean of the micrometer-readings in the different positions for the coincidence reading. The difference between this value and the reference reading of 10", being converted into arc, and applied to a Circle-reading taken before or after the coincidence is observed, gives the Circle-reading for the pointing of the telescope in the direction indicated by that coincidence, that is, the zenith point. If the telescope has a fixed wire for reference, the wire and its image might be brought into coincidence by movement of the Circle by clamp and tangent-screw, if such apparatus should be conveniently within reach, and the Circle-reading then taken would be at once the zenith point. But it would be a more certain method to obtain, for a given Circle-reading not differing much from the zenith point, the reading for coincidence of the micrometer-wire with its image in the way just stated, and then after finding the reading for coincidence of the micrometer-wire with the fixed wire, and converting the difference of the two readings into arc, to infer the zenith point by applying the difference to the above mentioned Circle-reading. For finding the reading for coincidence of the micrometer-wire with its image, I finally adopted the following method, considering it more accurate than those before mentioned; but it is only applicable in case there is a fixed wire. A Circle-reading differing little from the zenith point having been taken, the micrometer-wire was placed so that either its image, or that of the fixed wire, was mid-way between the two wires, a position which could be determined with nicety by the judgment of the eye. Then, the micrometer-reading for coincidence with the fixed wire being taken, if a be the reading for that wire whose image is mid-way between the two, and b the reading for the other, the reading r for coincidence of the micrometer-wire with its image is evidently given in all cases by the equation $r = a + \frac{1}{4}(b - a)$.

As the zenith point is a reference quantity, for the sake of accuracy it should be frequently observed, regard being had at the same time to changes of temperature, and on each occasion the Runs should be taken.

Plate II. represents the Troughton Mural Circle, which, in conjunction with a similar one by Jones, was used at the Greenwich Observatory till both were superseded by the erection of the present Transit Circle. An historic interest pertains to Troughton's instrument, it being the first of the kind. The Cambridge Mural Circle is very like this both as to construction and mounting, but differs in being eight feet, instead of six feet, in diameter, and in some minor details. The six microscopes (for the description of which see Art. 162) are represented as mounted on brass supports satisfying the condition spoken of in Articles 158 and 187. The designations of the microscopes, in the order from *North* over the highest part of the Circle to *South*, are A, C, E, B, D, F and the readings are recorded in this order in two rows A, C, E and B, D, F, the latter under the other, so that the mean of the readings of each set of opposite microscopes can be readily inferred, and any considerable change of value be detected. The object-end of the telescope is at a, and the eye-end at e, and the positions of the two Ramsden's Ghost Apparatus (described in Articles 168—170) are at f and g. The friction wheels nn are suspended for the counterpoise action by the rods mm (see Art. 165). For the purposes of clamping and slow motion, a circular flat ring, somewhat larger than the graduated rim of the Circle, is fastened to the wall, and on this ring a clamp-and-screw c for slow motion slides, and may be clamped to it at any point, a clamp b, which bites the Circle and gives it motion, being at

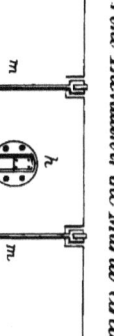

Front Elevation of the Mural Circle.

the other end of the moving screw. The necessity for this additional ring is superseded in the case of the Cambridge Circle by the construction of the clamp and tangent-screw, as explained in Art. 160, by reference to Figure 29. The letters P, Q, R mark the positions at which the three apparatus used with that Circle are attached to the wall, the position of P being about 30° below the microscope A. The small squares near the microscopes are sockets for inserting the support of the lamp by which the graduation read off by the microscopes is illumined. The lamp-supports of the Cambridge Circle are connected by hinges with attachments to the wall, and can be let down, or put up, as the purpose they are intended for requires.

III. METHODS OF OBSERVING WITH THE MURAL CIRCLE, AND CALCULATION OF APPARENT ZENITH DISTANCES.

192. It will be assumed, as in the case of taking observations with the transit-instrument, that in every instance the right ascension and polar distance of the object to be observed, and the error of a sidereal clock, suitably placed in the observing room, are at least approximately known. (See Art. 108.) Let us, first, suppose that the index at that end of the level of the setting-circle which is *northward* when the bubble is in mid-position, is made to point to the zero of the graduation when the telescope is directed to the zenith by any of the means described in Arts. 189—191, the graduation proceeding both ways from zero to 180°. Then another index attached to the level, and at a distance *below* the first equal to the colatitude λ, will point to the graduation-reading for polar distance of zenith, and, therefore, indicate the actual pointing of the telescope. The same index will serve for setting the telescope to any given polar distance $\lambda + z$, because if the setting-angle be increased by z, the telescope, in order to bring the bubble to mid-position, must be turned to point farther from the Pole by the same angle. This, in fact, is the usual method of setting for direct observations made with a Transit-instrument. But

with the Mural Circle, besides direct observations, many are taken by reflection at the surface of mercury, and it is, therefore, required to find the position of an index which may be used for setting in such observations.

193. If the telescope be moved northward from the zenith direction through an angle equal to the colatitude λ, it will be made to point to the Pole, and in order that the bubble may in that case be in mid-position, the level must be turned in the opposite direction through λ, by which operation the index for setting in polar distance (Art. 192) will be made to point to the zero of the graduation. If now the telescope be turned farther in the *same* direction through an angle equal to twice the latitude, it will point *below* the north horizon by an angle equal to the latitude, and, therefore, in the direction proper for observing by reflection an object at the Pole. Then in order to bring the bubble into mid-position the index and level must be moved farther in the same direction as before through twice the latitude. Hence, L being the latitude, if a *second* index be attached to the level so as to be farther on in the direction of the graduation than the first by $2L$, when that index points to zero, the bubble being in mid-position, the telescope will point in the direction proper for observing by reflection an object at the Pole. The same index will serve for setting for the reflection observation of a star of *any* polar distance; only it is to be noticed that the setting is to be performed as if the polar distance were *negative*. For if the telescope be moved to a depression below the north horizon greater than L by δ, that is, if it be set for the reflection observation of a star whose polar distance is δ, the new index will have to be moved from the zero through δ in the negative direction to put the bubble in mid-position. In accordance with the above explanations the Cambridge Mural Circle has a setting-circle furnished with two indices, movable with the level, and separated by an interval of $104° 26'$ ($= 2L$), and one is marked for use in direct observations, and the other for use in reflection observations.

194. But in the use of that instrument, a direct and a reflection observation of the same star, one quickly after the other,

THE MURAL CIRCLE. 189

are taken in the same meridian transit (by a method which will presently be indicated), and there is, consequently, not sufficient time for setting between the observations by means of the setting-circle. On this account the following process is adopted. (See Art. 164.) Let Z be the zenith point (inclusive of pointer reading), obtained as stated in Arts. 189—191, and in Figure 38, let $PZ\sigma$ ($=\delta$) be the star's polar distance, and

Fig. 38.

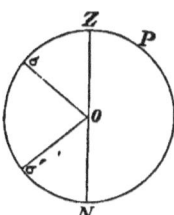

PZ ($=\lambda$) the colatitude of the Observatory. $O\sigma$ is the line of pointing of the telescope for the direct observation, and $O\sigma'$ that for the reflection observation, the angle $ZO\sigma$ being equal to the angle $NO\sigma'$ by the law of reflection. The pointer reading for the direct observation is greater than that for the zenith by the angle $ZO\sigma$, or $\delta - \lambda$, and is, therefore, equal to $Z + \delta - \lambda$, which, if A be put for the constant angle $Z - \lambda$, becomes $A + \delta$. The pointer reading for the reflection observation is *less* than that for nadir point by the angle $NO\sigma'$, or $\delta - \lambda$, and is, therefore, equal to $180° + Z - (\delta - \lambda)$, or $B - \delta$, the constant B being put for $180° + Z + \lambda$. In general the pointer reading for the direct observation is obtained by adding the approximate polar distance to the constant A just before proceeding to make the observation. But for bodies of the Solar system, the motions of which are given by ephemerides, and for stars which are to be observed directly a considerable number of times, it is advisable to tabulate for use the values of $A + \delta$; and in all cases of stars selected for frequent observation directly and by reflection at the same transit, the values both of $A + \delta$ and

$B - \delta$ should be calculated beforehand, and tabulated according to the approximate right ascensions.

195. According to what is stated in a Note to Art. 34 (p. 28), the use of a polar distance index may be dispensed with, and the zenith distance index be used in its place, if the setting-circle be turned relatively to the level with bubble in mid-position through an angle equal to the colatitude. For instance, let the telescope be pointed to the zenith, and the bubble be in mid-position, so that the zenith distance index points to zero and the polar distance index to the colatitude. To make the former index indicate polar distance when the bubble is in mid-position, it is only required to turn the zero of graduation through an angle equal to the colatitude from the northward horizontal direction towards the vertical direction. The mechanical construction allows of this being readily done. By this adjustment the radius drawn to the zero from the centre of the setting-circle makes an angle equal to the latitude with a radius drawn parallel to the axis of the telescope towards the object-end. After this alteration the setting is performed exclusively by the zenith distance index. In the case of the Cambridge Transit-instrument, which had two setting-circles, the change of the place of zero was made, in the above manner in each, when it was westward, after which, as before, either circle could be used for setting when on the *west* side of the telescope tube.

196. When, however, a meridian Instrument has two setting-circles, and, as in the instances of the Greenwich and Cambridge Transit-Circles, is *not reversed*, the second one may be made use of in one or the other of these two ways: either by the above-stated arrangements the setting for polar distance might be performed in exactly the same manner with both, in which case they could be conveniently used for setting for the direct observations of two objects coming in quick succession one after the other; or the second circle might be adapted for setting for *reflection* observations. The way in which the latter purpose is effected by the *eastern* setting-circle of the Cambridge Transit-Circle is such as follows. Suppose the telescope,

THE MURAL CIRCLE. 191

as before, to point at first to the zenith, the bubble being in mid-position, and of the two indices of the level that which is northward to point to the zero of the graduation. Then let the telescope be turned from the zenith northward through $180° - \lambda$, and the index be turned in the opposite direction till it points to $180° - \lambda$ of the graduation. The bubble will thus be brought to mid-position, and the telescope will be set for the reflection observation of a star at the Pole. After this there are two ways of proceeding. The circle may be shifted, every thing else remaining the same (see Art. 194), either so as to bring the zero of the graduation to the *north* index, or so as to bring it to the *south* index. In the former case the north index would be used for the setting, which, however, must be performed as for a *negative* polar distance, or for a star *sub polo*, because if that index be moved *upwards* from zero through an arc δ equal to the polar distance, the depression of the telescope below the north horizon, in order to bring the bubble to mid-position, has to be increased by δ, and the setting is, consequently, that required for the reflection observation of a star whose polar distance is δ. In the other case, the *south* index would have to be used for the setting, and, as in the ordinary setting for a direct observation, the movement of the index from zero will be *downwards*, so that the setting for a *reflection* observation by the *south* index of the *east* circle is performed in exactly the same manner as the setting for a *direct* observation by the *north* index of the *west* circle.

197. The rules for setting the telescope having been laid down, we may proceed to consider the different processes employed in taking Circle observations. For the *direct* observation of a *star*, the telescope is first to be set, according to the star's polar distance, either by the setting-circle (Art. 192), or by the pointer directed to the graduation-reading $A + \delta$ (Art. 194), and the Circle is then to be detained by the clamp of that tangent-screw which will be most conveniently reached when the telescope is pointed to the star. Knowing approximately the right ascension of the star and the error of the clock, the observer knows when the entrance of the star into the field is to be looked for, and the setting has secured that the direction

of the path shall be nearly across the middle of the field. Then, if no circumstances prevent, the star is bisected, just as it crosses the middle one of the five vertical wires, by the fixed wire, or, if there is no fixed wire, by the micrometer-wire in the reference position, the bisection being effected by moving the Circle with the tangent-screw. In order to satisfy the condition mentioned in Art. 160, of always performing the bisection by turning the screw in the direction indicated by the roughness of the milled head, if the star should not enter into the half of the field suitable for this operation, it has previously to be placed in that half by means of the tangent-screw. The Circle-reading is then to be taken (Arts. 180 and 181), and this completes the instrumental process.

198. If it should happen that the bisection is not made when the star is on or very near the middle wire, the place of bisection is recorded by stating its distance from that wire in integer wire-intervals and fraction of an interval, and prefixing a *minus* or *plus* sign according as the bisection was made before or after passing the mid-wire. This is done for the purpose of calculating a correction for the curvature of the path of the star in the field, to be applied to the Circle-reading to reduce it to what it would have been if the star had been bisected when on the meridian. This mode of indicating the place of bisection is sufficiently accurate if the plane of collimation of the telescope has been well adjusted mechanically to the plane of the meridian by the methods explained in Arts. 168—174, and if the curvature of the star's path be small. But for observations of stars near the Pole, as especially Polaris, δ Ursæ Minoris, and 51 (Hev.) Cephei, and generally for observing with precision a star, the curvature of whose path is considerable, the following process should be adopted. The clock-time of making the bisection should be set down, and then by knowing approximately the error of the clock and the star's right ascension, the hour angle from the meridian can be calculated. From this datum (which may be supposed to be accurate to within two or three seconds), and from the polar distance of the star, the correction for curvature of path can

be derived with accuracy by formulæ investigated in the next paragraph. If the R. A. of the stars be not known by contemporaneous transit-observations, they may be immediately inferred, in the instances of stars contained in the Catalogue of the Nautical Almanac, from apparent R. A. computed in that work, but in other cases they will have to be approximately deduced from mean R. A. given in standard catalogues.

199. The above-mentioned formulæ for calculating corrections for curvature of path are obtained as follows. In Figure 39, P is the pole, the arc PNS is in the plane of the

Fig. 39.

meridian, σ is the place of the star when its image is bisected, and the bisecting wire, being straight and horizontal, is projected into a great circle of which σN perpendicular to PS is a part. The dotted arc σS is part of the star's diurnal path, so that PS is equal to $P\sigma$ the polar distance. Hence, since the projection of the wire cuts the meridian at N instead of S, the error to be corrected is NS. Let $NS = e$, $P\sigma = \delta$, $\sigma N = D$, and $\angle \sigma PN$ (in time) $= t$, this hour-angle being obtained in the manner stated in Art. 198. Then since $PN = \delta - e$, the right-angled triangle $PN\sigma$ gives

$$\cos(\delta - e) = \cos \delta \sec D,$$
and $$\tan(\delta - e) = \tan \delta \cos 15t.$$

The first of these equations is applicable to observations of stars *not near the Pole*, D being the noted interval of the place of bisection from the mid-wire. Supposing, as is usually very nearly the case, the wire-intervals to be all equal, and the common equatorial interval in time to be τ', then if $m =$ the distance of the star from mid-wire, recorded as stated in Art.

198, we shall have D (in arc) $= 15m\tau$. As this is a small arc, $\sec D = 1 + \dfrac{D^2}{2R^2}$ nearly, R being the number of seconds in an arc equal to radius, and since e is very small,

$$\cos(\delta - e) = \cos\delta + \frac{e}{R}\sin\delta \text{ nearly.}$$

Putting, for the sake of convenience in the sequel, the complement of the declination Δ for δ, we obtain, with sufficient approximation,

$$\frac{e}{R} = \frac{D^2}{2R^2}\tan\Delta, \text{ or } e = \frac{(15m\tau)^2}{2R}\tan\Delta = [6{,}73672]\, m^2\tau^2 \tan\Delta.$$

In the instance of the Telescope of the Cambridge Circle $\tau = 16',6$, and the formula employed is

$$e = 0'',1503 \times m^2 \tan\Delta.$$

200. The second of the above equations is applicable to the case of those observations of stars *near the Pole* which are accompanied by determinations of the hour-angle t. This equation gives the exact result

$$\tan e = \frac{\sin 2\delta \sin^2 \dfrac{15t}{2}}{1 - 2\sin^2\delta \sin^2 \dfrac{15t}{2}}.$$

Hence, putting $\dfrac{e}{R}$ for $\tan e$, neglecting the second term in the denominator, which for circumpolar stars will always be very small compared to unity, and substituting $90° - \Delta$ for δ, there results the sufficiently approximate formula

$$e = R \sin 2\Delta \sin^2 \frac{15t}{2} = [5{,}31443]\sin 2\Delta \sin^2 \frac{15t}{2}.$$

Let us now suppose the declination Δ to be increased by the small arc n, and, in consequence, e to be changed to e'. Then, for a given value of t, we get by differentiation

$$e' = e + 2n\cos 2\Delta \sin^2 \frac{15t}{2} = (R\sin 2\Delta + 2n\cos 2\Delta)\sin^2\frac{15t}{2}.$$

For examples, let $\Delta = 88°. 40' + n''$ for Polaris, and $\Delta = 86°. 36' + n''$ for δ Ursæ Minoris. Then the formula gives

Cor. for Polaris............$\{[3,98212] - [0,30056] \times n\} \sin^2 \frac{15t}{2}$.

Cor. for δ Ursæ Minoris...$\{[4,38780] - [0,29796] \times n\} \sin^2 \frac{15t}{2}$.

These formulæ are calculated for the year 1879, and might, for hour angles not exceeding 6^m, be used for two years before and after that date.

201. The foregoing investigation gives formulæ for calculating the amount of the correction without respect to its *sign*. The proper sign is affixed according to rules derived as follows. The telescope, when pointed to an object between the pole and the equator at a little distance from the meridian, is turned, by reason of curvature of path, more towards the pole than when the object is on the meridian; the Circle-reading is consequently *too small*, and the sign of the correction is *plus*. If the object is either below the equator, or below the pole, the telescope for the same reason is turned farther from the pole than for bisection on the meridian, so that the Circle-reading is *too great*, and the sign of the correction is *minus*.

202. We have, next, to consider in what manner Circle-observations of the zenith distances of stars are made by means of *reflection* at the surface of mercury. As such an observation is almost always accompanied (for a reason that will be given afterwards) by the direct observation, I shall state at once the method of taking what is called *the double observation*. At the Greenwich Observatory under Pond's direction, two Circles exactly alike, named from their makers "Troughton" and "Jones," were handled by two observers for this purpose, one observing directly and the other by reflection, and the difference of the index errors of the two instruments had to be very exactly determined. The present Astronomer Royal introduced, first, at Cambridge in 1833, and adopted at Greenwich in 1839, the practice of taking both observations with a single Circle at the same meridian Transit by the following process. The

Circle is first set and clamped for the reflection observation, according to a prepared pointer reading (Art. 193), and the setting, if carefully performed, secures that the image of the star shall cross the field not far from the centre. Also the mercury-trough, which is usually of wood, and of rectangular form with the longer side north and south, has to be placed so that the light from the star falls near the middle part of the surface of mercury. This may be done in day-time by placing the trough where the light of the sky through the eye-piece is seen reflected at that part, while at night-time light from the illuminating lamp is sent back from the eye-end in sufficient quantity to be used for the same purpose. The bisections of the graduation-lines by the cross-wires of the microscopes are next made, but, as the observation may fail, are not usually read off; after which the star, as it crosses the field, is bisected by the *micrometer-wire* one or two intervals before passing the mid-wire, in order that the reflection observation may be made when the star is at about the same distance from the mid-wire on the opposite side. That this may be successfully done, the Circle is unclamped immediately after the reflection-observation, is then quickly set and clamped for the direct observation according to a prepared pointer reading, and lastly by means of the tangent-screw the star is bisected, if there is a fixed wire, either by the fixed wire or the micrometer-wire, otherwise of necessity by the micrometer-wire. The intervals between the places of the two bisections and the mid-wire, being fresh in memory, should next be recorded. The microscope-readings for the reflection observation are now written down, if omitted before, new bisections for the direct observation are made and recorded, and the micrometer is read off. This completes the operation. In the case of a Transit-Circle which has two setting-circles and is not reversed, the settings for the double observation would be performed by the two circles in the manner indicated in Art. 195, the telescope being pointed for the *reflection* observation by the *east* circle after preparing the *west* circle for the subsequent *direct* observation. It remains to remark respecting observations by reflection, that the trough ought to be placed on a solid basis of brickwork or masonry, and to be screened from wind, in order that the surface of mercury may be as free as possible from tremor; also that for obtaining satisfactory reflections the surface has frequently to be cleared of dross and dust. This is found to be effectually done by passing the mercury through a small hole in the side or bottom of a containing trough or vessel into the trough used for the observations.

203. The correction for curvature of path is plainly the same in amount for the reflection, as for the direct, observation, at the same distance from the mid-wire, but has always the *opposite* sign, the reason for which will be seen by remarking that the error of position of the telescope due to this cause is in opposite directions in the two observations relatively to the order of a graduation proceeding from 0° to 360°. (See Art. 201.)

204. The interval between the place of bisection and the mid-wire, which, except for small polar distances, may be taken to be the distance from the meridian, has to be noted in every instance of the bisection of a *moving* body out of the meridian, for calculating, as well as the correction for curvature of path, an additional correction for *change of polar distance*. As this change may in all cases be assumed to be proportional to the time-interval between the bisection and meridian-passage, let m be the space-interval between the place of bisection and the mid-wire, as recorded in wire-intervals, and τ' the common equatorial interval between the wires; then if Δ be the declination of the body, and Γ' its horary change of declination, derived either from the Nautical Almanac or from actual observation, the correction in *amount* will be $\dfrac{m\,\tau\,\sec\Delta}{3600} \times \Gamma'$. If the polar distance be *increasing*, or the value of Γ' in the Nautical Almanac be *negative*, the Circle-reading is *too small* if the bisection be made *before* meridian passage, and *too great* if made *after*, so that in the former case the sign of the correction is *plus*, and in the other, *minus*. If the polar distance be *decreasing*, or Γ' in the Nautical Almanac be *positive*, the sign is

minus before meridian passage and *plus* after. Consequently if m be the interval recorded with sign attached, as stated in Art. 198, and the sign of I' be taken as in the Nautical Almanac, the above expression gives the correction in all cases with its proper sign[1].

205. The Sun, Moon, and Planets, on account of the forms of their disks, require special modes of observation, the several descriptions of which are here subjoined. For observing the *Sun*, either the North, or the South Limb, is first to be set for according to a pointer reading derived, to the accuracy of about 1', from the Declination of centre and semi-diameter given in the Nautical Almanac; next, the Circle being (not forcibly) clamped, the microscope-bisections are made, but need not be read off; then the Limb is tangentially bisected by the *micrometer-wire* a little before the meridian passage of centre, immediately after which the Circle is unclamped and turned by hand till the image of the other Limb is placed near the middle of the field, and this Limb, after again clamping, is bisected tangentially, either by the fixed wire or the micrometer-wire, by means of the usual tangent-screw movement (Art. 160). The Circle-reading for the former observation, if not already recorded, is now written down, fresh microscope-bisections are made and read off for the second observation, and lastly the micrometer-reading is recorded. Generally the second observation, like the first, may be made when the Sun's centre is at a small distance from the meridian, and it might, therefore, not be worth while to state the distances. It would, however, be more exact always to bisect a Limb tangentially just when, according

---

[1] In Volume xvii. of the *Cambridge Observations*, for the years 1846—1848, pp. lxvi.—lxviii., are given, together with others, two Tables (iv. and v.), one of which is adapted for calculating by interpolation the correction for curvature of path for all declinations from 5° to 78°.40', and for every quarter of a wire-interval from $\frac{1}{4}$ to $4\frac{3}{4}$; and the other, which is computed to four places of decimals, for calculating by interpolation the change of declination, for *one* interval, for all horary variations from 1" to 10", and multiples or submultiples of the same by 10, and for all declinations from 0° to 40°, the equatorial interval being 16',6. Such tables shorten to an important amount the time occupied in reducing a continuous series of Circle observations.

to the judgment of the eye, it is cut symmetrically by one of the vertical wires, and to note the wire for the purpose of calculating corrections for reducing the observation to the meridian. The operations above detailed, together with the calculation of corrections for distances of the bisections from the meridian, and the reduction in arc, according to the rules given in Art. 184, of the micrometer-reading to the reference position, or to the fixed wire, furnish two Circle-readings, one applicable to the North Limb, and the other to the South Limb. The mean between these is the Circle-reading for the Sun's centre, and their difference is a measure of the Sun's diameter, both determinations involving, to an amount which will afterwards be taken into consideration, effects of *atmospheric refraction*.

206. From the above description of the mode of taking an observation of the Sun with the Mural Circle, it will be seen that on account of the Sun's large diameter it might be practicable for one observer, without assistance, to observe transits of the first and second Limbs, and bisect the North and South Limbs between the transits, at the *same* meridian passage if the Transit-instrument and Circle be in contiguous rooms. This, in fact, was occasionally done at the Cambridge Observatory. The two meridional observations may also be made simultaneously with a Transit-circle, but only with the aid of an assistant for making and recording the first set of microscope-bisections and making (without recording) the second set, in the interval between the transit-observations of the first and second Limbs. In the use of the Greenwich Transit-circle for the same purpose, the observer is aided by two assistants, who make and read off bisections with the microscope-micrometers, the observer himself reading off the telescope-micrometer. By these means, when, besides, "the transits are made by galvanic touch (Art. 131), it is found perfectly easy to take those of the first and second Limbs over nine wires, still leaving time enough for the observations of both Limbs in N. P. D." (See *Greenwich Observations* for 1874, p. ix.) It may here be remarked that according to Greenwich arrangements there are two clamps, acting one on the North side, and the other on the South side,

of a *clamping circle*, but they have *no slow motion*, all bisections of a limb or star being made with the micrometer-wire by turning its screw. By this process errors that might arise from deformation of the Circle are avoided.

207. In observations of the *Moon*, it is the practice, on account of the generally rapid change of polar distance, to apply the micrometer-wire to the Limb at each of the vertical wires, or at other positions, and to record the micrometer-reading and position immediately after each bisection. The several observations have then to be corrected for micrometer-reading, and for distance from the meridian according to rules already given (Arts. 184, 199 and 204), and the mean of the several results is the concluded Circle-reading. It is, however, to be remarked that although in the expression given in Art. 204 for the correction on account of change of declination, the time-interval between the bisection and meridian passage is with sufficient accuracy $m\tau$ for the Sun and Planets, in the case of the Moon this quantity has to be multiplied by two factors sensibly affecting the result, one on account of the retardation of the diurnal motion by the Moon's motion in her orbit, and the other on account of retardation due to parallax. The two factors, which have already been obtained in Arts. 123 and 124, are respectively

$$\frac{3600}{3600 - I} \text{ and } \frac{\sin(Z - \epsilon - p)}{\sin(Z - \epsilon)},$$

and consequently the complete expression for correction to the meridian in observations of the Moon is

$$m\tau \sec \Delta \times \frac{I'}{3600} \times \frac{3600}{3600 - I} \times \frac{\sin(Z - \epsilon - p)}{\sin(Z - \epsilon)},$$

I' as well as I being taken from the horary variations given in the Nautical Almanac under the head of Moon-culminating stars. The Local Zenith Distance Z, contained in this expression and in the formula obtained in Art. 124 for calculating p, may be readily deduced, to the accuracy of $1'$, from the difference between the Circle-readings for Zenith Point and the bisection of the Limb, corrected by a prepared Table of approxi-

THE MURAL CIRCLE. 201

mate refractions with zenith distance for argument. The above-mentioned formula, after putting p and P for their sines, and unity for ρ, becomes $p = P \sin(\chi - \epsilon)$, which is sufficiently approximate[1]. Also by expanding, and neglecting terms involving ϵ^2, ϵp, &c., it will be found that

$$\frac{\sin(Z-\epsilon-p)}{\sin(Z-\epsilon)} = \frac{\sin(Z-p)}{\sin Z}.$$ Hence finally

$$\text{Correction} = I' \times \frac{m\tau \sec \Delta}{3600 - I} \times \frac{\sin \text{ of } \mathrm{\rlap{)}{\cdot}\!\!\!D}\text{'s Geoc. Zen. Dist.}}{\sin \text{ of } \mathrm{\rlap{)}{\cdot}\!\!\!D}\text{'s Local Zen. Dist.}},$$

which formula is used at Greenwich and Cambridge, having been originally proposed by the Astronomer Royal when at Cambridge.

208. In the use of the Transit-circle, a transit of the Moon's *first* Limb may be observed, and different bisections of the full North or South Limb be made in the manner described in Art. 207, by a single observer at the same meridian passage. So also if the full North or South Limb be carefully set for, and the microscopes be read beforehand, the different micrometer-bisections of the Limb may be made and recorded, and a transit of the *second* Limb be taken, by one observer. When it happens either that the first and second Limbs, or that the North and South Limbs, are both sufficiently illumined for observation, in the former case transits of the two Limbs and a Circle-reading for one Limb, and in the latter a transit of one Limb and Circle-readings for the two Limbs, may be obtained with the aid of *one* assistant, as will be seen from the account given in Art. 206 of the method of observing the Sun with a Transit-circle.

209. The Circle-observation of a *Planet* which has a considerable disk is made by bisecting the North or South Limb with the fixed wire by tangent-screw movement, generally as it passes the mid-wire, and bisecting at the *same* time the other

---

[1] This is also true with respect to the calculation of parallax mentioned in Art. 124 (p. 115). It is incorrectly said there that p and P should not be substituted for their sines, this rule applying only when an exact value of the Moon's parallax is required.

Limb with the micrometer-wire. The Circle-readings for the two bisections, obtained according to rules already indicated, give by their mean the Circle-reading for bisection of the Planet's centre, and by their difference a measure of its diameter. The same results may be arrived at, if there should be no fixed wire, by bisecting alternately an equal number of times the North and South Limbs by the micrometer-wire, and recording the several places of bisection and micrometer-readings. From these data the Circle-reading for each bisection has to be reduced in the usual way to meridian passage, and then, after inferring the mean of the results for each Limb, the concluded Circle-reading for centre is the mean between the two means, and their difference measures the diameter. The Circle-observations of *Jupiter, the globe of Saturn, Mars,* and *Venus* are conducted as above stated; those of the other planets, inclusive generally of Mercury, are made by simple bisections of their estimated centres. When the observation is made by bisections of opposite Limbs, it is often required to correct the inferred Circle-reading for centre, and the measure of the diameter, for form of the illuminated disk, both when the disk is horned and when it is gibbous. We have, therefore, in the next place to enquire how such correction is calculated.

210. The investigation contained in Art. 113, and applied in obtaining a formula for correcting for defect of illumination in transit-observations of a gibbous disk, will serve for the present purpose, whether the disk be horned or gibbous. It will, first, be supposed that the disk is *horned*, which can be the case only in the instances of Venus, Mercury, and the Moon. The observation of Venus is not unfrequently made by bisections of a Limb and the opposite cusp; but this process is rarely employed in observing Mercury (see Art. 209) and the Moon. The following investigation of the correction required in the case of a horned disk applies equally to the three bodies. In Figure 26 (1), the arcs PS ($=\Delta$) and PV ($=d$) are respectively the polar distances of the Sun and the observed body, and the angle SPV ($=P$) is the difference of their Right Ascensions, at the given time of observation. These three

THE MURAL CIRCLE.

quantities may be supposed to be derived from the Nautical Almanac. The angle θ between the arc of the meridian through the centre and the line joining the cusps (Fig. 26 (2)) is equal

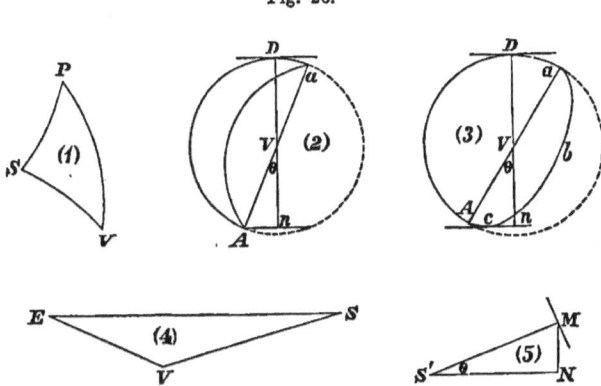

Fig. 26.

to $90° - \angle PVS$. Hence from a known formula of Spherical Trigonometry, which, according to the usual notation, is

$$\cot a \sin b = \cot A \sin C + \cos b \cos C,$$

we derive, *mutatis mutandis*,

$$\tan \theta = \cot \Delta \sin \delta \operatorname{cosec} P - \cos \delta \cot P,$$

from which equation θ is to be calculated. Then if $2D_0$ be the full diameter of the body, and D be the difference of zenith distance measured by the observation (inclusive in the case of the Moon of a small correction for refraction), we shall have $D = DV + Vn$ (by Fig.) $= D_0 + D_0 \cos \theta = 2D_0 \cos^2 \frac{\theta}{2}$. Hence $D_0 = \frac{D}{2} \sec^2 \frac{\theta}{2}$, which is the semi-diameter as deduced from the observation, and the required amount of the correction of D is $2D_0 - D$, or $D \tan^2 \frac{\theta}{2}$. Also the local zenith distance of centre is obtained from that of the full Limb (calculated as will be subsequently shewn), by applying the correction $\pm \frac{D}{2} \sec^2 \frac{\theta}{2}$, according as the Limb is the upper or lower one.

211. In Art. 113 it has been shewn that if R and r be respectively the tabular distances of the Earth and the observed body from the Sun, and e be the eccentricity of the elliptic boundary of illumination in the case of a *gibbous* disk, the value of e is $\frac{R}{r} \sin SV$. But by the spherical triangle SPV,

$$\frac{\sin SV}{\sin PS} = \frac{\sin SPV}{\sin PVS}; \text{ or } \sin SV = \frac{\sin \Delta \sin P}{\cos \theta}.$$

Hence
$$e = \frac{R \sin \Delta \sin P}{r \cos \theta}.$$

Now suppose the angle θ' to be derived from the equation $\sin \theta' = e \sin \theta$, which equation, by substituting the above value of e, and putting for $\tan \theta$ the expression obtained in Art. 210, becomes $\sin \theta' = \frac{R}{r} (\cos \Delta \sin \delta - \cos \delta \sin \Delta \cos P)$, giving the value of θ' in known quantities. Hence the value of the perpendicular Vn on the tangent cn (Fig. 26 (3)), which, by Conic Sections, is equal to $D_0 \cos \theta'$, can be calculated. Then, as θ' simply takes the place of θ, the inferences relating to a gibbous disk are precisely analogous to those obtained at the end of Art. 210 relative to the horned disk.

212. When, as is most frequently the case, the Moon is gibbous, the calculation of θ' may be simplified in a manner like that indicated in Art. 115 with reference to a transit observation of opposite Limbs. As there shewn, $\frac{R}{r} = 1$, and $e = \sin SV$ nearly. Hence, if $S'M$ be the arc joining M the Moon's centre and S' the point opposite the Sun's centre, so that $SV = 180° - S'M$, it follows that $e = \sin(180° - S'M) = \sin S'M$, and $\sin \theta' = \sin S'M \sin \theta$. Since, therefore, in this kind of observation $S'M$ is always a small arc, we have nearly $\theta' = S'M \sin \theta = MN$ in Fig. 26 (5). Thus to obtain θ' with sufficient approximation, it is only required to find the difference of the polar

distances of the Moon and the point opposite the Sun at the time of the Moon's transit. The *upper* or *lower* Limb is defective according as the polar distance of the Moon's centre is *greater* or *less* than that of the point opposite the Sun's centre.

213. It is evident from Figure 26, (1) (2) and (3), that the *lower* or *upper* Limb is defective according as the angle PVS is *less* or *greater* than 90°, or according as θ is *positive* or *negative*, and that this rule is applicable both to a horned and a gibbous disk. Hence from the expression for tan θ in Art. 210 we may infer that the lower or upper limb is defective according as cot Δ tan δ is greater or less than cos P.

214. For observing with the Cambridge Circle asteroids and small stars, when they were too faint to admit of illuminating the field sufficiently for bisections with a fine wire, a broad rectangular metal bar was attached (in September 1853) to the east side of the frame carrying the micrometer-wire, with an edge parallel to this wire and distant from it nearly 10^r in the direction *from* the micrometer-head, and a vertical edge nearly coincident with the middle vertical wire. A faint object could then be bisected by the edge of the bar with very little, and sometimes without any, illumination of the field by the lamp. Under these circumstances, suppose the micrometer to be read for a bisection by the bar just as if the micrometer-wire were in the place of the bar-edge, and, M being such reading, let B and W be respectively the readings for coincidence of the bar and wire with the *fixed* wire at an interval from mid-wire equal to that of the object when bisected. Of these readings W is usually known already (see Art. 184), and B, if not known, should be taken immediately after the observation. Now the micrometer-reading for the actual position of the micrometer-wire when the bar-bisection was made, is plainly $M + 10^r$, and, by inference, the reading for bisection of the object, if made by the micrometer-wire, would have been $M + 10^r - (B - W)$. Hence the correction to be added to the recorded reading M, for reduction to the wire-bisection, is the algebraic excess of 10^r above the measured interval $B - W$ between the wire and bar-edge. After applying this correction, the reduction to the fixed wire for

micrometer-reading is made in the usual way. It may be observed that the use of the fixed wire for the purpose above mentioned might be superseded by measures of the interval between the bar-edge and micrometer-wire obtained from time to time, at the vertical wires, by bisecting with each a meridian mark, or point of a collimator. The interval at other positions might then be obtained by interpolation.

Finding that the bisection of a faint object by the edge of a bar is rendered uncertain by the diffraction of the light, I adopted the method of placing the edge, by micrometer-movement, as exactly as possible in a line with the path of the object just *before* its arrival at the mid-wire, or the vertical edge. By noting simultaneously the clock-time of the transit at the edge, an approximate Right Ascension of the object might also be obtained, which, in taking Circle-observations of faint objects, as the asteroids, is often required for the purpose of identifying them.

Calculation of Local Zenith and Polar Distances.

215. The difference between the Circle-reading for the direct observation of any object and that for Zenith Point, gives the apparent Zenith Distance of the object northward or southward. As inferred from a reflection observation, the same zenith distance is the difference between the Circle-reading for the observation and that for the Nadir Point. The arc so measured is called *apparent* zenith distance, because its value involves the amount by which the zenith distance is diminished by atmospheric refraction. The calculation of this effect of refraction is usually performed according to Bessel's Refraction Tables, which are contained in pages 538—542 of the *Tabulæ Regiomontanæ*, and are founded on an investigation resting on Laplace's solution of the Problem, given in pages lix.—lxiii. of the same work. The data required in using these Tables are, the height of the Barometer, the temperature of the external air, the temperature of the mercury of the Barometer, and the above-defined apparent zenith distance as general argument.

The readings of the Barometer and Thermometers are recorded, if not at each observation, frequently enough to take account of all changes that might sensibly affect the calculation of the refractions. A zenith point given immediately by the collimating eye-piece (Art. 191) may be used for inferring the apparent zenith distances, as it will only differ slightly from a finally adopted value derived (as will afterwards be shewn) from direct and reflection-observations of stars. Bessel's Tables have been adopted both at Cambridge and Greenwich, altered in form and expanded, as they are given in an Appendix to the Volume of Greenwich Observations for 1836, and more recently at Greenwich, as given in an Appendix to the Volume for 1853, with a small correction of Bessel's mean refraction. In these computations the external thermometer only is used, unless the Zenith Distance exceeds 85°.

216. After adding the amount of refraction to the apparent zenith distance obtained as above stated, the result is the zenith distance of the object measured from the astronomical zenith of the place of observation, or from the local direction of the action of gravity. This is the angle which is designated by Z in Arts. 124 and 207, and in the latter Article is called 'local zenith distance.' In order to deduce from this angle local polar distance, or the angle which the direction of the object, as seen from the place of observation, makes with a parallel to the earth's axis, it is required to find the zenith distance of the Pole. This is done by taking Circle-observations of the *same* star above and below Pole, and correcting the observations for refraction, the mean between the results for the upper and lower culminations being the zenith distance of the Pole. But before using zenith distances of stars for this purpose, account has to be taken of an instrumental irregularity which is named the *Discordance of Zenith Points*. It is found, in fact, that the mean between the Circle-readings for the direct and reflection observations of a star corrected for refraction, which mean should be the reading for the horizontal direction northward or southward, is somewhat different as determined by different stars, and, consequently, that the zenith point,

which is the reading for the horizontal direction ± 90°, varies with the zenith distance. It becomes necessary, therefore, to ascertain by what process the calculation of polar distances may be performed so as to be free from effects of this irregularity, that is, to obtain *corrections for discordance of zenith points*. The following method, described in pages xxxi. and xxxii. of Vol. xx. of the Cambridge Observations (for 1855—1860), is for the most part the same as that which was adopted by Professor Airy for the observations of 1833, the year in which the Circle was first brought into use.

217. The discordance is such that the Circle-reading for zenith is in general less by a star observed to the south of zenith than by a star observed to the north of zenith, the telescope, whether directed to the heavens or the trough of mercury, requiring, when the object is to the south of zenith, to be turned for bisecting it a little farther in the direction of the graduation than if the inequality did not exist, and when the object is to the north of zenith a little in the contrary direction. Whatever may be the cause of the discordance, the *law* of the error it produces may be presumed to be approximately represented by the differences between an ascertained zenith point for a *given* zenith distance and the zenith points corresponding to other zenith distances, if these be sufficiently consecutive and numerous. It will be assumed that the collimating eye-piece gives correctly the zenith point corresponding to the zenith direction, and that the double observation gives zenith points that require corrections. Let, therefore, M be a zenith point given by the eye-piece, and Z_1 a zenith point obtained nearly contemporaneously by a double observation. Since, if there were no discordance Z_1 would be equal to M, it follows from the foregoing principles that the Circle-reading for the direct observation and that for the reflection-observation both require to be corrected by $M - Z_1$, it being assumed that they are alike affected by the inequality both in amount and as respects the direction of the graduation. In order, therefore, to obtain the true zenith distance, or that which, except for the discordance, would be given by adopting the zenith point M,

the quantity $M-Z$ has to be *added* to the Circle-reading for the *direct* observation, and *subtracted* from that for the *reflection* observation. Hence the algebraic excess of the latter Circle-reading above the other is twice $M-Z$. This quantity is also equal to the algebraic excess of the polar distance derived from the reflection observation above that derived from the direct observation, before taking account of the discordance of zenith points, polar distances below Pole being considered negative. In short, the values of $M-Z$ are most conveniently obtained by halving the excesses thus deduced from polar distances provisionally calculated with the values of M.

218. This being understood, suppose the values of $M-Z$ derived from all the double observations made in a year to be collected, the mean, in each instance of several being given by the same star, to be taken, and the separate results, whether a single value or the mean of several, to be divided into groups such that the stars in each group shall not greatly differ in zenith distance. Then let each value in a group be multiplied by the number of observations by which it was determined, and the corresponding zenith distance by the same number. After dividing the sum of each series of products by the whole number of observations in the group, the resulting value of $M-Z$ may be considered to belong to the resulting zenith distance. Conceive these zenith distances to be set off on a straight line of abscissæ, and ordinates proportional, according to an arbitrary scale, to the values of $M-Z$ corresponding to the several abscissæ, to be drawn at right angles to the line. The extremities of the ordinates will thus mark a certain number of positions, the probable error of each of which may be supposed to be smaller the greater the number of the observations by which it was determined. This having been done, the next step is to trace by hand among the points laid down a continuous *curved* line, making it at the same time approach nearer to any position the less the probable error. Now let intervals of 5° be marked off on the line of abscissæ, commencing at the zero of zenith distance, and proceeding both in the positive and negative directions, and let ordinates to the curve be raised at

the points of division. Then these ordinates, their scale-measure being known, will give the values of $M-Z$ for every fifth degree of zenith distance, from which the value for any other zenith distance may be found by interpolation. It appears from the foregoing reasoning, that the required *correction* is $M-Z$ with its proper sign, or with the sign changed according as the zenith distance is inferred from a direct or a reflection observation. A *Table* of the corrections for discordance of zenith points, both for direct and reflection observations, for every fifth degree of zenith distance, or polar distance, should be formed to facilitate the interpolations.

219. We may now proceed to consider how the colatitude of the Observatory may be exactly determined by observations of circumpolar stars above and below Pole. The mean between the zenith distances of the same star at the upper and lower culminations, after correcting both for refraction and discordance of zenith points, is the exact zenith distance of the Pole, or the Colatitude. As this is an essential element in calculating polar distances, the adopted value should be the mean result of a large number of observations; and on account of the uncertainties of refraction, stars should be selected that are not far from the Pole. Polaris is especially adapted for this purpose, both because of its nearness to the Pole, and because it is visible in day-time in telescopes usually attached to a Mural Circle, or a Transit-circle. In calculating the mean result, less weights should be given to individual results in proportion as the stars rendering them are more distant from the Pole. I proceed now to exemplify the above method by stating in some detail the processes by which the colatitude of the Cambridge Observatory at present adopted was obtained.

220. The first accurate determination was made by Professor Airy in 1833 from observations, taken in that year, of the polar distances of ten circumpolar stars, extending from the polar distance of Polaris ($=1°.35'$) to that of a Cassiopeiæ ($=34°.23'$). The data for the calculations are given in p. xxxv of Volume VI. (for 1833) of the *Cambridge Observations*. The whole number of observations employed was 917, consisting of

observations above and below Pole in unequal portions, and also of unequal portions of direct and reflection observations. The weight attached to the result given by each star was proportional to the product of the number of observations from which it was deduced by an estimated factor varying with the polar distance. The value of the colatitude finally obtained was $37°.47'.8''{,}43$, inclusive of a correction $+0''{,}15$ required in consequence of its being subsequently found that the barometer reading was $0^{in}{,}1$ too small. By similar means I calculated (taking into account the barometer-correction) the colatitude from observations made in 1836, 1837, and 1838, consisting of very nearly equal numbers of direct and reflection observations, but of observations above and below Pole in unequal numbers. The data and results of the calculations are contained in pages liii—lviii of Volume XI. (for 1838). The weight given to the value deduced from the observations of any star was calculated in the following manner. If the number of observations of the star above and below pole be respectively a and b, the weight, so far as it depends only on the number of the observations, was taken to be proportional to $\dfrac{ab}{a+b}$, the reciprocal of $\dfrac{1}{a}+\dfrac{1}{b}$. It was also considered that an observation below pole of a star of more than $40°$ polar distance would be of no value for determining the exact colatitude; and as the correction for refraction is less uncertain the less the zenith distance, it was assumed that the factor by which $\dfrac{ab}{a+b}$ should be multiplied to take account of the increment of the weight due to greater certainty in the amount of correction for refraction, might be increased by 2 for every $1°$ decrease of polar distance, beginning with zero at $40°$ of P.D. Accordingly, Δ being the star's P.D., the adopted weight is equal to $2(40-\Delta)\dfrac{ab}{a+b}$. Suppose now the corrected zenith distance of the star to be z_1 above pole and z_2 below pole, and $\lambda+x$ to be the true colatitude, λ being put for $37°.47'.8''{,}43$. Then the polar distance is equal to $\lambda+x-z_1$ and also to $z_2-\lambda-x$; whence it follows that $(\lambda-z_1)+(\lambda-z_2)$

$= -2x$, and that x is obtained by halving, and changing the sign of, the sum of the calculated polar distances above and below pole, the latter being negative. Every such determination was multiplied by a weight computed according to the foregoing formula, and for each of the three years the sum of the products was divided by the sum of the weights. In this manner the value of x was found to be $+0'',01$ by 354 observations in 1836, $-0'',18$ by 446 observations in 1837, and $-0'',12$ by 392 in 1838. By giving to these results weights proportional to the sums of the weights of the respective years, the final value of x was found to be $-0'',11$, and the corrected colatitude 37°. 47'. 8'',32. This result depends on 1190 observations, and that obtained in 1833 on 917. If the weights of the two results be taken to be proportional to these numbers, the colatitude concluded from their combination is 37°. 47'. 8'',37, which is the value I have adopted for all the observations made in the years 1838—1860, and contained in Volumes XI.—XX.

To shorten the calculation of Circle observations the colatitude printed in the Calculation Books is 37°. 47'. 8'',00, and the correction $+0'',37$ for error of assumed colatitude is incorporated in the Table of corrections for discordance of zenith points mentioned in Art. 218.

221. I propose to conclude this *Division* III. by considerations respecting the effect of the *flexure* of the telescope of the Cambridge Circle, and the extent to which it might have given rise to the discordance of zenith points. For conducting the enquiry as to the amount of flexure, I employed the two collimators mentioned in Art. 172, the construction of which, and the mode of using them for finding collimation error, are described in Arts. 53—55. For the present purpose they were mounted in such manner that the Circle-telescope could be made to collimate with either, both horizontally and in directions inclined at *any* angles to the horizontal direction. This was effected by firmly attaching the collimators to a wooden plank 11,5 feet long, 12 inches wide, and one inch and a half thick, with 6 feet of the length at the middle part of double thickness, then fastening symmetrically the plank, thus loaded and

THE MURAL CIRCLE.

strengthened, to the end of a stout axle 52 inches long, having two cylindrical pivots, and finally placing this apparatus on a braced wooden stage, so as to be capable of being moved about an axis nearly in a line with the axis of motion of the Circle. The collimators were fixed to the vertical face of the plank near its ends, and pointed in opposite ways in the direction of its length, and their axes could, by means of screws provided for this purpose, be adjusted so as to revolve very nearly in the plane of the motion of the telescope-axis. Thus for any angular elevation the Circle-telescope could collimate with both, one after the other, and their collimation with each other could be effected through the aperture made at the middle part of the telescope-tube (Art. 172). For steadying the collimators' support while a set of collimations was being taken, the ends of the two arms were clamped to the pier by means of brass rods and appropiate clamping gear. The wooden stage supporting the revolving part of the apparatus rested at first on the floor of the Circle-room. I made trial of these arrangements for measuring the effects of flexure on various occasions from November 16, 1855, to February 13, 1856, and obtained values for every tenth degree of zenith distance South from 20° to 90°. The results of this preliminary investigation shewed that the floor was not a sufficiently firm support of the apparatus, and that the brass clamping rods had an injurious effect, owing to variations of temperature. To get rid of these sources of error, the brass rods were replaced by wooden ones, and the stage was placed on a mass of brick-work surmounted by a thick stone slab. After these changes the measures of flexure were much more consistent with each other, and seemed to be sufficiently trustworthy. The instrumental operations for determining the flexure, and the calculation of its amount, were conducted in the following manner.

222. It will be assumed that the plane of collimation of the Circle-telescope has been placed very nearly in the plane of the meridian, and that its wire-frame has been adjusted to geometrical focus by the method described in Art. 83. Under these circumstances the wires of the collimators are placed exactly at the geometrical

foci by adjusting the wire-frames, one after the other, so that the images of the wires may be seen distinctly at the geometrical focus of the Circle-telescope; for when this condition is satisfied the images are formed by parallel rays issuing from the collimators' object-glasses. Supposing these adjustments to be made in positions of the telescope which are approximately opposite, or differ, according to the graduation of the Circle, by nearly $180°$, the collimators will by this means be made to point towards each other, so that the images of the wires of either will be seen distinctly at the geometrical focus of the other. If, for instance, the centre of the field of each be marked by the intersection of a vertical and a horizontal wire (see Fig. 13 in pp. 44 and 215, and Arts. 54 and 55), the two lines joining the points of intersection and the respective optical centres of the object-glasses will be very nearly parallel, and the image of the point of intersection of the wires of one collimator will be formed near the point of intersection of the wires of the other. (I provided screw-adjustments for correcting, if there was need, defect of parallelism of these optical axes, and also for having the means of placing both in a line with the optical axis of the telescope, which latter condition, although not optically necessary, it was considered desirable to be able to fulfil approximately.) At the same time that these adjustments are made, the horizontal wires of the collimators might be placed, according to judgment of the eye, parallel to the image of the horizontal wire of the telescope, by being slightly rotated by hand and then clamped. If, however, all bisections be made near the centres of the collimators' fields, small deviations of the wires from horizontality will have no sensible effect. After the foregoing arrangements the operation for measuring the flexure proceeds as follows.

223. The collimators were first made to collimate with each other through the aperture in the telescope-tube. In order to make this adjustment with precision the double micrometer-wire of B (Fig. 13) was moved to a position such that the image of the horizontal wire of A was seen, near the middle of the field, exactly between the two wires, and the micrometer-reading of B was recorded. This was done six or more times, and after calculating the mean of all the readings the micrometer was set to this mean. The telescope having been turned to collimate with B and the Circle clamped, the telescope-micrometer-wire was placed six or more times between the images of the two wires of B, the several micrometer-readings were recorded,

THE MURAL CIRCLE. 215

and the mean being computed was taken to be the concluded micrometer-reading (x_1). The telescope was then put in collimation with A by placing its micrometer-wire in coincidence with the image of the horizontal wire of A, the coincidence reading (x_2) being ascertained by the method of alternate contacts (Art. 191). The Circle-reading having been taken in the usual manner for each position of the

Fig. 13.

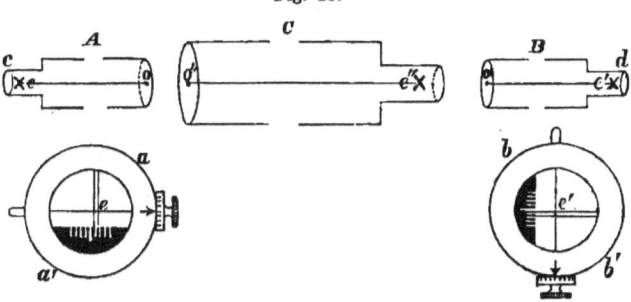

telescope, suppose the reading to be C_1 for the first position, the telescope pointing *northward* from Zenith, and C_2 for the second position, the telescope pointing *southward*. Then if a be either the micrometer-reading for coincidence with the fixed wire, or the reference micrometer-reading, and m the value in arc of one micrometer revolution, the concluded Circle-reading for the first bisection is $C_1 - m(x_1 - a)$, and that for the other $C_2 - m(x_2 - a)$. Now if there should be no effect of flexure, the *excess* of the second reading above the first would, by reason of the collimating process, be exactly 180°. Also it is to be considered that the effect of flexure might be such as to cause the optical centre of the object-glass and the reference point in the field of the telescope to be equally displaced in parallel directions, in which case the flexure would produce no angular displacement of the optical axis, and the difference of the Circle-readings would still be 180°. But if the arc $(C_2 - mx_2) - (C_1 - mx_1)$ through which the telescope-axis is turned differs from 180°, this will shew that sensible effect is produced by the bending of the telescope-tube, and it may readily be seen that according as that arc is *less* or *greater* than 180°, the *drooping* of the object-end of the tube is *greater* or *less* than that of the eye-end. As the difference of the droopings may be assumed to be the same in the two positions of the telescope, the error of the Circle-reading due to flexure is in amount equal to *half* the difference between $180°$ and the arc $C_2 - C_1 - m(x_2 - x_1)$. Also since, for a given pointing of the optical axis *southward*, the greater drooping of the object-end would *diminish* the Circle-reading, the correction of the error is a *positive* or *negative* quantity, according as the object-end droops *more* or *less* than the eye-end, that is, by what is argued above, according as $180°$ is greater or less than $C_2 - C_1 - m(x_2 - x_1)$. From mechanical considerations we may conclude that the correction is equal in amount with *opposite* sign for a pointing equally distant from the zenith *northward*. Hence, generally, flexure-correction $= \pm \frac{1}{2}\{180° - (C_2 - C_1) + m(x_2 - x_1)\}$ according as the telescope points *southward* or *northward* from the zenith. The following examples will serve to illustrate the process of calculation.

224. On September 10, 1856, at $3\frac{1}{2}^h$, I took measures as follows for ascertaining the flexure-correction of the Cambridge Mural Circle for the horizontal pointing of the telescope *southward*.

Object-glass.	Circle-reading.	Micr. reading.
N.	246° 40′ 38″,95 (C_1)	10,451 (x_1).
S.	426 40 39,67 (C_2)	10,486 (x_2).
N.	246 40 38,81 (C_3)	10,563 (x_3).

The value of m, the micrometer revolution, being $20″,850$, $m(x_2 - x_1) = 20″,85 \times 0^r,035 = +0″,73$; and $C_2 - C_1 = 180° + 0″,72$.

Hence $+\frac{1}{2}\{180° - (C_2 - C_1) + m(x_2 - x_1)\} = \frac{1}{2}(-0″,72 + 0″,73)$

$= +0″,005 =$ flexure-correction by the first and second measures.

Also $+\frac{1}{2}\{180° - (C_2 - C_3) + m(r_2 - r_3)\} = \frac{1}{2}\{180° - (180° + 0″,86)$

$+ 20″85 \times -0^r,77\} = \frac{1}{2}(-0″,86 - 1″,63) = -1″,245$, which is the flexure-correction by the second and third measures. The mean between the two results is $-0″,62$, which is the concluded correction as given by these measures. It will afterwards appear, notwithstanding the discrepancy between the separate results, that this mean is entitled to considerable weight. The sign of the correction thus obtained proves that the *eye-end* droops most. See Art. 223.

225. As the above-mentioned discrepancy appeared to be due to gradual change of the reading of the B-micrometer for coincidence with the wire of A, the measures were repeated on September 12, and the coincidence reading was taken three times. The telescope was pointed horizontally and the B collimator was *northward*. After making trial of different methods of conducting the collimations it appeared to be most satisfactory to put *one* of the wires of B in collimation with the wire of A, by means either of alternate contacts or of actual coincidences, and then to place the images of these two wires, one after the other, mid-way between the telescope's fixed and micrometer wires separated by an arbitrary small interval. Suppose that for such collimation northward the Circle-reading and micrometer-reading are C_1 and x_1, and for the southward collimation C_2 and x_2, and that the reference micrometer-reading is a. Then the reduced Circle-reading for the northward collimation is $C_1 - \frac{m}{2}(x_1 - a)$, and that for the other is $C_2 - \frac{m}{2}(x_2 - a)$. Hence by the same argument as that in Art. 223, the flexure-correction is $\pm \frac{1}{2}\{180° - (C_2 - C_1)$ $+ \frac{m}{2}(x_2 - x_1)\}$ according as the telescope is pointed *southward* or *northward*, $\frac{m}{2}$ simply taking the place of m in the former expression. The following data were obtained by such measures on September 12.

Pointing of telescope.	Reading of B-micrometer.	Circle-reading.	Reading of Circle-micrometer.
N.	$10^{r},012$	$246°\ 40'\ 3,43''\ (C_1)$	$10^{r},559\ (x_1)$
S.	$10,018$	$426\ 40\ 4,18\ (C_2)$	$10,582\ (x_2)$
N.	$10,033$	$246\ 40\ 3,72\ (C_3)$	$10,703\ (x_3)$

As the readings $10^{r},018$ and $10^{r},033$ were taken before and after the south pointing of the telescope, the coincidence reading at the time of pointing is assumed to be the mean between these, or $10^{r},025$. Compared with this the reading $10^{r},012$ is *too small* by $0^{r},013$, and consequently, as the micrometer-head was *downwards*, the wire of B was *too high*, and the Circle-reading *too great*, by this quantity. It was found by taking the difference of the Circle-readings for collimations of the fixed wire of the telescope with the double-wire of B at positions separated by 24^{r}, that $1^{r} = 58'',07$. Hence the required correction of $C_1 = -0,013 \times 58'',07 = -0'',76$, the corrected seconds of $C_1 = 2'',67$, and $C_2 - C_1 = 180° + 4'',18 - 2'',67 = 180° + 1'',51$. Also $\frac{m}{2}(x_2 - x_1) = \frac{1}{2} \times 20'',85 \times 0,023 = +0'',24$. Hence the flexure-correction $= +\frac{1}{2}(-1'',51 + 0'',24) = -0'',63$. Similarly, as compared with $10^{r'},025$ the reading $10''',033$ is too great by $0^{r'},008$, and the correction of $C_3 = +0,008 \times 58'',07 = +0'',46$. Hence the corrected seconds of C_3 are $3'',72 + 0'',46 = 4'',18$, and the flexure-correction

$$= +\frac{1}{2}\left\{180° - (C_3 - C_2) + \frac{m}{2}(x_3 - x_2)\right\} = \frac{1}{2}\left(0'',00 + \frac{1}{2} \times 20'',85 \times -0,121\right)$$

$= -0'',63$. Hence the mean result from the two sets of measures is $-0'',63$, agreeing nearly with that of Sept. 10. If the calculation had been made without reference to the change of reading of the *B*-micrometer, the result would have been $-0'',26$ by the first and second measures, and $-0'',86$ by the second and third, the mean being $-0'',56$. This result shews that the effect of that change is nearly eliminated by adopting the principle of *alternate* measures.

226. October 23, 1856, $3\frac{1}{2}^h$—5^h, the following measures were taken for determining the flexure-correction when the telescope is horizontal and looks *northward*.

Pointing of telescope.	Reading of B-micrometer.	Circle-reading.	Reading of Circle-micrometer.
S.	$9,916$ (x_1')	$66° 43' 57'',10$ (C_1)	$9,659$ (x_1).
N.	$9,928$ (x_2')	$246\ 43\ 55,57$ (C_2)	$9,585$ (x_2).
S.	$9,935$ (x_3')	$66\ 43\ 57,05$ (C_3)	$9,634$ (x_3).
N.	$9,938$ (x_4')	$246\ 43\ 56,52$ (C_4)	$9,669$ (x_4).

Calculating, first, without regard to the change of the micrometer-reading of B, the flexure-correction by the first and second measures

$$= -\frac{1}{2}\left\{180° - (C_1 - C_2) + \frac{m}{2}(x_1 - x_2)\right\} = -\frac{1}{2}(-1'',53 + 0'',77) = +0'',38;$$

by the second and third measures $= -\frac{1}{2}\left\{180° - (C_3 - C_2) + \frac{m}{2}(x_3 - x_2)\right\}$

$= -\frac{1}{2}(-1'',48 + 0'',51) = +0'',48;$ and by the third and fourth measures

$$= -\frac{1}{2}\left\{180° - (C_3 - C_4) + \frac{m}{2}(x_3 - x_4)\right\} = -\frac{1}{2}(-0'',53 - 0'',36) = +0'',45.$$

The mean of the first and second results is $+0'',43$, and that of the second and third is $+0'',47$, and the concluded mean is $+0'',45$.

For taking account of the change of the micrometer-reading of B,

THE MURAL CIRCLE. 219

the times of taking the Circle readings and those of both micrometers were noted. Hence it was found by interpolating that the reading of B-micrometer was $9^r,920$ corresponding to C_1, and $9^r,936$ corresponding to C_3. Hence x_2' is too great by $9^r928 - 9^r,920$, which in arc $= 58'',07 \times 0,008 = + 0'',46$. Consequently, since the B-micrometer was *northward*, the wire was *too low* by $0'',46$ and the corrected seconds of $C_2 = 55'',57 + 0'',46 = 56'',03$. Thus by the formula
$-\frac{1}{2}\left\{180° - (C_1 - C_2) + \frac{m}{2}(x_1 - x_2)\right\}$, the flexure-correction $= -\frac{1}{2}(-1'',07 + 0'',77) = +0'',15$. So in comparing the second and third measures, x_2' is too small by $9^r,936 - 9^r,928$, or in arc $0,008 \times 58'',07 (= 0'',46)$, and the corrected seconds of $C_2 = 55'',57 - 0'',46 = 55'',11$. Hence

flexure-correction $= -\frac{1}{2}\left\{180° - (C_2 - C_3) + \frac{m}{2}(x_2 - x_3)\right\} = -\frac{1}{2}(-1'',94 + 0'',51) = +0'',71$, and the mean of the two results $= \frac{1}{2}(0'',15 + 0'',71) = +0'',43$. Again, in comparing the third and fourth measures x_4' is *too great* by $9^r,938 - 9^r,936$, or in arc $0,002 \times 58'',07 (= 0'',12)$. Hence corrected seconds of $C_4 = 56'',52 + 0'',12 = 56'',64$, and flexure-correction
$= -\frac{1}{2}\left\{180° - (C_3 - C_4) + \frac{m}{2}(x_3 - x_4)\right\} = -\frac{1}{2}(-0'',41 - 0'',36) = +0'',39$.

Hence the mean of the two results by the second and third measures and by the third and fourth $= \frac{1}{2}(0'',71 + 0'',39) = +0'',55$. Consequently concluded mean $= \frac{1}{2}(0'',43 + 0'',55) = +0'',49$, which differs little from $+0'',45$ obtained by the other calculation.

227. I propose now to give two examples of the calculation of the flexure-correction for *oblique* positions of the telescope, one for a southward, and the other for a northward, zenith distance. In the first example it was required to find the flexure-correction when the telescope is pointed $70°$ from the zenith *southward;* for calculating which, on September 13, 1856, at 3^h, the following measures were taken:

Pointing of telescope.	Reading of B-micrometer.	Circle-reading.	Reading of Circle-micrometer.
N.	$10^r,188\ (x_1')$	$226°\ 39'\ 49'',93\ (C_1)$	$10^r,684\ (x_1)$
S.	$10,186\ (x_2')$	$46\ \ 39\ \ 50,22\ (C_2)$	$10,648\ (x_2)$
N.	$10,211\ (x_3')$	$226\ \ 39\ \ 49,50\ (C_3)$	$10,624\ (x_3)$

The readings of B-micrometer being left out of consideration, we have $C_2 - C_1 = 180° + 0'',29$, $\frac{m}{2}(x_2 - x_1) = \frac{1}{2} \times 20'',85 \times -0,036 = -0'',37$, and flexure-correction $= \frac{1}{2}(-0'',29 - 0'',37) = -0'',33$. Also $C_3 - C_2$ $= 180° + 0'',72$, $\frac{m}{2}(x_3 - x_2) = \frac{1}{2} \times 20'',85 \times 0,024 = +0'',25$, and flexure-correction $= +\frac{1}{2}(-0'',72 + 0'',25) = -0'',23$. Concluded value $= \frac{1}{2}(-0'',33 - 0'',23) = -0'',28$. Taking account, now, of the readings x_1', x_2', x_3', since the Circle reading for the south pointing of the telescope was obtained between x_2' and x_3', the corresponding reading of the B-micrometer will be assumed to be the mean of these, or $10'',198$. The value of x_1' being *less* than this by $0'',010$, and the collimator B being *northward* and its micrometer-head *downwards*, the wire was *too high* by $0,010 \times 58'',07$, or $0'',58$, and the Circle reading C_1 requires the correction $-0'',58$. Similarly, C_3 requires the correction $+0,013 \times 58'',07$ or $+0'',75$. These corrections being included, it will be found by calculating as before, that the flexure-correction is $-0'',33 - 0'',29$, or $-0'',62$ by the first and second measures, and $-0'',23 + 0'',37$, or $+0'',14$ by the second and third. The mean of the two results is $\frac{1}{2}(+0'',14 - 0'',62)$, or $-0'',24$, agreeing well with the value obtained above. I am not sure that the former method of calculation is not the best.

In the second example, the following measures were taken on October 21, 1856, at 22^h—$23\frac{1}{2}^h$, for calculating the flexure-correction for the pointing of the telescope $70°$ from the zenith *northward*.

Pointing of telescope.	Reading of B-micrometer.	Circle-reading.	Reading of Circle-micrometer.
S.	$9'',765$ (x_1')	$86° 42' 55'',43$ (C_1)	$9'',659$ (x_1).
N.	$9'',760$ (x_2')	$266\ 42\ 54,95$ (C_2)	$9'',701$ (x_2).
S.	$9'',760$ (x_3')	$86\ 42\ 55,90$ (C_3)	$9'',724$ (x_3).
N.	$9'',758$ (x_4')	$266\ 42\ 55,35$ (C_4)	$9'',677$ (x_4).

Calculating without reference to the readings of the B-micrometer, the following values of the flexure-correction are obtained from the first and second measures, the second and third, and the third and fourth, respectively:

$$-\frac{1}{2}\left\{180° - (C_1 - C_2) + \frac{m}{2}(x_1 - x_2)\right\} = -\frac{1}{2}(-0'',48 - 0'',44) = +0'',46$$

$$-\frac{1}{2}\left\{180° - (C_2 - C_3) + \frac{m}{2}(x_2 - x_3)\right\} = -\frac{1}{2}(-0'',95 + 0'',24) = +0'',36$$

$$-\frac{1}{2}\left\{180° - (C_3 - C_4) + \frac{m}{2}(x_3 - x_4)\right\} = -\frac{1}{2}(-0'',55 + 0'',49) = +0'',03.$$

The mean of the first and second results is $+0'',410$, and the mean of the second and third $+0'',195$, and the concluded mean $= +0'',30$, which is the required flexure-correction. By calculating according to the other method I obtained $+0'',31$, differing very little from the previous result, because, in fact, the B-micrometer readings differ little from each other. I have not thought it worth while to give the details of the second calculation, which was conducted exactly as in like cases already adduced.

228. In the interval from September 10 to October 24, 1856, I took thirty-three different sets of measures to determine the flexure-correction for zenith distances separated in some instances by 5°, but generally by 10°, and extending from 90° south to 90° north. The subjoined Table contains the results for every tenth degree of zenith distance, as deduced from these measures. It is to be understood that in all the calculations micrometer-readings of the collimator B were taken into account. The corrections for discordance of zenith points (Arts. 217 and 218) adopted in the years 1837 and 1856 are inserted in the Table for the purpose of drawing inferences from a comparison of the law and amount of the two classes of corrections.

Z. D. South.	Flexure-corrⁿ.	Discordance-corrⁿ.		Z. D. North.	Flexure-corrⁿ.	Discordance-corrⁿ.	
		1837	1856			1837	1856
°	″	″	″	°	″	″	″
90	−0,50			0	−0,08	0,00	+0,02
80	−0,58			10	+0,13	−0,46	−0,57
70	−0,33	+0,44	+1,25	20	+0,27	−0,56	−0,83
60	+0,15	+0,30	+0,88	30	+0,29	−0,51	−0,51
50	+0,24	−0,08	+0,53	40	+0,11	−0,32	+0,02
40	+0,32	+0,23	+0,51	50	+0,16	+0,15	+0,35
30	+0,18	+0,55	+0,84	60	+0,43	−0,14	
20	+0,34	+0,55	+1,16	70	+0,31	−0,34	.
10	−0,14	+0,37	+0,71	80	+0,38		
0	−0,08	0,00	+0,02	90	+0,49		

The measures from which the four values of the flexure-correction for the zenith distances 20°—50° S. were calculated, were taken with all the shutters of the room closed, and lamps were suspended at the ends of the arms supporting the collimators for the purpose of throwing light on the reflectors of *diagonal* eye-pieces whereby the wires were made to appear dark on bright fields for collimation with the telescope. It was found, however, that the heat from the lamps had a perceptible effect on the reading of B for collimation with A, producing apparently anomalous values of the correction in the above-mentioned instances; on which account the use of the lamps was discontinued. The other measures were taken with the shutters open, the illumination of the collimators' fields being effected by light from the sky. That they might be the better comparable with each other all the measures were taken by myself, often, however, under considerable difficulties arising from effects of wind and changes of temperature. The results for the north zenith distances, as depending on measures taken after I had had experience in overcoming the difficulties incident to the experiment, are more trustworthy than those for the south zenith distances, and, I believe, may be considered to be a pretty close approximation to the law and amount of the flexure.

The corrections for discordance of zenith points in 1837 were interpolated from the values for the direct observations given in the Table in page lxi of the Volume of Cambridge Observations for that year. The adopted corrections of 1856 were formed by first taking the means of the values obtained for 1852, 1853 and 1854, the forms of the discordance-curves for the three years being very nearly the same, and then correcting these means by $-0'',32$. This correction was applied because the discordance-correction for 0° Z.D. deduced from the three years' observations was $+0'',32$, whereas by the results of numerous observations made in 1856 near the zenith, as given in page lvi of Vol. xx., the correction was found to be zero. Discordance-corrections for zenith distances greater than 70°, as depending on few observations, were considered too uncertain for the present purpose, and in the years 1852—1856 none were obtained for north zenith distances greater than 42°. I proceed now to state the inferences which appear to be deducible from the foregoing discussion.

229. It may, in the first place, be concluded that the

discordance of zenith points cannot be accounted for by flexure of the telescope-tube, inasmuch as the flexure-correction agrees neither in law nor amount with the discordance-correction. Hence, since from mechanical considerations the flexure must under any circumstances produce its effect, there must be some cause which, by operating according to a different law and in greater degree, *veils* this effect. From the consideration I have given to the question I incline to the opinion that the sudden and somewhat violent movement of the instrument from the position for the reflection observation to that for the direct observation gives rise to a *temporary deformation* of the Circle, of the nature of flexure, from which it does not recover in the short interval between the observations, especially if it be clamped for the direct observation. This view receives support from the following facts. (1) The change is most rapid at small zenith distances, where the difference between the pointings of the telescope for the reflection and direct observations is greatest. (2) The discordance is greater for the Cambridge 8-feet Mural Circle than for the Greenwich 6-feet Transit-circle, and follows different laws in the two instruments. (This will be seen by comparing the values for 1856 in the above Table with corresponding values in page lvii of the Volume of Greenwich Observations for that year.) These facts seem to indicate that the discordance is due as to law and amount to the construction of the instrument, or to mode of using it, or to these conditions combined. (3) As far as regards the Cambridge Circle, the discordance increased in amount from 1837 to 1856, the law remaining nearly the same; which might be the result of an increasing liability of the Circle to deformation consequent upon prolonged use. (4) The discordance-curves from which the corrections in the above Table for the years 1837 and 1856 were deduced, as well as all the curves that I formed from the observations of other years, exhibited a *minimum* value at about $50°$ Z. D. south, and a *maximum* value at the same Z. D. north. Possibly at these positions the effect of the flexure due to the action of gravity, and the temporary effect due to the movement impressed on the instrument, might most nearly counterbalance each other. From the aggregate of the values of the flexure-correction in the above Table, with the exception of those for 20°—60° south, all of which appear to have been affected by disturbing causes (see Art. 228), it may be inferred that the law and amount of the flexure are very nearly the same on the opposite sides of zenith (as from mechanical considerations might be expected), and that they may be supposed to be approximately expressed by the formula, horizontal flexure × sin Z. D. This formula is, in fact, assumed in the reduction of the Greenwich observations, the zenith points obtained after the application of the flexure-correction being supposed to depend on some unknown cause different from flexure. It is plain that according to the supposed law of flexure, the correction is the same in amount, and in direction as regards the order of the graduation, for a reflection as for a direct observation, and consequently that its effect on the zenith point is taken into account in the usual treatment of the double observation, when, as in Cambridge practice, it has not been previously eliminated. If it had been eliminated from the results of the observations of 1856, the residual discordance would have been much greater for the Cambridge instrument than any hitherto obtained at Greenwich, proving, I think, that the 8-feet Mural Circle was too large, and that a decided advantage has been gained by substituting for it a 3-feet Transit-circle. If the foregoing theory be accepted, it will follow that in the case of a double observation, the Z. D. deduced from the reflection observation, corrected only for flexure, is likely to be more accurate than that similarly deduced from the direct observation, and that the uncertainty of the direct observation increases with the size of the Circle[1].

[1] As far as I am aware the experiment of which I have here given an account is the only one in which a determination of the flexure-corrections for *oblique* positions of the Telescope has been attempted to be made. I intended originally to offer a communication of all the details for publication in the *Memoirs of the Royal Astronomical Society*, instead of inserting them in a Volume of the *Cambridge Observations*, but press of other occupation prevented my carrying out this intention. The details and explanations, as far as they are given in the present work, may suffice to make known the character of

230. No account has been taken in the preceding explanations of any effect on the Circle-reading that may be due to *errors of graduation*, which at Greenwich are carefully determined, and allowed for in the reduction of the observations. Respecting this source of error as affecting observations with the Cambridge Circle, I am only able to say, that since the corrections of the assumed colatitude obtained in the years 1836, 1837, and 1838 by observations made with the telescope in three different positions on the Circle were $+0'',01, -0'',18$, and $-0'',12$ respectively (see Art. 220), the close agreement of these results may be regarded as evidence that the Circle-readings were very little, if at all sensibly, affected by errors of graduation (see Art. 158), considering that a difference of latitude equal to $0'',1$ corresponds to only 10 feet on the earth's surface.

IV. CALCULATION OF THE MEAN NORTH POLAR DISTANCES OF STARS, AND OF THE GEOCENTRIC NORTH POLAR DISTANCES OF THE SUN, MOON, AND PLANETS.

231. After correcting the observed zenith distance of any object for refraction, and adding an *assumed* value of the colatitude (Art. 220), the result is its apparent North Polar Distance subject to farther correction for the error of the assumed colatitude and the discordance of zenith points. As these two corrections are always of very small amount, instead of taking account of them for each observation, it is preferable, when the object is a *Star*, to determine the star's mean polar distance for the beginning of the year by applying to the uncorrected apparent polar distance the value of $+(\delta' - \delta)$ calculated by the formulæ given in Arts. 146 and 147, and afterwards to obtain the mean

---

the experiment, and so much of its results as may possess scientific value. Nothing relating to them has been previously published, except a brief abstract of results in a Letter inserted in Vol. v., No. 100, page 28, of *The Astronomical Journal*, and addressed to Dr Gould the Editor. In the same letter I took occasion to state that the experiment gave the means of determining the law of flexure of a loaded beam for different angular elevations, and that this law was shewn to be approximately expressed by the formula $A \sin Z$, but more exactly by the formula $A \sin Z + B \sin^2 Z$, Z being the angular distance from the vertical.

of all such determinations made in the course of the year. It will then suffice to apply to this mean a correction for error of assumed colatitude and discordance of zenith points interpolated from a Table including both kinds of correction, with polar distance for argument, the discordance-correction having been deduced from all the double observations of the year. (See Arts. 217 and 218.) The values of mean polar distance thus derived from the observations of different years can subsequently be referred to January 1 of a given year, by applying annual variations calculated by the formula, $-n \cos \alpha$, in Art. 145, and thus a concluded mean N. P. D. be obtained depending on a large number of observations. This process, combined with that for calculating concluded mean R. A, described in Art. 154, gives the means of forming a *Catalogue* of the mean places of stars for a given epoch.

232. In observations of the *Sun*, *Moon*, and *Planets*, the corrections for discordance of zenith points have to be applied to the polar distances severally, and may either be deduced, in conjunction with the correction for error of assumed colatitude, from the above-mentioned Table, or be added to that correction after being taken by measurement with compasses and scale from the discordance-curve (Art. 218). By these means the local polar distance of a bisected north or south limb of any of these bodies is accurately determined, and to obtain the *geocentric* polar distance, it is only required to subtract therefrom the *parallax*, which is the angle subtended at the observed limb by the radius from the earth's centre to the place of observation. Let p be this angle, P the equatorial horizontal parallax of the body's centre, $P + \alpha$ that of the limb, Z the zenith distance by the observation, ρ the local terrestrial radius, and ϵ the local angle of the vertex ; then (Art. 124),

$$\sin p = \rho \sin (P + \alpha) \sin (Z - \epsilon).$$

In using this formula for the *Moon*, the value of P, as obtained by interpolation to second differences from the Nautical Almanac, may at present be adopted without correction ; but it would not be sufficiently exact to substitute p and P for their sines, on which account it is convenient to construct a Table from which, within the limits of the values of P, the arc may be readily inferred from the sine, and the sine from the arc. (Such a Table is given in the *Greenwich Observations* of 1874, p. cxxix.) The small quantity a, required to be added to P to obtain exactly the parallax of the Limb, may be derived by interpolation from the subjoined values, which have been extracted from a general Table contained in the *Cambridge Observations*, Vol. IV. for 1831, p. 147, having been constructed by Professor Airy to be used in the calculation of Occultations by the Moon for obtaining the parallax of any point of her periphery:

Zenith Distance...	30°	40°	50°	60°	70°	80°
Cor^n. for N.L...	$-0'',03$	$-0'',05$	$-0'',06$	$-0'',08$	$-0'',09$	$-0'',09$
Cor^n. for S.L...	$+0'',10$	$+0'',12$	$+0'',13$	$+0'',15$	$+0'',16$	$+0'',16$

For all the bodies except the Moon the formula $p = P \sin(Z - \epsilon)$ is sufficiently approximate. The Equatorial Horizontal Parallaxes may either be adopted from the Nautical Almanac, or computed from that of the Sun and the Log. of true distance from the Earth given in Ephemerides. {For the Cambridge Observatory $\rho = [9,9990916]$ and $\epsilon = 11'.12''$.}

233. Parallax having been taken into account in the manner above stated, the *geocentric* polar distance of the *centre* of the body, being the mean between the polar distances of the limbs, may be at once inferred when the north and south limbs have been both observed, and corrections have been applied for defect of illumination. The difference between these polar distances is a measure of the *geocentric diameter*. If only one limb be bisected, the polar distance of centre is obtained by adding the value of the geocentric semidiameter to the polar distance of the north limb, and subtracting it from that of the south limb. This value, if not adopted from the Nautical Almanac, might be computed either from a sufficient number of Circle measures of diameter, such as that just mentioned, or from the mean result of measures taken expressly for the purpose with an Equatorial instrument, it being supposed that the Log. distance of the body from the Earth, which is required for such computation, is

supplied by the theory of gravity. For instance, from 68 measures of the Moon's diameter obtained in the years 1836—1842, partly by differences of transits of limbs taken with the transit-instrument and five-feet Equatorial, and partly by Circle measures of diameter, I found that the semidiameter adopted at that time in the Naut. Alm. required the correction $+2'',21$. (*Cambridge Observations*, Vol. XIV. for 1842, p. xxxviii) It will be proper to mention here that the corrections of Mr Baldrey's transits of limbs spoken of in Art. 135, were deduced by taking as standards my own transits of limbs, which, in the case of the Sun, gave values of diameters agreeing very nearly with those of the Naut. Alm., while the latter, according to the results of Mr Glaisher's Circle-measures in the 5 years 1836—1840, required a correction not larger than $-0'',16$. (See Vol. XIII. pp. xxii and xxiii.)

234. In the reduction of meridional observations of the Moon I have taken account of the above correction, $+2'',21$, of semidiameter; but for reduction to centre, in the cases of observations of a single Limb of the Sun, I adopted without change the semidiameters, and sidereal intervals occupied by transits of semidiameters, given in the Nautical Almanac. At the same time yearly comparisons of numerous diameters derived from observation with tabular values, have furnished data for ascertaining what amount of correction the assumed value of the Sun's semidiameter at the earth's mean distance may require. The tabular semidiameters (as also the parallaxes) in the Naut. Alm., are at present, in consequence of corrections applied from time to time for reasons drawn both from theory and observation, more likely to be trustworthy than at the period referred to in Art. 233. In the Greenwich calculations of the Apparent R. A., and Geocentric N. P. D., of centre from observations with the Transit Circle made in 1874, the adopted parallaxes (those of the Moon corrected as stated in Art. 232) and the semidiameters are all interpolated from the Naut. Alm., with the following exceptions pertaining to cases in which only one Limb was observed. For these the tabular semidiameter of the Sun is corrected by $-0'',53$, to agree with the value derived

from the observations of several previous years; those of Mars, Jupiter, and Saturn are corrected from the results of neighbouring observations; and to the semidiameter of Venus a special correction is applied, the amount of which, as given in the *Greenwich Observations* of 1874, page lxxii, is $+0'',392 + 0,027 \times$ tabular semidiameter. This expression was deduced by Mr Stone from 370 vertical measures of diameter taken with the Greenwich Transit Circle from 1851 to 1862. (See *Monthly Notices* of the R. Ast. Society, Vol. xxv., No. 3, p. 57.) The reasons for applying such a correction may be gathered from the following investigation, which, however, is founded on observations of Venus made with the Cambridge Mural Circle.

235. It being understood that each vertical measure of the Planet's diameter is compared with the diameter computed in the Naut. Alm., let e be the excess, in any particular instance, of the tabular above the observed value. Then if the latter should be affected by a constant error such that the observed always exceeds the true value by the constant E, the excess of the tabular above the true diameter will be $e + E$. Again, suppose the tabular diameter to be Δ, and the true diameter, by reason of error in the value assumed for the unit of distance, to be $\Delta (1 - \epsilon)$. Then $\epsilon\Delta$, being the excess of the tabular diameter above the true, is equal to $e + E$; or, $e - \epsilon\Delta + E = 0$. Hence if values of e and Δ, extracted from a Table of the above-mentioned comparisons, be substituted in this equation, as many different equations will be formed as there are observations for comparison, and if the number be large, by dividing them into two groups, values of ϵ and E may be deduced, the probable errors of which may be presumed to be small. For example, I formed 78 such equations from the Circle measures of diameter in 1838, and as the Planet passed through conjunction in the first half of the year and through opposition in the other, the equations pertaining to the two halves were put in separate groups, the mean result from which gave, $\epsilon = -0,0267$ and $E = +0'',94$. From 74 equations derived from measures in 1839, and similarly treated, it was found that $\epsilon = -0,0375$ and $E = +0'',40$. Hence the mean results of the 152 measures are, $\epsilon = -0,0321$, and $E = +0'',67$. As the tabular diameter for unit of distance was $8'',25$, the true diameter according to this investigation is $8'',25 (1 + 0,032) = 8'',51$. Also the half of E, $0'',34$, is the

apparent enlargement of the semidiameter. This effect I supposed to be due to *irradiation*[1]. (See *Cambridge Observations*, Vol. XI. for 1838, p. li, and Vol. XII. for 1839, p. xx.)

236. It only remains to make mention of a correction of the observed N. P. D. which may be required on account of *error of position of the plane of collimation*. It has already been stated in Art. 174 that transits of known stars are usually taken from time to time with the Circle-telescope in order to determine the deviation (in time) of its pointing from the meridian for any polar distance. A correction for reduction to the meridian is in strictness required on account of such deviation; but after adjusting the Circle in the manner described in Arts. 167—172, the amount can be sensible only in observations of the Moon, and in case the change of her N. P. D. is rapid. (Instances of data from observations of stars for calculating this correction are given in the *Cambridge Observations*, Vol. XVIII. p. xlix, and Vol. XIX. p. xxx.)

237. Having now gone through the treatment of Circle observations under the *four Divisions*, I propose to add examples of the complete reduction of observations taken with the Cambridge Mural Circle, both when the object is a star, and when it is a moving body. The subjoined instances will serve to exemplify the application of most of the various calculations that have been under consideration.

1856, April 23, 10^h, δ Ursæ Majoris was observed by reflection (R) at the mid-wire, and directly (D) at three intervals from the mid-wire. The micrometer-reading both for R and D was $8^r,176$, and the reference reading $9^r,000$. One micrometer revolution $= 20'',869$. Barometer, $29^{in}, 880$; External Thermometer, $44°,3$; Approximate

---

[1] The investigation by Mr Stone cited in Art. 234 is similar to the above. From 219 observations made with the Greenwich Mural Circle in 1839—1850, and 370 made with the Transit Circle in 1851—1862, he obtains for the semidiameter at the unit of distance, $8'',47$. He also finds that the apparent enlargement of semidiameter is $0'',729$ by the former instrument, and $0'',392$ by the other, and remarks that as these quantities are nearly in the inverse ratio of the diameters (4 inches and 8 inches) of the object-glasses, they might be attributable to irradiation.

THE MURAL CIRCLE.

R. A. of ✶, 12ʰ. 8ᵐ; Approximate Decl., + 57°. 50'. For the direct observation "Negative Correction for Runs" was noted.

	δ Ursæ Maj. (R) ° ′ ″	δ Ursæ Maj. (D) ° ′ ″
Pointer Reading	162 15	331 0
Microscope A.	3 63,1	4 52,6
,, B.	63,6	48,8
,, C.	64,1	50,8
,, D.	59,2	43,4
,, E.	60,6	49,3
,, F.	63,7	50,0
Correction for Run of + 0″,9	+ 0,7	(N. C.) 0,0
Mean of Microscope Readings	162 19 2,50	331 4 49,15
Micrometer correction	+ 17,20	+ 17,20
Reduction to Meridian	0,00	+ 2,14
Concluded Circle Reading	162 19 19,70	331 5 8,49
Adopted Zenith Point	336 42 12,73	336 42 12,73
Apparent Zen. Dist. South	− 5 37 6,97	− 5 37 4,24
Refraction-correction	− 5,80	− 5,80
Assumed Colatitude	37 47 8,00	37 47 8,00
Apparent North Polar Distance	32 9 55,23	32 9 57,96
Reduction to Mean N. P. D.	+ 2,31	+ 2,31
Mean N. P. D., 1856,0	32 9 57,54	32 10 0,27
Corⁿ. for discordance of Z. P.	+ 0,35	− 0,35
Corⁿ. of assumed Colatitude	+ 0,37	+ 0,37
Concluded Mean N. P. D., 1856,0	32 9 58,26	32 10 0,29

The micrometer correction = $(9,000 - 8,176) \times 20″,869$. The correction for Runs applied in the reflection observation = $\frac{4}{5} \times + 0″,9$; for the other observation the correction is 0″,0 because "Negative correction" was recorded (see Art. 179). The reduction to meridian for three intervals in the direct observation = $0″,1503 \times 3^s \times \tan 57°.50'$; in the other the reduction is zero because the mid-wire was nearly in the meridian (see Art. 198). The observer noted that the reflected image of the star was "much diffused," and that the reflection observation was "worth little." I find, in fact, by comparison with other observations of the star, both at Greenwich and Cambridge, that this observation is discordant.

1856, October 2, both Limbs of the Sun were observed, N. L. at two intervals before, and S. L. at two intervals after, mid-wire. One

232   PRACTICAL ASTRONOMY.

equatorial interval $= 16^s,6$. The micrometer reading $= 16^r,482$ for each Limb; reference reading $= 9^r,000$. One micrometer revolution $= 20'',869$. Approximate declination of the Sun's centre, $- 3°. 45'$. Horary variation of declination $= - 58'',12$. Barometer, $29^{in},766$; External Thermometer, $61°,3$.

	Sun N. L. ° ′ ″	Sun S. L. ° ′ ″
Pointer Reading	32 25	32 55
Microscope A.	0 33,4	2 31,8
,, B.	37,0	37,0
,, C.	33,6	30,4
,, D.	27,4	24,4
,, E.	30,0	27,3
,, F.	33,2	29,8
Correction for Run of $+2'',3$	+ 0,2	+ 1,1
Mean of Microscope readings	32 25 32,47	32 57 30,22
Micrometer correction	− 2 36,14	− 2 36,14
Cor^n. for curvature of path	− 0,04	− 0,04
Cor^n. for change of declination	+ 0,54	− 0,54
Concluded Circle Reading	32 22 56,83	32 54 53,50
Adopted Zenith Point	336 42 15,68	336 42 15,68
Apparent Zen. Dist. south	+55 40 41,15	+56 12 37,82
Refraction-correction	+ 1 22,88	+ 1 24,54
Assumed Colatitude	37 47 8,00	37 47 8,00
Apparent N. P. D. of N. L.	93 29 12,03	
,, ,, S. L.	94 1 10,36	App^t. Diam^r. 31 58,33
		Parallax cor^n. − 0,04
Apparent N. P. D. of centre	93 45 11,20	
Correction for Parallax	− 7,08	Geoc. Diam^r. by } 31 58,29
Cor^n. for discordance of Z. P.	+ 0,68	observation
Cor^n. of assumed Colatitude	+ 0,37	Tabular 32 1,76
Concluded Geocentric N. P. D.	93 45 5,17	Excess of Ta- } + 3,47
Seconds of Tabular N. P. D.	4,53	bular Diam^r.
Excess of Tabular N. P. D.	− 0,64	

The corrections for Runs are $+ 2'',3 \times \dfrac{0,5}{5}$ and $+ 2'',3 \times \dfrac{2,5}{5}$. The micrometer correction $= (9,000 - 16,482) \times 20'',869$. Correction for curvature of path $= - 0'',1503 \times 2^s \times \tan 3°.45'$ (Arts. 199 and 201), correction for change of decl. $= \dfrac{2 \times 16,6 \times \sec. 3°. 45'}{3600} \times 58'',12$, and is *plus* for N. L. and *minus* for S. L. (Art. 204). For these, as for all other observations, the refractions were calculated by Bessel's Tables (see Art. 215). The Parallax, taking $8'',5776$ to be the Sun's Equat. Hor. Parallax, is $[0,9325] \times \sin(Z - \epsilon)$ (see Art. 232), which gives $7'',06$ for N. L., $7'',10$ for S. L., $7'',08$ for centre, and $-0'',04$ for the parallactic correction of measured diameter. The corrections for discordance of zenith points applied to the observations both of April 23 and October 2, were derived from a Table calculated for the year 1856, and contained in page lvi. of Vol. xx. of the *Cambridge Observations*.

THE TRANSIT CIRCLE.

238. Although in treating of the Transit Instrument and Mural Circle separately all explanations have been given that are essential for understanding the taking and reducing of meridian observations, it will be proper to add some account of the construction and use of a Transit Circle, this instrument being adaptable by itself to every kind of meridional observation, and at the present time very generally employed in large Observatories. In speaking of the latest improvements that have been made in the instrumental means of taking meridian observations, I have already had occasion to advert to various particulars relating to the observational apparatus of the Greenwich and Cambridge Transit Circles[1]: my object at present is to give a description in detail of the latter instrument, making use for that purpose of Plate III. (taken from a Photographic Picture), with accompanying explanations and remarks. [The Greenwich Transit Circle is described in full detail in Appendix I. of the *Greenwich Observations* of 1852.]

239. The Cambridge Transit Circle was constructed and mounted by Mr Simms, and in December 1870 was completely

[1] Notices respecting the parts and uses of a Transit Circle, as distinguished from those of the Transit Instrument and Mural Circle, are contained in the following Articles: 6, 7, 29, 34 (note p. 28), 37, 38, 43 (note p. 36), 47, 48 (note p. 40), 33—56, 58, 64, 86, 102 (note p. 98), 112, 116, 128, 137 (note p. 148), 164 (note p. 159), 173, 196, 202, 206, 208.

ready for use. The following account of it is extracted from the Report of the Council of the Royal Astronomical Society, February 10, 1871, under the head of *The Cambridge Observatory.* (Monthly Notices, Vol. XXXI. p. 106.) "Professor Adams believes that this instrument, for which the Observatory is indebted to the munificence of Miss Sheepshanks[1], will prove to be one of the finest of its class. The object-glass of the telescope is an excellent one, by Cooke, of 8 inches aperture [and about 9 feet focal length]. There are two divided circles of 3 feet in diameter, one of them being fixed relatively to the axis of the telescope, and the other moveable and capable of being clamped to the axis in any position. Each of the divided circles is read off by means of four micrometer microscopes, and additional microscopes, if required, may be readily applied. There are two collimating telescopes each of 6 inches aperture, which can be directed upon each other through an opening in the central cube. A powerful apparatus is likewise provided by which the instrument may be readily and safely reversed[2]. There are no screw adjustments for azimuth and level, but the axis of the instrument is brought into its definitive position by scraping the Y supports[3]."

The following additional particulars are given in the Monthly Notices, Vol. XXXII. p. 151. "Small buildings have been erected over the collimators, which are now, therefore, virtually in the same room with the Transit Circle, and the consequent steadiness in the images of the collimator wires is very remarkable."

"Mr Graham [First Assistant at the Cambridge Observatory] has adopted a method of determining the intervals of the right-ascension wires which has been found to be very accurate

---

[1] The donation in 1863 of £2000 by Miss Sheepshanks, to be employed for the benefit of the Cambridge Observatory, with special preference for its application in the purchase of one or more large instruments, and the legacy in 1858 of £10,000, left by her brother, the Rev. R. Sheepshanks, for the promotion of the science of Astronomy in the University, are recorded in the yearly issues of the *Cambridge Calendar* under the head of The Observatory.

[2] See Note 2 to Art. 48, p. 40.

[3] See at the end of Art. 47.

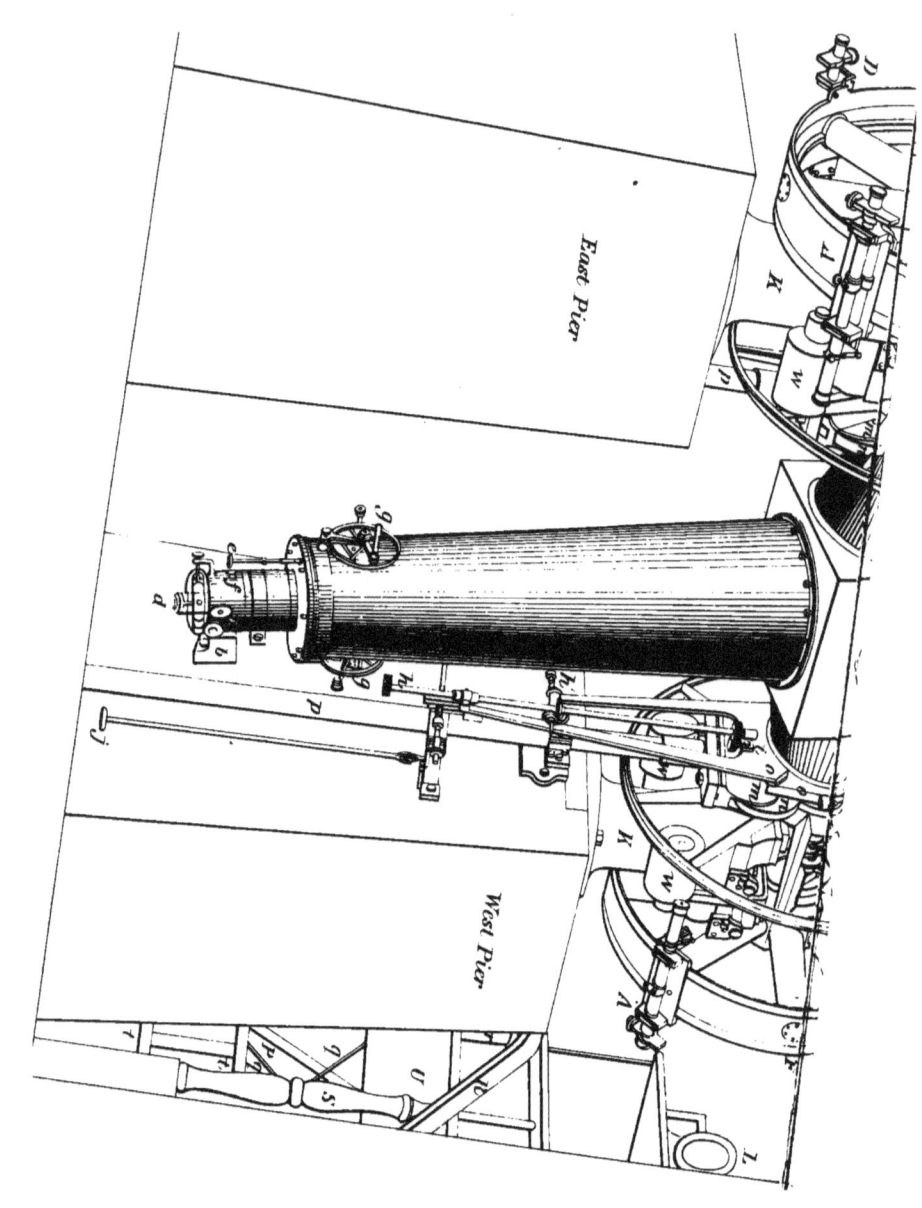

and convenient. The eye-piece is turned through 90°, and each of the right-ascension wires, now become horizontal, is in turn directed to the horizontal wires of either collimator, and the corresponding readings of both circles are taken. The values of the revolutions of the micrometer-screws have been found in a similar way with great accuracy."

240. In Plate III., O is the object-end of the telescope, and a, b, c, d, e, f are the positions of mechanical arrangements at the eye-end adapted to the following purposes: at a are two graduated micrometer-heads with fixed indices, one for reading off the integral micrometer revolutions, and the other the parts of a revolution, whereby the position of the wire-frame in the vertical direction is recorded. The latter micrometer-head is turned by hand, and by tooth-and-wheel action turns the other in the required ratio. Apparatus of exactly the same kind is provided for moving the wire-frame in the horizontal direction, and both are covered by the angular plate b, which is attached for preventing accidental disturbances of the micrometer-heads, and may be readily detached as occasion requires. In taking an observation the eye-piece is shifted horizontally by turning the milled head c, and by turning that at d it may be shifted vertically and independently of the horizontal movement, so that by the two movements the position of the eye-piece, within certain limits, may be changed *ad libitum*. Turning the milled-head at e regulates the illumination of the field, or of the wires. (See Arts. 37 and 38.) At f there are two sets of antagonistic screws, acting in transverse directions, one for turning the eye-end bodily about the axis of the telescope, and the other for changing its distance from the object-glass. (See Arts. 36 and 83.) The former action is explained in Art. 163, and in p. 157 is represented by the part of Figure 31 indicated by the letter C. At g, g are the setting-circles, respecting which and the mode of using them, in the case of non-reversion of the instrument, see what is said in Art. 196.

241. Relatively to the wires used in taking transit observations it is requisite to state, that as there is no vertical micrometer-wire, the abbreviated method of observing transits

of stars near the Pole, described in Art. 112, cannot be conveniently made use of; on which account six wires, in addition to the usual number of seven, have been inserted and arranged as follows. Supposing the whole system of wires to be $A, B, C, \alpha, \beta, \gamma, D, \delta, \epsilon, \zeta, E, F, G$, the intervals between A, B, C, D, E, F, G are *quam proxime* equal and of usual magnitude, while those between C, α, β, γ and those between $\delta, \epsilon, \zeta, E$ are all equal but much smaller, and considerably smaller than the remaining intervals between γ, D, δ. Polaris and other stars in proximity to the Pole are observed at the wires $C, \alpha, \beta, \gamma, D, \delta, \epsilon, \zeta, E$, and the larger intervals from γ to D and from D to δ allow of observing them both by reflection and directly at the same meridian transit.

242. After what has been said in Arts. 53—56 respecting the construction and use of collimators, it will suffice to mention the following particulars respecting those of the Cambridge Transit Circle. The focal length of each is very nearly $6\frac{1}{2}$ feet. In Plate III., I is the opening in the central cube through which the collimators are pointed towards each other (Art. 239). At the geometrical focus of the *south* collimator, there are two parallel wires, separated by a small interval, and crossing the middle of the field *vertically*, and also two other wires intersecting near the middle of the field, and inclined by equal small angles to a horizontal direction passing through their point of intersection. These two sets of wires are attached to the same frame, which is moveable in the horizontal direction by turning a graduated micrometer-head. At the geometrical focus of the *north* collimator there are two like sets, but the parallel wires are horizontal and the crossing wires are inclined by equal small angles to the vertical direction. For determining collimation error, first, the parallel wires of the *south* collimator are moved by turning the micrometer-head into a position such that the image of the point of intersection of the cross wires of the *north* collimator is exactly mid-way between them, and at the same time near the place of their intersection by the image of the north collimator's parallel wires. Then, after pointing the telescope to the *south* collimator, the part of the wire D

near the centre of the field is placed by the telescope-micrometer mid-way between the images of the collimator's parallel wires, and upon turning the telescope to the *north* collimator, the *same* part of D is made by the micrometer-movement to bisect the image of the cross wires of that collimator. These operations having been performed, and the micrometer-readings recorded (six or more times for each pointing of the telescope), from the difference of the means of the readings the error of collimation may be calculated, as shewn in Art. 55, and if of excessive amount might be mechanically corrected. The parallel horizontal wires of the north collimator, and the vertical cross wires of the south collimator, might be employed in an analogous manner for determining by the process explained in Arts. 222 and 223 the amount of horizontal flexure.

243. The East and West Piers, which are those of the old Transit-instrument reduced in height, are found to give sufficiently steady support to the much heavier Transit Circle. They are surmounted by strong iron pillars K, with which the Ys in which the pivots rest have rigid connection, so as not to be susceptible of any screw adjustment (see Art. 47). The arrangements relative to the graduations and microscopes are the same on the East and West sides of the telescope, being, in fact, mutually convertible by a reversion of the instrument. (The reference letters are generally the same for both sides.) G, G are the graduated circles, turning with the instrument about the horizontal axis, that on the West side having a permanent position relative to the axis, and that on the East side capable of being shifted rotationally into any position relative to the graduation of the other. To the wheels F with axle and spokes are attached four micrometer-microscopes A, B, C, D, pointing to the graduations on the circles G. The weight of F is counterpoised by two weights W, one on each side of the axis, the fulcrum being the pillar K. H is an additional microscope also directed to the graduation, but only used as a pointer, or for more exact setting, its reading for the setting being previously calculated to the nearest minute. For instance, to set for a reflection observation so as to secure that the star shall

cross very near the middle of the field, the setting would first be performed in the usual manner by the setting circle, and then, if on looking into microscope H it should be found to be not sufficiently exact, it might by the tangent-screw movement be readily changed; after which the bisections of the graduations by the four microscopes would be proceeded with. (See note to Art. 164 in p. 159.) The rods p, p acting upwards by weights and leverage, as already described at the end of Art. 29, sustain, by intervention of the friction wheels m, n, the greater part of the weight of the instrument, and thus lessen the wear of the pivots by friction. Relative to the apparatus for the Circle-readings it only remains to mention, that the parts marked x, x furnish means of adjusting the axes of the wheels F so that they shall be nearly in a line with the axis of motion of the instrument.

244. The apparatus for slow-motion about the horizontal axis is attached to the West Pier. At i there is a small screw the end of which abuts on a collar of the axis, and by turning the milled-head at h this end is made to act on the collar, and thereby brings down the parts o, o, o, the uppermost of which is a half-ring fitting into an angular groove, the pressure on the faces of which clamps the instrument. After this the slow movement is effected by turning the handle j, which by screw-and-wheel action with Hooke's joint, gives horizontal motion to the lower end of the apparatus o, o, o. On unclamping by turning h, the two lower parts o, o are upheld by the upper half-ring by means of intermediate screws with capstan heads. The handle j is usually supported by a moveable upright stand (not represented in the Figure), which is furnished with a row of hooks for placing the handle in a position convenient to the observer. Similar apparatus at h', with handle and Hooke's joint, was added, at Mr Graham's suggestion, for being occasionally used by an assistant, especially for clamping the instrument in different positions in taking zone observations of stars, the changes of setting being quickly made by the observer by simply pulling a string tied to the Circle and passing over a fixed pulley.

245. At L a large lens, placed on a stage, not only throws light from a paraffin lamp on a lens near the end of the transverse axis for the illumination of the field and wires[1] (Art. 240), but also, by means of transmission of light through glass prisms suitably formed and placed, illumines the graduations towards which the five microscopes are directed. S, R, r are parts of a rail and balustrade for the observer to ascend by steps t, t to take his position for reading off the microscopes; p is part of the bracing of the stage for supporting the lamp; q, q are iron wires for propping the flap U, on which, as there is occasion, the observer places his memorandum book and hand-lamp. The arrangements for illumining the field of view and the graduations, and reading off the microscopes, are precisely the same on the eastern, as on the western side, the interior plate which reflects light towards the field being adaptable, by turning the screw e, to the incidence of the light from either the western or eastern lamp. At X is the clock-face, which is illumined by a large lamp moveable by a hinged support, and can be looked at in most of the positions occupied by the observer.

246. With respect to the means employed for obtaining corrections for forms of the pivots of the Transit Circle, little need be said in addition to the statements made in the Note to Art. 102 in page 98. The dot is marked in the centre of a small metallic disk, which is affixed to a piece of plate-glass covering the end of the opening for the passage of the rays that illumine the field of view. After removing the lens at the end of the axis (Art. 245), a powerful microscope-micrometer is placed so that an image, or ghost, of the dot (suitably illuminated) is formed where its horizontal and vertical coordinates can be measured by the micrometer-wires. The arrangements are precisely the same at the two ends of the transverse axis, two microscope-micrometers being made use of. For adjusting the

[1] Mr Graham informs me that he has found the visibility of very faint stars to be sensibly improved by interposing between the lens at L and that at the extremity of the axis a plate of red glass, which by means of an attached string can be readily put in place or removed.

micrometer-wires horizontally and vertically, the spirit-level apparatus mentioned in clause (3) of the Note in page 98 is had recourse to, but not till its indication has been corrected or verified by a small plumb-line used in the manner described in Art. 95 with reference to the Transit-instrument.

247. After the explanations that have now been given of the processes of taking and reducing meridian observations, it only remains, for the complete determination of celestial positions, to deduce from observations of the Sun the *exact* position, at any epoch, of the origin of right ascension, assumed to be the point of intersection of the Equator and the Ecliptic, or *First Point of Aries*, and also to find exactly, for the same epoch, the inclination of the Ecliptic to the Equator, or *The Obliquity of the Ecliptic*. In the determination of these two *Elements* by Flamsteed's Method (given in Arts. 137 and 138), only observations of the Sun taken near the times of the Equinoxes are employed; by which means provisional values may be obtained sufficiently accurate for adoption by the theoretical astronomer for the purpose of expressing his theoretical places in right ascension and polar distance for immediate comparison with the results of observation. It has already been intimated (Art. 141) that it is usual for the *practical* astronomer to employ such comparisons for obtaining mean errors of the Sun's Tabular right ascension and polar distance from *all* the meridian observations of the Sun in the course of a year, and thence to infer the error of the assumed Obliquity, and the mean error of the assumed right ascensions of the fundamental stars. These corrected values may still admit of being farther corrected by a continuation of the series of observations. But to effect these corrections so as to obtain the most accurate values of the Solar Elements that observation is capable of giving, it is necessary to have recourse to the Solar Tables themselves; which is done on principles that are quite logical, as will be seen from the following explanations, which are applicable to the method employed in Vols. X—XX. of the *Cambridge Observations*. (See Vol. XX. pp. xxxvi. and 96.)

THE TRANSIT CIRCLE. 241

248. The true longitude λ and true Polar Distance Δ of the Sun's centre, and the true obliquity of the Ecliptic I, are related to each other at any epoch by the equation
$$\cos \Delta = \sin \lambda \sin I;$$
and the Tabular Longitude $\lambda + \delta\lambda$, the Tabular Polar Distance $\Delta + \delta\Delta$, and the Obliquity $I + \delta I$ assumed in the Nautical Almanac, for the same epoch, by the equation
$$\cos (\Delta + \delta\Delta) = \sin (\lambda + \delta\lambda) \sin (I + \delta I).$$
Hence, assuming that the Tabular errors $\delta\Delta$, $\delta\lambda$, δI are already so small that powers of these quantities higher than the first may be neglected, we have by expanding,
$$\delta\Delta + \frac{\cos \lambda \sin I}{\sin \Delta} \delta\lambda + \frac{\sin \lambda \cos I}{\sin \Delta} \delta I = 0 \ldots\ldots\ldots(A).$$
The reasoning now proceeds on an assumption which is justified by the existing state of the Solar Theory, namely, that the *changes* of the values of λ and I in the course of a year are in agreement with the indications of theory, and consequently that the values themselves as given in the Nautical Almanac are affected, if by any, by constant errors, which it is proposed to find.

249. The actual errors of the Solar Tables in Polar Distance cannot be immediately derived from comparisons of the calculated with the observed values, because, although mere errors of observation may be supposed to be eliminated in the mean result of a large number of observations, there may still remain uncorrected instrumental errors and errors of reduction. Representing, therefore, by α any excess of the tabular above the observed polar distance, and by p the unknown excess of the observed above the true polar distance, we shall have

Excess of tabular above true polar distance $(= \delta\Delta) = \alpha + p$.

As we are ignorant of the causes to which p may be owing, we can only undertake to deduce from observation the mean of its values within the limits of the Tropics. (See what is said in Art. 229.) By putting m for $\sin I \times \delta\lambda$, n for $\cos I \times \delta I$, and $\alpha + p$ for $\delta\Delta$, the equation (A) becomes
$$\alpha + m \operatorname{cosec} \Delta \cos \lambda + n \operatorname{cosec} \Delta \sin \lambda + p = 0.$$

C. 16

Instead of forming a separate equation from this formula for each value of a, let the whole number of observations be divided into twelve groups, the mean of the values of a in each group be calculated, and considered to correspond to the day nearest the numerical mean of the days of observation in the group, and let λ and Δ be taken for the mean noon of the mean day from the Nautical Almanac. In this manner twelve different equations will be formed, which may be divided into four groups of three, corresponding roughly with four quarters of the year. Then three equations for determining the three unknown quantities m, n, p may, as respects probable errors, be advantageously obtained as follows:

First Quarter + Second + Third + Fourth = 0,
First Quarter + Second − Third − Fourth = 0,
First Quarter − Second − Third + Fourth = 0.

250. For the purpose of exemplifying the results to which the foregoing reasoning conducts, those obtained from the Cambridge Meridian Observations of the Sun in the year 1855 are here subjoined. The total number of the Circle observations was 125, inclusive of 24 observations of single Limbs. These were divided into four quarter-groups containing severally $28\frac{1}{2}$, $27\frac{1}{2}$, $28\frac{1}{2}$ and $28\frac{1}{2}$ observations, one of a single Limb being taken to be half an observation. Each of these groups was subdivided into three in such manner that the twelve groups all contained nearly the same number of observations. In calculating the value of a for any group, half-weight was given to an observation of a single Limb. The rest of the reasoning was conducted by the process explained in Arts. 247 and 248, and the following results were obtained by the solution of the three final equations:

$$m = -1'',239, \quad n = -0'',233, \quad p = +0'',329.$$

Hence $\delta\lambda \;(= m \operatorname{cosec} I) = -1'',239 \operatorname{cosec} 23°.\;27',5 = -3'',112,$
which is the mean excess for the year of the Tabular, above the true, Longitude of the Sun;

$$\delta I \;(= n \sec I) = -0,'233 \sec 23°.\;27',5 = -0'',254,$$

which is the excess of the Obliquity assumed in the Tables above the true Obliquity; and $p=$ the mean excess within the Tropics of the N.P.D. determined by the reduction of the Circle observations of 1855 above the true N.P.D. (See Art. 229.)

If δR represent the *mean* excess for the year of the Tabular R.A. above the true R.A. in arc, we may assume that $\delta R = \delta \lambda$. Hence in the present instance δR (in time) $= -0^{s}.207$. The comparisons of all the observations of the year with the Tabular values of the Nautical Almanac gave $-0^{s}.187$ for the mean excess (A) of the Tabular above the observed R.A. by 128 comparisons, and $-0''.364$ for the mean excess (D) of the Tabular above the observed N.P.D., by 113 comparisons. Hence, putting q for the mean excess of the assumed R.A. of the fundamental stars above the true, we have the following results:

$\delta \Delta = $ (Tab. N.P.D. $-$ Ob. N.P.D.) $+$ (Ob. N.P.D. $-$ True N.P.D.)

$= D + p = -0''.364 + 0''.329 = -0''.035$,

$q = $ (Tab. R.A. $-$ True R.A.) $-$ (Tab. R.A. $-$ Ob. R.A.)

$= \delta R - A = -0^{s}.207 + 0^{s}.187 = -0^{s}.020$.

Hence the observed R.A. were *too small* by the mean quantity $0^{s}.020$; or the assumed position of the first point of Aries was *in advance* of the true by this quantity.

251. By 127 observations of the Sun in R.A., and 117 observations in N.P.D. in the year 1856, it was found that $m = -0''.254$, $n = -0''.199$, $p = +0''.451$, $D = -0''.447$, $\delta \Delta = +0''.004$, $\delta I = -0''.217$, $\delta R = -0^{s}.043$, $A = -0^{s}.155$, and $q = +0^{s}.112$. Hence the assumed R.A. of the fundamental stars were *too great* by $0^{s}.112$. (See *Camb. Obs.* Vol. xx. pp. 204 and 205.) It may be noticed that these results, with the exception of the value of m and the consequent values of δR and q, agree closely with those given by the observations of 1855. On account of the discrepancy between the values of δR ($= m \csc I$), the corrections of the assumed R.A. of the fundamental stars are different for the two years, being $+0^{s}.020$ for 1855 and $-0^{s}.112$ for 1856, although the mean excess of the assumed R.A. above those of the Naut. Alm. was almost exactly the same for both. (See Introduction of Vol. xx., p. xlix.)

Hence as δR is the excess of the *tabular* above the *true* R.A. of the Sun, it might be inferred that the R.A. of the Tables, as calculated for the years 1855 and 1856, require the respective corrections $-0^s,207$ and $-0^s,043$. Such an inference, however, cannot be safely drawn, in the existing state of practical and physical astronomy, unless the amount and variation of the error be determined by the mean results of the observations of a large number of years. The foregoing explanations will, however, have indicated by what kind of process small residual errors of the Tables are ascertained.

252. The values of the corrections p and q having been determined by observations of the *Sun*, those of $\delta R \, (= A + q)$ and $\delta \Delta \, (= D + p)$ are readily deducible, inasmuch as A and D are respectively the excesses of the tabular above the observed R.A. and N.P.D. usually computed by the observer. From the values of the errors δR and $\delta \Delta$ thus obtained, mean errors may be derived from arbitrary groups of the individual errors, and may severally be supposed to correspond in epoch to the mean of the times of the observations furnishing the group of errors. These rules for calculating mean tabular errors are applicable alike to meridian observations of the Sun, Moon, and the *older* planets, in all cases in which a continuous series of such observations have been taken. For the sake of the theoretical astronomer, the mean errors in R.A. and N.P.D. are converted into errors in Longitude and Ecliptic Polar Distance by known formulæ[1], the application of which is much facilitated by making use of the factors contained in Appendix II. of the Volume of Greenwich Observations for 1836. In the same volume (p. lxii) are given formulæ for calculating errors of Heliocentric Longitude and Ecliptic Polar Distance from the errors of Geocentric Longitude and Ecliptic Polar Distance obtained in the manner just indicated. The results of such calculation in the cases of the minor planets Ceres, Pallas, Juno, and Vesta, and all the larger planets, are given in the Volume just named, and in all succeeding Volumes, of the Greenwich Observations.

[1] See the Rev. R. Main's *Practical and Spherical Astronomy*, Chap. III. Art. 7 (p. 74).

253. The practical astronomer is required to give, together with his observations, the *times* at which they were taken. With respect to fixed stars, when their places are observed with a Transit-circle, it generally suffices to name the year and day of the observation; but for the moving bodies the precise time of taking each observation has to be calculated. The science of *Time*, as consisting of measurements of time-intervals, and determinations of epochs, depends essentially on data furnished by Practical Astronomy. This will appear from the following discussion, the purpose of which is to shew how to calculate the *epochs* of observations.

254. Bessel, by a comparison of his own observations of the Sun in 1820—1825, and those of Bradley in 1753 and 1754, with Carlini's Solar Tables, obtained the Sun's mean longitude at a given epoch; from which the following fundamental quantity[1], used in the computations of the Nautical Almanac for the years 1836—1863, was derived:

At the Greenwich Mean Noon of January 1, of the year $1800 + t$, the Sun's Mean Longitude (M) is

$$283°.53'.32'',71 + t.27'',605844 + t^2.0'',0001221805 - f.14'.47'',083,$$

where $14'.47'',083$ is the change of Mean Longitude in 6^h of mean time, f denotes, *for the nineteenth century*, the number of years from the year immediately preceding $1800 + t$ which is divisible by 4 without remainder. It is to be observed that this value of the Mean Longitude includes the effect of aberration.

A *sidereal day* is the interval between two consecutive transits of the first point of Aries across the meridian of any place. A *mean solar day* is the interval between two consecutive transits of a fictitious Sun supposed to move in the Equator with the Sun's mean motion in Longitude, or Right Ascension.

It has been agreed by astronomers to call *the sidereal time*

---

[1] See the *Astronomische Nachrichten*, No. 133; Bessel's *Tabulæ Regiomontanæ*, p. xxiv; and the Nautical Almanac for 1863, p. vi.

at any place, the arc intercepted between the *actual* first point of Aries and the point of the Equator which is on the meridian of the place, converted into time at the rate of 15° to an hour. But the first point of Aries, since its position is determined by the intersection of the plane of the earth's Equator with the plane of the Ecliptic, moves relatively to stars on the plane of the Ecliptic, and, on account of nutation, by an irregular motion. Hence according to the adopted reckoning astronomical sidereal time is not strictly uniform. No sensible error, however, arises from this circumstance, because the irregular motion takes place relatively to a *mean* position which has a uniform motion, and the fluctuations of position (called the Equation of the Equinoxes) are very slow, and much slower than those to which the rate of the best constructed time-piece is liable from incidental causes. The particular advantage of this conventional reckoning is that the sidereal time at which a celestial object passes the meridian becomes identical with its apparent Right Ascension. The calculated apparent R. A. of the fundamental stars are accordingly referred to the true Equinox. Hence, since the error of the clock is the difference between its indication and the calculated Apparent Right Ascension of a fundamental star, it follows that the time-piece is virtually regulated to point to 0^h whenever the first point of Aries (affected by aberration as a star) is apparently on the meridian.

255. From the definition above given of Sidereal Time it follows that the Sidereal Time at Mean Noon = Sun's Mean R. A. + Nutation in R. A., the addition of the Nutation giving to the sidereal time the fluctuating value it is required to have by being referred to the *true* Equinox. By means of this equation the relation between the sidereal time and the mean solar time of the same epoch is ascertained by the following process. From the calculations of Bessel already referred to, it was found that the mean motion of the Sun in $365\frac{1}{4}$ mean solar days was less than 360° by 22″,617656; whence it follows that the sidereal year, or complete revolution of the Sun with respect to fixed space, is $365^d . 6^h . 9^m . 10^s,7496$, or

365,256374417 mean solar days. Taking, according to the same authority, the mean amount of the precession of the equinoxes in the t years succeeding 1800 to be

$$50'',22350t + 0'',0001221805t^2,$$

the mean length of the tropical year between $1800+t$ and $1800+t+1$ is

$$365^d.5^h.48^m.47^s,8091 - 0^s,00595t,$$

or $\qquad 365^d,242220013 - 0^d,0000000686t.$

Dividing $360°$ by the length of the tropical year, the mean motion of the Sun in Longitude in a mean solar day will be found to be $59'.8'',3302$, and consequently the mean daily motion in R.A. expressed in time, $3^m.56^s,555348$. Hence by the foregoing equation we have at the Greenwich mean noon of any day (n) of the year $1800+t$,

$$\text{Sidereal time} = \frac{M}{15} + 3^m.56^s,555348\,n + \text{Nutation in R.A.}$$

It thus appears that from one mean noon to the next following the sidereal time increases by the mean quantity $3^m.56^s,555348$, and consequently that 24^h of mean time are equivalent to $24^h.3^m.56^s,555348$ of sidereal time. The above formulæ for M and for sidereal time are adopted in the Nautical Almanacs for the 29 years 1835—1863.

256. In the Nautical Almanacs for 1864 and following years the tabular places of the Sun are taken from Leverrier's Tables contained in Vol. IV. of the Annals of the Imperial Observatory of Paris. According to results deduced by Leverrier from 8911 observations of R.A. of the Sun made in the years 1750—1850, the mean sidereal motion of the Sun in a Julian year of $365\frac{1}{4}$ days is $1295977'',38234$, or $22'',61766$ less than $360°$. Also he obtains $50'',23572t + 0'',00011289t^2$ for the general precession in the interval t reckoned in Julian years of $365\frac{1}{4}$ days, and commencing at the Paris mean noon of January 1, 1850. Hence he finds (by a process equivalent to that indicated in Art. 255) that the mean day is equal to the sidereal day multiplied by $1,002737909$, or $24^h.3^m.56^s,555338$,

nearly agreeing with Bessel's value above. If therefore M' be the Sun's Mean Longitude for the meridian of Greenwich for the mean noon of January 1, $1850 + t$, as inferred from Leverrier's Tables of the Sun, allowance being made for the difference between the meridians of Paris and Greenwich, we shall have

$$\text{Sidereal time} = \frac{M'}{15} + 3^m . 56^s . 555338 \times n + \text{Nutation in R. A.}$$

257. After shewing how the foregoing two sets of formulæ were obtained, I propose now to give an example of the application of each of them for calculating the sidereal time of mean noon. Supposing that it is required to calculate the sidereal time of the Greenwich mean noon of January 24, 1863, the following is the process according to Bessel's formulæ contained in Arts. 254 and 255. In this example $t = 63$, and therefore $f = 3$. Hence it will be found that

$27'',605844 \times t = 28'.59'',168, \quad 0'',0001221805 \times t^2 = 0'',485,$
and $\qquad -14'.47'',083 \times f = -44'.21'',249.$

Consequently

$M = 280°. 53'. 32'',71 + 28'. 59'',168 + 0'',485 - 44'. 21'',249$
$\quad = 280°. 38'. 11'',11 = \text{(in time)} \ 18^h. 42^m. 32^s,741.$

Also since mean noon was *earlier* than apparent noon, the day before is to be taken, so that $n = 23$. Hence

$$3^m . 56^s,555348 \times n = 1^h. 30^m. 40^s,773 ;$$

and since the nutation-correction, as given in the Naut. Alm. (p. 242), was $+ 1^s,09$, it follows that the sidereal time

$= 18^h. 42^m. 32^s,74 + 1^h. 30^m. 40^s,77 + 1^s,09 = 20^h. 13^m. 14^s,60,$

which agrees with the value in page 3 of the Naut. Alm. for 1863.

To exemplify the application of Leverrier's formulæ, let it be required to find the sidereal time at mean Greenwich noon of 1879 January 20. The Sun's Mean Longitude (M'), as given in p. 102 of Vol. IV. of the Annals of the Paris Observatory, is

$$280°.\ 46'.\ 43'',51 + 1296027'',6784t + 0'',00011073t^2,$$

the time t being reckoned in Julian years of $365\frac{1}{4}$ days, and commencing at the Paris mean noon of January 1, 1850. Hence $t = 29\ (= 4 \times 7 + 1)$, and $f = 1$. These values give

$$1296027'',6784 \times 29 = 10440°.\ 13'.\ 22'',736,$$
and $\qquad\qquad 0'',00011073 \times 29^2 = 0'',093.$

Hence, rejecting $29 \times 360°$, we have

$M' = 280°.\ 46'.\ 43'',51 + 13'.\ 22'',736 + 0'',093 - 1 \times 14'.\ 47'',083$
$\quad = 280°.\ 45'.\ 19'',194 = $ (in time) $18^h.\ 43^m.\ 1^s,279.$

Since the mean noon, at the epoch, is earlier than apparent noon, the day before is to be taken. Hence $n = 19$ and

$$236^s,555338 \times n = 1^h.\ 84^m.\ 54^s,551.$$

As Paris is $9^m.\ 20^s,63$ East of Greenwich, the correction of M' for difference of meridians is $+23'',033\ (= +1^s,535)$. Also correction for nutation in R. A., derived from $15'',16$, the equation of the equinoxes in Longitude is $0^s,927$. Consequently the sidereal time

$= 18^h.\ 43^m.\ 1^s,279 + 1^h.\ 14^m.\ 54^s,551 + 1^s,535 + 0^s,927$
$= 19^h.\ 57^m.\ 58^s,292,$

which agrees with the time given in page 3 of the Naut. Alm. for 1879. In calculating sidereal times for the formation of Ephemerides, much assistance would be derived from the Tables arranged by Leverrier for this purpose in the before cited Volume. The object of the foregoing discussion has been to indicate processes, and furnish data, whereby the sidereal time of any epoch might be calculated independently of such aid.

258. For calculating epochs in mean time, it is required to know *the mean time of the transit of the first point of Aries;* which may be inferred as follows from the sidereal time of

mean noon. Let σ be this sidereal time; that is, the horary angle, at mean noon of any day, of the first point of Aries *westward*. This point is, accordingly, on the meridian at an interval before noon equal to σ reckoned in sidereal time, and therefore equal to $\mu\sigma$ reckoned in mean time, μ being the ratio of the Sun's mean diurnal motion to that of a star. Hence, as mean time is reckoned from mean noon, the required mean time is $24^h - \mu\sigma$. Selecting, for example, the sidereal time at mean noon of January 20, 1879, namely, $19^h. 57^m. 58^s,292$, and either taking 1,002737909 for the ratio of the mean day to the sidereal day, as given in Art. 256, or employing the Table of equivalents in pages 490 and 491 of the Naut. Alm. for 1879, it will be found that $\mu\sigma = 19^h. 54^m. 42^s,030$. Hence $24^h - \mu\sigma$, or the mean time required, is $4^h. 5^m. 17^s,970$. As, however, this mean time is *less* than the sidereal time at mean noon of January 20, it must be the mean time of transit of the first point of Aries on January 19. It agrees, in fact, with the value given in the Naut. Alm. for this date.

259. By knowing the sidereal time of mean noon, and the mean time of transit of the first point of Aries, which may be called *sidereal noon*, mean time and sidereal time are mutually convertible by calculating according to the following rules:—
Sidereal time *required* = that at the preceding mean noon + the equivalent in sidereal time of the *given* mean time. Mean time *required* = that at the preceding sidereal noon + the equivalent in mean time of the *given* sidereal time.

To convert, for example, $2^h. 22^m. 25^s,62$ Greenwich Mean Time, January 20, 1879, into sidereal time, the process is as follows. The sidereal time at the mean noon of January 20, 1879, (calculated as is shewn in Art. 257) $= 19^h. 57^m. 58^s,292$, and the sidereal equivalent of the mean time

$$2^h. 22^m. 25^s,62 = 2^h. 22^m. 49^s,018,$$

as calculated by the Tables in pages 488 and 489 of the Naut. Alm. for 1879. Hence the sidereal time required

$$= 19^h. 57^m. 58^s,292 + 2^h. 22^m. 49^s,018 = 22^h. 20^m. 47^s,310.$$

To obtain, conversely, the mean time of January 20, 1879,

corresponding to the sidereal time $22^h.20^m.47^s{,}310$ on that day, the mean time of sidereal noon on January 19 is to be taken, in accordance with a rule which, to assist the memory, has been expressed in the terms, "If the sidereal time be more, take the day before" (see Art. 257); the result of the computation in Art. 257, as also the contents of pages 3 and 4 of the Naut. Alm., shewing that for January 19 the sidereal time at mean noon exceeds the mean time of sidereal noon. Now the latter, either as obtained independently by the calculation in Art. 258, or taken immediately from the Naut. Alm., is $4^h.5^m.17^s{,}970$. Also the equivalent of $22^h.20^m.47^s{,}310$ in mean time is

$$22^h.17^m.7^s{,}654$$

by the Tables in pages 490 and 491. Hence the mean time required $= 4^h.5^m.17^s{,}970 + 22^h.17^m.7^s{,}654 - 24^h = 2^h.22^m.25^s{,}624$, being for January 20 the excess above 24^h.

260. For astronomical purposes it is convenient to express the epochs of all events in mean times of a fixed meridian, and the English astronomer selects, of course, the meridian of Greenwich. If the longitude of any observatory *eastward* from the meridian of Greenwich be L (in time), a given star, or the first point of Aries, passes the meridian of that Observatory before it passes the meridian of Greenwich by a sidereal interval equal to L. Hence if S be the sidereal time of any event, as determined by transit observations at the Observatory, that is, the *local* sidereal time, the Greenwich sidereal time of the event is $S - L$. With this sidereal time the Greenwich mean time may be calculated by the process indicated in the preceding paragraph. Or, if with the local sidereal time S the mean time be calculated just as for Greenwich, the correction for difference of meridians is $-\mu L$, μ being the factor (the reciprocal of $1{,}002737909$) for converting sidereal time into mean time. For the Cambridge Observatory, the longitude is $22^s{,}75$ East, and the correction $-\mu L$ is equal to $-22^s{,}69$.

261. It will thus be seen that in order to deduce from the sidereal times of events, as obtained by observations made at any Observatory, the epochs of the same events in Greenwich

mean time, it is necessary to know the *Longitude* of the Observatory with reference to the meridian of Greenwich. The same element, together with *Latitude,* is necessary for indicating the locality of the Observatory on the earth's surface, which the practical astronomer is required to do. Having already had under consideration (in Arts. 219—220) a method of obtaining exactly the Latitude of an Observatory, we have now to enquire by what means its Longitude may be accurately determined. Terrestrial Longitude is the arc between the meridian of any place and that of Greenwich measured on the earth's Equator, and is called East or West Longitude, or is designated by the signs — or +, according as the place is eastward or westward from Greenwich. This arc is measured *astronomically* by the interval between the transits of a given *star* across the two meridians, because the star is apparently carried from one meridian to the other in that interval by the diurnal motion of the heavens due to *the earth's rotation about its axis,* to which rotation, assumed to be uniform[1], all astronomical measures of time are referred. The *sidereal day,* which is the interval from the passage of a star, or the first point of Aries, across a meridian to its return to the same, being supposed to consist of 24 hours, it follows that the *astronomical measure* of longitude has to 24 hours the same ratio as the *arc of terrestrial longitude* has to $360°$, or that one hour corresponds to $15°$. Hence it may be seen that the longitude of a place is measured by the difference between the sidereal time at the place, and the sidereal time at Greenwich, of the *same* event, because the difference between such times measures the arc between the two meridians. This is the general principle on which terrestrial longitude is ascertained by astronomical means.

262. There are various ways of comparing the simul-

[1] It was suggested by M. Delaunay, late Astronomer Imperial at the Paris Observatory, and has since been maintained by other physicists, that the uniformity of the earth's rotation might be disturbed to a sensible amount by the effect of the attractions of the Sun and Moon on the tidal waves. If this be the case, the calculation of all time-intervals and epochs will be in some degree altered. I am not aware that the amount, or even the actuality, of disturbance from this cause has been strictly deduced from hydrodynamical principles.

taneous sidereal times of two Observatories. The simplest appears to be that by *transfer of Chronometers*, according to which, when several chronometers are used, each is first timed and rated by the transit-clock of one of the Observatories, is then transferred to the other Observatory, and after being timed, and rated by a clock and transit-instrument there, is conveyed back to the first for another timing and rating, and so on as often as may be thought desirable. In this way, after taking account of the meridional adjustments of the instruments, and the recorded clock and chronometer times and the calculated rates, the simultaneous sidereal times of the two positions are repeatedly brought into comparison; and by taking the mean of the several differences given by a single chronometer, and then the mean of the means given by the several chronometers, the concluded difference of longitude is obtained. By a process such as this, the details of which are contained in the *Cambridge Philosophical Transactions* (Vol. III. p. 168), Professor Airy found the Longitude of the Cambridge Observatory to be 23°,54 eastward from Greenwich.

263. Since the invention of the Electric Telegraph, astronomers have taken advantage of the transmission of galvanic signals for the determination of the difference of longitude between two Observatories. When it can be so arranged that the signals can be sent and received at both Observatories, and the operator can also take his time from the transit clock as in an ordinary transit observation, the sender notes the time of sending a signal and the receiver that of receiving it, according to the indications of the respective clocks. The signal is usually the movement of a galvanic needle produced by contact being made by the sender for completing the galvanic circuit, which movement the receiver has to watch for. The signals are transmitted a certain number of times, previously agreed upon, alternately from the two stations. After applying corrections for the errors of position of the transit-instruments, and the errors of the clocks on the true sidereal times of the Observatories, and also eliminating as far as may be errors which might be due to difference of personal equations of the observers,

whether as regards the taking of transits or noting the signal-times, the mean of all the values of the difference of the sidereal times measures the longitude, if the transmission of the signals be absolutely instantaneous. If there should be a sensible rate of transmission, its effect on the determination of the longitude is got rid of by sending signals in opposite directions, as may be thus shewn. Supposing the passage of a signal between the two stations to occupy the interval τ, the time recorded at the receiving station will be *too late* by this interval. Hence, since the measure of longitude is the excess of the sidereal time at the east station above the simultaneous sidereal time at the west station, according as the receiving station is to the east or the west of the other the measure of longitude will be too great or too small by τ. Hence the true longitude is the mean between two measures obtained by signalling from the two stations in opposite directions. Consequently to eliminate this source of error the number of signals should be the same from both stations. In that case a very probable value of 2τ would result from subtracting the mean of all the measures deduced from signals made at the *west* station from the mean of those deduced from the signals made at the *east* station.

264. If the signal-times, whether of sending or receiving, are not immediately taken from the transit-clock, but are recorded by means of transportable chronometers at any position more or less distant from the clock, the noted times have to be reduced to sidereal times of the Observatory by means of comparisons of the chronometers with the clock, made both before and after the series of signals, in order to have data for making allowance for the rates of the chronometers. The comparison of a sidereal chronometer with the transit-clock should, for greater certainty, be made by the intervention of a mean-time chronometer, because this time-piece may be accurately compared with each of the other two by *coincidence of beats*. In making, according to this process, a determination of the longitude of the Cambridge Observatory by galvanic signals on May 17 and 18, 1853 (the particulars of which are given in

the *Transactions of the Cambridge Philosophical Society*, Vol. IX., pages 487—514), the Greenwich signals were received and the Cambridge signals transmitted at the Cambridge Telegraph Office of the Great Eastern Railway, and the Observatory sidereal times were inferred in the manner just stated by comparisons of chronometers. The mean result, deduced from 281 signals, gives for the longitude of the Cambridge Observatory, 22ˢ,75 East of Greenwich. Also from all the signal-times, reduced to sidereal times of the Observatories, the interval (τ) occupied by the transmission of the signals was found to be 0ˢ,020. On account of the comparatively small length of the conducting wire between the two stations (about 55 miles), this value is not trustworthy for calculating the rate of transmission, but may, I think, be regarded as a limit which the interval of transmission did not exceed.

265. Having now gone through the explanation of the means actually employed at the present time for determining with as much accuracy as possible by meridian instruments the *places* in Right Ascension and Polar Distance of fixed stars and of the centres of the moving bodies, and also the processes for calculating the *epochs* at which they have the assigned places, we may be said to have taken account of all that is of fundamental importance in practical astronomy. Accurate determinations of celestial positions by instruments adapted for observations out of the meridian, depend ultimately on the results of observations made on the meridian. Again, as respects *physical astronomy*, suppose the elements of the orbit of a moving body to have been deduced from observations (by processes such as those given in Gauss's *Theoria Motus Corp. Cel.*), till they are known with sufficient accuracy for being employed by the theoretical astronomer for calculating *Tables* whereby the body's place at a given time may be computed. Then the farther corrections which the elements may require are ascertained by means of continued observations of the body made either actually on the meridian, or by the intervention of places of stars determined by meridian observations, the Right Ascensions and Polar Distances given by such observations

being subsequently compared with theoretical values computed with the assumed elements. Suppose, for illustration, A and Δ to be the tabular R.A. and P.D. of a moving body at the assigned epoch of an observation from which the values, as observed, are found to be a and δ. The practical astronomer, after obtaining such observational values with all attainable accuracy, gives values of $A-a$ and $\Delta-\delta$, as deduced from a series of observations, to the theoretical computer, calling them *Errors of the Tables* on the assumption that the observations are exact. Each such error is put on one side of an equation, the other side of which consists of a *variation* of the function from which A, or Δ, was computed, with respect only to the letters it contains representing the elements, and these variations of the elements are consequently to be taken as algebraic excesses of the values of the elements employed in constructing the Tables above their true values. From a large number of such equations, divided into groups and treated according to the rules of the *method of least squares*, values of the variations of the elements are obtained, which, *with signs changed*, are the *corrections* to be applied to the assumed elements. In proportion as the number of the equations is larger, these corrections may be considered to be more free from the effects of incidental errors of observation. An exemplification in part of the investigation above indicated is given by the process described in Arts. 247—252, whereby the errors of the assumed obliquity of the Ecliptic and the assumed R.A. of the fundamental stars are corrected. I conclude this first and principal Part of the Lectures with making the remark, which the foregoing statement of the relation between practical and physical astronomy justifies, that however far the theorist may carry his analytical investigations, the accuracy thereby attainable is limited by the accuracy of the data furnished by instrumental observation.

The remaining portion of the Lectures will treat of (1) the construction and use of instruments employed for obtaining celestial places by observations out of the meridian, especially the Equatorial and the Altazimuth Instrument; (2) other fixed instruments, particularly those adapted to finding exact values of the constants contained in the formulæ for calculating star-corrections given in Arts. 140—147; (3) uses of movable and portable instruments, as the Zenith Sector and the Theodolite, for determining geographical positions, and of the Sextant for nautical purposes; (4) various applications of instruments for special purposes, such as finding longitudes by land and sea, measures of small differences of R. A. and P. D. between the places of two objects, measures of the relative positions of the components of double and multiple stars, &c., together with miscellaneous subjects not treated of in the previous sections.

THE EQUATORIAL.

266. An Equatorial Instrument is made suitable by its construction and mounting for determining Right Ascensions and Polar Distances of objects at whatever part of the heavens they are visible. For this purpose it has an axis of revolution placed parallel to the earth's axis. Like the Transit-circle, the Equatorial is a complete instrument for determining celestial places; but as being liable to unsteadiness on account of the oblique position of its axis, it is not as trustworthy for determining absolute R. A. and P. D. as a meridian instrument. If, however, it be used only for *differential* observations, that is, for comparing the place of an unknown object with that of a known star, or one whose exact place is afterwards ascertained by meridian observations, very accurate results are obtainable. The Equatorial is employed in observing Comets, newly discovered small Planets, Eclipses of the Sun, Moon, and Jupiter's Satellites, occultations of stars and Planets by the Moon, double and multiple stars, and star-groups. It is also much used for securing approximate places of unknown stars, and of stars selected for comparison in differential observations, in preparation for afterwards obtaining their exact places by meridian observations. As in the instances of the Transit Instrument

258   PRACTICAL ASTRONOMY.

and Mural Circle, the Equatorial will be treated of under four Divisions.

I. DESCRIPTION OF THE PARTS AND MODE OF MOUNTING.

267. Figure 40 delineates a wooden model exhibited in my Lectures for indicating the parts, mounting, and uses of an Equatorial. The different parts correspond as to position, and in most respects as to proportion, to those of the five-feet Equatorial of the Cambridge Observatory, to which instrument the following descriptions will more especially apply. Q and R

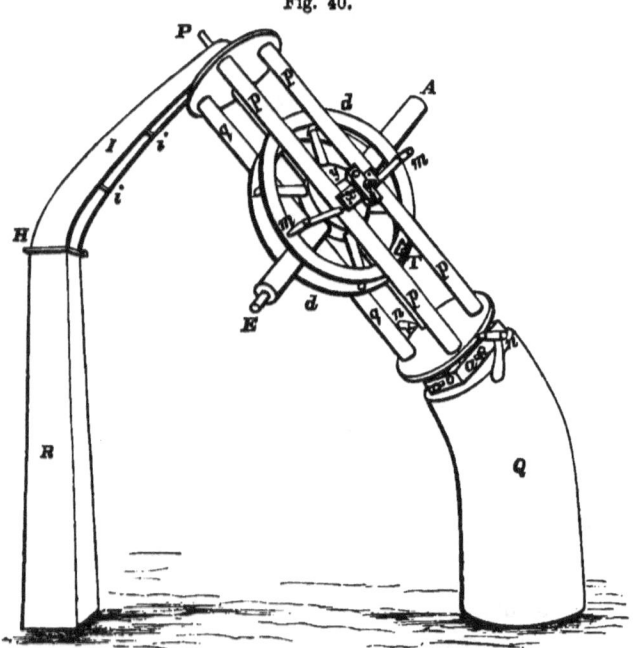

Fig. 40.

are solid stone piers placed, the former southward and the other northward, so as to be cut symmetrically by the plane of the meridian, on a massive stone pillar rising about 20 feet from the level of the ground, on which also a sidereal clock (Graham

THE EQUATORIAL.

with Harrison's gridiron pendulum) is posited for convenience in taking observations. I is an iron frame attached at H to the support R, and having at its upper extremity a Y in which the pivot P at the higher end of the axis of motion rests. The plane faces of the frame I and of the bracing pieces i, i, are made to be parallel to the axis of the Telescope when pointed towards them, so that they only intercept a quantity of light corresponding to their narrow edges. The lower pivot of the polar axis turns in a cup fixed in a plate which is adjustible in the plane of the meridian by two opposing screws, one of which is represented at b, and in the transverse direction by the screw represented at a and one opposite. Above this plate there is another which is pushed from beneath by strong springs against four small wheels at the ends of four arms joined to the polar frame, by which means the weight of the instrument is in great part taken off, and the friction in the cup diminished, without impeding the movement of rotation about the axis.

268. The two terminals of the polar frame (represented in the Figure as circular plates) are in great part open frames, the upper one being such for the admission of light from objects near the Pole. The lower one carries the hour-circle graduation. These terminals are joined by the columns pp, pp, and the opposite columns qq, qq, so that the Telescope EA, with the connected circle of declination dd, may be turned between the two sets of columns about a transverse axis supported by them. The pivot at one end of this axis (that not seen in the figure) rests in a Y fixed in a cross-piece joining the columns qq, qq and that at the other end in a Y adjustible by being moved transversely to the cross-piece x, and fixed in position by the screw s. E is the eye-end and A the object-end of the Telescope; m, m are the positions of two opposite micrometer-microscopes by which the declination-circle is read off, which are of the same construction as the microscopes of the Mural Circle; and n, n are the positions of two opposite microscopes for reading off the Hour-circle. To one of the microscopes of each set a pointer is attached for setting the instrument in declination and hour-angle. Slow motion is given to the declination-

17—2

circle by tangent-screw-and-clamp apparatus attached to the polar frame at T, and worked by a handle at the command of the observer. Analogous arrangements are made for slow motion of the Hour-circle.

269. The graduation-lines of the declination-circle are 5' apart, and, for convenience, the pointer is so placed with reference to the zero of graduation, that its reading is the approximate declination or polar distance of the object. The upper face of the Hour-circle is graduated in hours, which are subdivided into minutes, and the graduation-lines on the edge, which are looked at with the microscopes, are 20" apart, one micrometer revolution is 4", or one-fifth of a graduation-interval, and each second is divided on the micrometer-head into 10 parts. Thus the hour-circle reading, which gives the hour-angle of the object from the meridian, consists of a pointer reading suitably adjusted, the number of integral revolutions of 4" indicated by indentures of the comb, and the fraction of a revolution in seconds and tenths of a second, read from the micrometer-head. In the eye-piece of the Telescope, which is an achromatic refractor of the usual construction, are a double-wire micrometer, one equatorial fixed wire, and five vertical wires. The required adjustments of the wires and wire-frame are effected by screws nearly as in the case of a Transit-circle. [For additional particulars respecting the Cambridge five-feet Equatorial, see p. xxi. of the Introduction to Vol. v. of the Cambridge Observations (for 1832)].

270. To assist in pointing the Telescope to objects, an Equatorial, especially if it be of large size, requires to be furnished with a *Finder*, which is a telescope of small aperture and large field, having its axis parallel to that of the large one, and attached either to the telescope-tube, or to the connected declination-circle. The Finder is sufficiently adjusted if an object seen at the centre of its field is seen at the same time at or near the centre of the field of the telescope.

271. The *Illumination* of the Telescope's field of view is effected by suspending a lamp in gimbals opposite one end of the transverse axis, at which a lens is fixed for throwing light from

THE EQUATORIAL. 261

the lamp on a ring reflector placed at the middle of this axis, and inclined by an angle of 45° to the axis of the telescope. Before incidence on the lens the light passes through an adjustible opening for regulating the quantity.

272. In case there should not be room for the observer's head when looking through the telescope at an object in the neighbourhood of the Pole, a diagonal eye piece is substituted for the ordinary Rasmden eye-piece, to give the means of looking at the image in a direction at right angles to the axis of the telescope. The simplest form of the diagonal eye-piece is that with two lenses described in Art. 23. The form usually adopted for an Equatorial has four glasses, the shapes and arrangement of which are the same as those of the four-glass erecting eye-piece mentioned in Art. 22. As no description of *the erecting eye-piece* was there given, I shall now describe it in detail, before proceeding to give an account of the four-glass diagonal eye-piece. Fig 41 exhibits the course, through the four glasses, of a

Fig. 41.

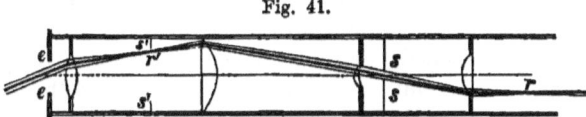

pencil of rays having its origin at r, which may either be a point of an image formed at the geometrical focus of the object-glass of the telescope, or any point of a wire at the same focal distance. The pencil, after passage through the first lens, crosses the axis of the eye-piece at ss, where it is limited by having to pass through a small circular hole in a diaphragm in that position. Then after passing through the second lens, by which it is made a convergent pencil, it is refracted by the third lens, so as to form an image of r at the point r' of a field which is bounded by the diaphragm $s's'$. This image is seen by the eye at ee by rays that are made suitable for vision by passing through the fourth glass. As the image of the point r below the axis is formed at the point r' above the axis, the inverted image formed at the focus of the object-glass is made erect by this eye-piece.

PRACTICAL ASTRONOMY.

273. The construction of the four-glass diagonal eye-piece may be explained as follows by reference to Figure 42. The first lens is at that end of the broken tube GFE which is inserted in an outer tube terminating at G in a screw by means

Fig. 42.

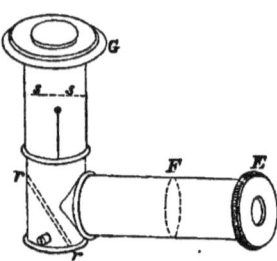

of which the eye-piece is attached to the telescope. At the position ss (corresponding to that marked by the same letter in Fig. 41), there is a diaphragm with a small circular hole for limiting all the pencils, its centre being on the axis. A pencil of rays from any point of an image at the focus of the object-glass, after passing through the first lens, the aperture in the diaphragm at ss, and the second lens a little further on (see Fig. 41), is incident on the reflector rr inclined at an angle of 45° to the axis, and being thereby turned through 90° from its original direction, passes through the two-glass eye-piece FE, forming an image between the glasses which is viewed through the eye-glass at E. The adjustment to distinct vision of the telescope-wires is effected by shifting the eye-piece bodily from or towards the wire-frame by a movement of the inner tube within the outer tube with reference to the fixed position G, and by moving the eye-piece FE inwards or outwards in the direction of the axis.

Ramsden's Refraction-piece, an example of which is attached to the eye-end of the Telescope of the Cambridge five-feet Equatorial, will be most suitably described when the corrections for refraction out of the meridian are under consideration.

II. The Adjustments of an Equatorial.

274. If an Equatorial be only used for taking differential observations (Art. 266), it will generally suffice to make an initial determination of the instrumental errors, and thereby correct them *mechanically*, without repeating the operations unless there be reason to suppose that the residual errors have considerably changed in course of time, or through incidental causes. If the object and the star with which it is compared do not differ greatly in R.A. and P.D., it may be presumed that the positions of both are for a long time affected to the same amount by *changes* of the instrumental errors, if the errors be originally small, and the instrument be steadily mounted. It is, however, an essential condition that it should never be largely out of adjustment; on which account the values of the errors ought to be determined from time to time. Such determinations can *all* be made by means of observations of *stars*, and the errors so ascertained might be corrected by mechanical means. If small instrumental errors should be allowed to remain uncorrected, their effects on the observations of R.A. and P.D. will have to be got rid of by applying corrections calculated according to appropriate formulæ. In this way celestial places may be independently determined by Equatorial observations. I propose to obtain formulæ proper for thus correcting instrumental errors by calculation, as well as to shew how to correct them mechanically. I shall also take occasion to indicate processes by which certain of the errors may be corrected exclusively by mechanical means.

275. The adjustments of an Equatorial by means of *stars* will first be considered, and the different processes will have to be indicated in a certain *logical* order.

(1) The first operation is to correct the *Index Error* of the declination-circle. In Figure 43 (1), $C\Pi$ represents the position of the axis of motion of the instrument, and C is a point through which the axis of the declination-circle may be

264   PRACTICAL ASTRONOMY.

supposed to pass; for if the two axes do not actually intersect, the reasoning will hold good if the declination-circle be conceived to be transferred bodily to a position in which its axis will pass through C. Also this axis may be assumed to be at right

Fig. 43.

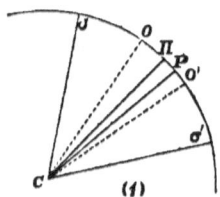
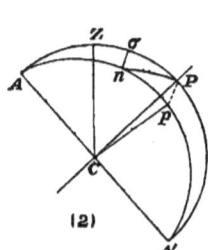

angles to $C\Pi$ on the principle that a small uncorrected deviation from that direction will not affect the determination of the index error. This being understood, suppose the circular arc $\sigma O\Pi$, which is in the plane containing $C\Pi$ and the axis $C\sigma$ of the Telescope when directed to a certain star, to be a portion of a complete circle marked by exactly the same graduation as that of the declination-circle. Then, if the mean of the two microscope readings be either zero, or 180° from zero, when the Telescope points in the direction CO, the index error to be determined is the arc $O\Pi$, supposing that the circle-reading is required to be 0°, or 180°, when the pointing is in the direction $C\Pi$ parallel to the axis of motion. In the case of the Cambridge instrument, the graduation proceeds from 0° to 360°, and the Circle-reading is 180° when the graduated face of the declination-circle is westward and the Telescope points in CO, the reading increasing as the Telescope is moved from that position *southward*. To obtain a measure of $O\Pi$ the Telescope is first directed to a star (which should be near the meridian), and the circle is read off; and then, after turning the instrument through half a revolution by the hour-circle-graduation in order to reverse the declination axis, the Telescope is pointed to the same star, and the circle is again read off.

Let O' be the position to which O is brought by the reversion, so that $O\Pi = O'\Pi$; and let the measure of the arc σO, inferred from the first reading, be D, and that of the arc $\sigma O'$, inferred from the second reading be D'. Then the measure of the index error $O\Pi$ is $\frac{1}{2}(D' - D)$. By this quantity *both* microscope readings require to be corrected to get rid of index error, the sign of the correction being determined by the relative magnitudes of D and D'. The correction is made mechanically by altering the readings of the microscopes to the required amount by screws provided for that purpose (see Arts. 175 and 176). The microscope readings are farther to be corrected so that in some *one* position of the instrument they are made to bisect opposite lines of graduation. To do this, take the microscope readings for the selected position, and supposing $180° + e$ to be the *algebraic* excess of the reading of that microscope which in the order of the graduation is in advance of the other, correct (by the screws above-mentioned) the former reading by $-\frac{e}{2}$, and the latter by $+\frac{e}{2}$. Then as the mean microscope reading will not thereby be changed, the index error will remain corrected as before, and the microscope readings will be just $180°$ apart. This will generally not be the case for other positions of the instrument; but as the effect of a displacement of the circle is *quam proxime* to increase one microscope reading just as much as it diminishes the other, the mean microscope-reading will always be nearly free from instrumental error.

When the graduation of the Circle proceeds (as in the instance of the Cambridge Equatorial) from $0°$ to $360°$, the index error might be left uncorrected and of any amount, and be corrected in the reduction of the observations as in the case of a Transit-circle. For the meridian instrument such index errors are eliminated by means of determinations of the "zenith point," whereas in the use of an Equatorial, they would have to be corrected by observations of *known* stars. It is usual, however, for greater convenience, to correct the index error mechanically, so that a *Pointer* attached to one of the microscopes, and agreeing in its indication with the reading of that microscope to the accuracy of 1' or 2', may be used for *setting* immediately for an object according either to declination or polar distance.

(2) The next step is to make the angular elevation of the axis of motion equal to the altitude of the Pole of the heavens. For simplicity it will be supposed that after actual or virtual correction of index error, the mean Circle-reading is 0° when the Telescope is pointed in a direction parallel to that axis. Then, if a *known* star be bisected when it is on, or very near, the meridian, as indicated by the sidereal clock (the error of which has been ascertained by comparisons with the transit-clock), the mean Circle-reading will be the apparent polar distance of the star affected only by error of the angular elevation of the polar axis, and the amount of refraction. The true polar distance of the star and the colatitude of the place of observation being known, the refraction can be calculated (as subsequently shewn) with *true* zenith distance for argument, and the Circle-reading be corrected accordingly. The algebraic excess of the true polar distance above the result thus obtained, which may be called the instrumental polar distance, is the correction to be applied to the latter for error of elevation of the polar axis. For greater certainty, the adopted correction might be the mean of values obtained by observations of several known stars. The mechanical correction would be effected by turning the lower screw b (Fig. 40) on which the instrument bears, in the direction indicated by the sign of the correction, and to the required amount, and then fixing the position of the pivot by bringing the opposite screw into bearing. The amount of revolution to be given to the screw for correcting a given error might be calculated by a simple proportion after measuring the interval, parallel to the axis of the screw, occupied by an integral number of threads, and the distance between the two pivots. (See Art. 171.)

(3) For finding the deviation of the axis of revolution from the plane of the meridian, the polar distance of a *known* star is to be observed when the R.A. differs from the corrected

sidereal time of the clock by 6^h, or its hour-angle from the meridian is $90°$. In that case the observation gives the instrumental polar distance with sufficient exactness whether or not the error of position of the axis in altitude be mechanically corrected. But the measurement is affected by the apparent displacement of the star in a vertical plane caused by refraction, and the correction in polar distance required on that account has to be calculated according to a formula the investigation of which will be given in a subsequent article. The algebraic excess of the true above the instrumental polar distance thus corrected is the quantity which measures the angular deviation of the axis of revolution from the plane of the meridian, the sign of the excess indicating the direction of the deviation. The error in amount and direction being thus known, the mechanical process by which the polar axis is brought into the plane of the meridian and fixed by means of the screw a (Fig. 40) and its antagonist, is exactly analogous to that indicated above for the correction in altitude.

(4) The next operation is to determine the error of collimation of the Telescope. For this purpose equatorial stars are to be selected, as the determination would otherwise be affected by error of position of the declination-axis, the mode of correcting which has yet to be indicated. The transit of an equatorial star having been taken and recorded for an arbitrary position of the plane of collimation (not far from the meridian), the Hour Circle is to be read off. Then after reversing the Telescope by changing the graduated face of the declination circle with respect to eastward and westward, and clamping the instrument, a transit of the same star is to be taken and the Hour Circle again read off. Let T_1 be the interval between the transit-times, and T_1' the difference of the Hour Circle readings. Then supposing the plane of collimation to deviate too much towards the *east* in the first position, the times recorded before the reversion will be *too early* by a certain quantity, and that recorded after the reversion will be as much *too late*. Hence T_1 will be greater than T_1', the latter being the hour-angle through which the instrument is turned from one position to the other,

and the difference $T_1 - T_1'$ will measure *double* the error of collimation. Hence the required *correction* in arc is

$$\mp \frac{15}{2}(T_1 - T_1'),$$

whether T_1 be greater or less than T_1', if the upper sign be taken to apply to a position the same as that which the instrument had before the reversion, and the lower sign to the reverse position. This error is to be corrected by moving the wire-frame.

(5) The method of correcting by star-observations the position of the declination-axis, may be explained by reference to Figure 43 (2). Let C be the intersection of the corrected plane of collimation with the axis of motion of the declination-circle; through C conceive to be drawn a plane parallel to the earth's equator, cutting the plane of collimation in ACA'; and through the same point C draw CP parallel to the earth's axis. Then the plane of collimation, being required to pass through the pole of the heavens, should coincide with the plane of the arc APA'. Suppose that it actually coincides with the plane of the arc ApA'. Then the error of position of the declination-circle which has to be corrected is the small spherical angle at A; or, if a plane through C at right angles to ACA' cut the plane ApA' in Cp, it is the small angle PCp, and is measured by the arc Pp. The value of the angle is obtained by observing as follows. Transits of a star are taken, and the Hour Circle is read off, just as for determination of collimation-error. (See (4) above.) Let T_2 be the interval between the two transits, and T_2' the difference of the Hour Circle readings. Then σn representing a small portion of the diurnal path of the star, the angle σPn, due to the error of position of the declination-axis, is equal in arc to $\frac{15}{2}(T_2 - T_2')$. Let the required arc Pp be α, and the star's polar distance $P\sigma$ be δ. Then

$$\angle \sigma Pn = \sigma n \operatorname{cosec} \delta = Pp \cos \delta \times \operatorname{cosec} \delta = \alpha \cot \delta.$$

Hence $\quad \alpha = \angle \sigma Pn \times \tan \delta = \dfrac{15}{2}(T_2 - T_2') \tan \delta.$

A star should be selected for which the factor tan δ is not large, otherwise the result may be injuriously affected by errors in the observed values of T_2 and T_2'. In general the value of δ should be less than the colatitude of the Observatory, or the transits be taken between the zenith Z and the pole P. Just as in the case of correction of collimation error, the required correction is $\mp \dfrac{15}{2} (T_2 - T_2')$ tan δ, according as the position of the instrument is the same, or not the same, as that it had before the reversion. The amount and sign of the correction being thus found, the correction may be effected mechanically by means of a screw acting transversely to the declination-axis at one end of it, as indicated by the screw s in Figure 40.

(6) The effect of the foregoing adjustments being to make the plane of collimation pass through the pole of the heavens, the last adjustment is to correct the index-error of the Hour Circle. The two microscopes (n, n in Fig. 40, p. 258) point to the opposite ends of the horizontal diameter of the Circle, the graduation proceeds from 0^h to 24^h, and the microscope-readings are supposed to *increase* as the instrument is turned *eastward*. Let the reading of the East microscope be set to its zero, as indicated by the hole of the comb and the graduated micrometer-head, and let the instrument be turned about the polar axis by the slow-motion handle till the graduation-line corresponding to 0^h is bisected by the cross-wires of the microscope. It will be assumed that the position and pointing of the microscope has already been so adjusted, that the plane of collimation is thus made to be nearly coincident with the plane of the meridian. Under these circumstances, let the transit of a *known* equatorial star be taken, and after obtaining the error of the clock by means of chronometer reference to the transit-clock, let the sidereal time of the transit be calculated. The difference between this time and the apparent R.A. of the star measures the distance of the plane of collimation from the meridian. For instance, if the sidereal time be *less* than the star's R.A. by the interval β^s, the plane of collimation requires to be turned *westward* through β^s to be placed in coincidence

with the meridian. This may readily be done with either microscope by first bisecting a graduation-mark by the cross-wires, and after diminishing the reading by turning the micrometer-head through β, causing by the slow-motion handle the same mark to be again bisected. Then, supposing the graduation of the declination-circle to be *westward*, after setting each microscope to its zero-reading, the east one is to be pointed exactly to 0^h, and the west one exactly to 12^h, by means of the screws provided for adjusting the pointings of the microscopes. By these processes the index error is corrected, and at the same time the microscope-readings are adjusted. After this the mean microscope-reading measures the deviation of the plane of collimation from the meridian *eastward*.

Mechanical Adjustments of an Equatorial.

276. The adjustment of the direction of the axis of revolution of the instrument, as having reference to the Pole of the heavens, is necessarily performed by means of observations of stars. The other adjustments have reference either to the earth or the instrument, and can consequently be effected either by stars, as already shewn, or by mechanical means. It is convenient to be furnished with means of applying the latter method in order to be readily able to secure that the instrument shall at no time be largely out of adjustment.

(1) The axis of the Declination-circle has to be placed so as to be coincident with, or parallel to, a straight line cutting the polar axis at right angles.

(a) To satisfy this condition use is made of a *swinging*, or *hanging*, spirit-level, which is held in position by two parallel supports fixed to the *nave* of the declination-circle (see Fig. 40, p. 258). Hitherto we have only had under consideration the *striding* level (Arts. 67—79); it is now required to describe the swinging spirit-level, and the arrangements whereby it is applied for adjusting the declination-axis of an equatorial. This I proceed to do by reference to Figure 44, (1), (2), (3). In (1) is

THE EQUATORIAL. 271

represented the brass tube ab, containing the level and movable about two cylindrical pivots e and e', which are fixed to the two supports so as to have their axes in coincident directions. The tube has at its two circular ends two Y's, which rest on the pivots in the manner shewn in (3). The positions of these Y's

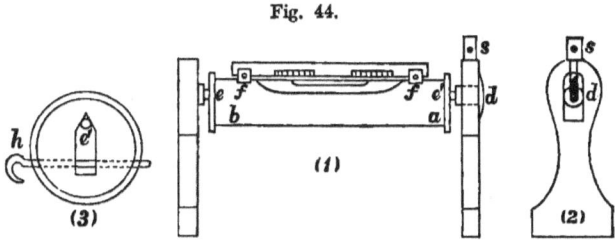

Fig. 44.

are *eccentric* with respect to the ends of the tube, so that the part ba is always downwards. A scale for reading off the bubble-ends is attached to the brass tube by means of screws ff, which fix the scale in position after it has been shifted in the direction of the axis of the tube for adjusting the level-readings. For making this adjustment it is required to be able to *reverse* the level. This can be done after drawing out the rod h (3), which detains the pivot e' within the limits of the aperture when the instrument is being turned about its axis. The tube of the level may then be reversed, and the readings of the bubble-ends for an arbitrary position of the scale may be determined for the case in which the axis of motion of the tube is *horizontal*, in the manner described in Arts. 69 and 70. It will be convenient to adjust the scale in that case so as to make the two readings equal, care having been previously taken to place the instrument so that the bubble is not far from the mid-position relative to the exposed part of the glass.

(β) After this adjustment, turn the Telescope till the Level is immediately *above* the declination-axis, move the Polar Frame till the axis of motion of the Level is placed horizontally by means of the scale-readings of the bubble-ends, and clamp the Hour Circle. That axis being now in the position AB [Figure 45 (1)], bring it to $A'B'$ by turning the Telescope in declination through 180°. Then AB and $A'B'$ are in a vertical plane and make equal angles with the declination-axis Oc, and the movement of the bubble measures the angle AOA', which is double the angle of deviation of Oc from the horizontal direction. Hence the bubble must be brought half-way back by the screw s [Figure 44 (1) and (2)], which changes the height of the pivot e' [as shewn in (2)], by which means the

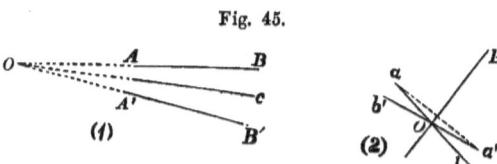

Fig. 45.

level-axis is placed parallel to the declination-axis. The aperture in which the pivot is caused to move is covered by a movable plate d.

(γ) The Level-axis and Declination-axis being now parallel, place the latter in a horizontal position by the Level, and read off the Hour Circle. Turn the Polar Frame half round till the declination-axis is again horizontal, read off the Hour Circle again, and then make the Hour Circle reading to differ just 12^h from the first reading. Suppose that by these operations the declination-axis is moved from its original position ab to $a'b'$ [Figure 45 (2)]. Then since the two positions are thus made to incline equally in opposite directions to the polar axis OP, the dotted line aa' joining the two places of any point a of the declination-axis is perpendicular to OP, and the error of position of the axis in the second position is the angle $aa'b'$, which is the half of the angle aob'. Hence the requisite correction is made by placing the instrument in a position intermediate to the two last, as inferred from the Hour Circle readings for these positions, and then putting the bubble in mid-position by the adjusting screw s (Fig. 40) of the declination-axis. Any future want of adjustment will then be indicated by the scale-readings.

(2) The Collimation Error may be corrected by means of a *mark* near the meridian as follows. The declination-axis being adjusted, bisect the mark by the middle wire, and read off the Hour Circle. Reverse the Telescope, by reversing the instrument from graduation West to graduation East or the contrary, bisect the mark again, and read off the Hour Circle. Then by means of the difference of the Hour Circle readings, point the Telescope in a direction intermediate to the two former directions, and by the wire-frame adjustment make the middle wire bisect the mark. The error of collimation of the middle wire is then corrected.

(3) Lastly, the Index Error of the Hour Circle may be corrected, after all the foregoing adjustments have been performed, by first placing the declination-axis horizontal by the swinging Level, whereby the plane of collimation is brought into the plane of the meridian, and then by the proper adjusting screws (Art. 272 (6)) making the Hour Circle reading zero.

277. Supposing all the adjustments to have been made as exactly as possible by means of stars, if the plane of collimation be placed in the meridian, and the Telescope be pointed by the declination-circle to a polar distance equal to the colatitude $+ 90°$, the line of collimation will be directed due southward, and will be horizontal. The Cambridge Five-feet Equatorial has a swinging Level attached to the Telescope-tube, with its axis of motion approximately parallel to the axis of the tube. As this level, like the one pertaining to the declination-circle, is capable of reversion (Art. 276 (1) (*a*)), let the adjustment of its axis to horizontality be performed when the Telescope's line of collimation points, as just mentioned, horizontally southward. The level-axis will thus be placed *parallel* to the line of collimation. After making this arrangement, the level-scale might be adjusted so that the readings for the bubble-ends are equal, or the bubble is put in mid-position. Then if at any future time the line of collimation be pointed horizontally southward by the graduation of the declination-circle, and the bubble be seen to have shifted from its mid-position, this will

indicate that the angular elevation of the *polar axis* has changed, and may require re-adjustment.

278. The determination of the position of the polar axis by the method given in Art. 275 (2) (3) requires the use of a graduated declination-circle. If the Equatorial has no declination-circle, and only a graduated sector (which is the case of the Northumberland Equatorial, as there will be occasion to mention subsequently), the error of position of the polar axis may be found by the following process, by means of an equatorially adjusted micrometer-wire. In Figure 46, Z is the zenith, P the

Fig. 46.

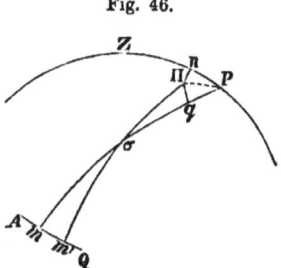

pole of the heavens, Π the instrumental pole, σ the place of a star, $ZP\sigma(\theta)$ its hour-angle reckoned positive towards the *west*, and Πn, Πq are respectively perpendicular to ZP and $P\sigma$. It is required to find the coordinates $Pn(x)$ and $\Pi n(y)$ of Π. Let $P\Pi = \rho$ and $\angle n P\Pi = \alpha$. Then $Pq = P\Pi \cos \angle \Pi Pq = \rho \cos (\theta - \alpha) = \rho \cos \alpha \cos \theta + \rho \sin \alpha \sin \theta = x \cos \theta + y \sin \theta$. For another hour-angle (θ') of the same star, the Telescope being clamped so as to prevent change of instrumental polar distance $Pq' = x \cos \theta' + y \sin \theta'$, and for a third hour-angle (θ''), $Pq'' = x \cos \theta'' + y \sin \theta''$. Then, Πq being drawn perpendicular to $P\sigma$, $q\sigma$ is very approximately the instrumental polar distance, and $Pq' - Pq$ is the difference of the instrumental polar distances in the first and second positions. This difference, which would vanish (but for refraction) if Π coincided with P, is measured by the difference of the micrometer-readings for those positions. Let, therefore, the values of $Pq' - Pq$ and $Pq'' - Pq$ when converted into arc, and corrected for refraction (by a process to be given subsequently), be respectively α and β. Then we have

$$\alpha = x (\cos \theta' - \cos \theta) + y (\sin \theta' - \sin \theta),$$
$$\beta = x (\cos \theta'' - \cos \theta) + y (\sin \theta'' - \sin \theta).$$

These two equations give for calculating x and y,

$$x = \frac{\beta \cos \frac{1}{2} (\theta' + \theta) \sin \frac{1}{2} (\theta' - \theta) - \alpha \cos \frac{1}{2} (\theta'' + \theta) \sin \frac{1}{2} (\theta'' - \theta)}{2 \sin \frac{1}{2} \sin (\theta'' - \theta) \sin \frac{1}{2} (\theta'' - \theta') \sin \frac{1}{2} (\theta' - \theta)},$$

$$y = \frac{\beta \sin \frac{1}{2} (\theta' + \theta) \sin \frac{1}{2} (\theta' - \theta) - \alpha \sin \frac{1}{2} (\theta'' + \theta) \sin \frac{1}{2} (\theta'' - \theta)}{2 \sin \frac{1}{2} \sin (\theta'' - \theta) \sin \frac{1}{2} (\theta'' - \theta') \sin \frac{1}{2} (\theta' - \theta)},$$

x being the angular deviation of the pole of the instrument from the Pole towards the *south*, and y that towards the *west*.

279. For taking observations of Right Ascension and Polar Distance with an Equatorial, the following adjustments and determinations are also required. (1) To adjust the fixed transverse wire equatorially, small stars near the equator are made, by screw-adjustment, to traverse the wire so as to be bisected by it from end to end. (2) Readings of the two micrometer-wires for coincidence with the fixed wire are taken at the five vertical wires to correct for defect of their parallelism to that wire. (3) The equatorial intervals between the vertical wires are determined by transits of Polaris, after bringing the plane of collimation to coincide nearly with the plane of the meridian. (4) The value in arc of one revolution of the eye-piece double micrometer is found by means of transits of Polaris. The modes of performing these operations need not be stated here, since they have been given in full detail in treating of the Transit and Mural Circle, or the Transit-Circle. (For examples of numerical calculations of equatorial adjustments, and determinations of micrometer-revolutions, see the *Cambridge Observations*, Vol. XI. pp. lxxiii—lxxvi. and Vol. XII. pp. xxix—xxxvi.)

III. METHODS OF OBSERVING WITH THE EQUATORIAL, AND ELIMINATION OF INSTRUMENTAL ERRORS.

280. The Apparent Right Ascensions and Polar Distances of celestial objects may be obtained by the use of an Equatorial either directly, or by the intervention of *stars of comparison*. I propose to adduce both methods, giving, first, the processes for determining the places of objects independently of known places of stars. First, the instrument has to be *set* for observing the given object. It will be supposed that the right ascension and polar distance of the object are, at least, approximately known; which will be the case except under circumstances that will be subsequently mentioned. The setting for observing in P.D. is effected by the graduation of the declination-circle and the index coincident in position with *A*, one of the two microscopes. The setting for an observation in R.A. to be taken at a chosen epoch of mean time is performed by means of the *eastern* microscope of the hour-circle, and the known error of the clock on sidereal time. By subtracting the calculated sidereal time of epoch from the approximate R.A. of the object, its approximate hour-angle *eastward* at that time is obtained. If the hour-circle index be set to this hour-angle, the setting in declination secures that the object shall pass nearly across the middle of the field, and the observation can be taken both in R.A. and P.D. if the observer has been careful to set the instrument a convenient interval before the object enters the field. The process of taking the observation is just the same as that for an observation with a transit-circle. The times of transit across the vertical wires are recorded, and the object is bisected by the fixed wire, by slow movement of the declination-circle, or optionally by one of the two micrometer-wires. By use of the three wires several objects following in quick succession, and of approximately the same P.D., may be observed without move-

ment of the declination-circle for a fresh setting. The readings of the two declination-microscopes are then recorded, just as in Circle-observations, as are also the micrometer readings (if any), and the readings of the two hour-circle microscopes.

281. The Runs of the declination-microscopes have to be taken from time to time, as well as, on the same principles, those of the hour-circle microscopes, in order to apply in both cases corrections for Runs in the reduction of the observations. Also errors of hour-angle and polar distance, due to ascertained error of position of the polar axis, have to be corrected in the reduction of the observations. Formulæ proper for calculating the amounts of these corrections may be obtained as follows. In Figure 46, σ is the place of the object, Pq is error of position of the polar axis in P.D., being equal to $P\sigma - \Pi\sigma$; and if $\Pi\sigma$ and $P\sigma$ be produced so as to cut the Equator AQ in m' and m, mm' is the error of hour-angle. As in Art. 278,

$$Pq = x \cos \theta + y \sin \theta,$$

x and y being the ascertained errors of the position of Π in altitude and azimuth. The value of Pq given by this formula is the quantity to be *added* to the observed P.D. to obtain the true P.D. Also

$$\Pi q = \rho \sin (\theta - a) = \rho \cos a \sin \theta - \rho \sin a \cos \theta = x \sin \theta - y \cos \theta.$$

Let $P\sigma = \delta$. Then the error in hour-angle, mm', is equal to $\angle m\sigma m' \cos \delta$, or $\Pi q \cot \delta$. Hence mm' (in time)

$$= \frac{\cot \delta}{15} (x \sin \theta - y \cos \theta).$$

Since when x and y are both positive, and the hour-angle is positive, or σ is on the *west* side of the meridian, the point m' is more *westward* than m, it follows that the above quantity is to be *added* to the R.A. by the observation to obtain the true R.A.

282. After applying the above corrections for the error of position of the polar axis, in case there may still be residual error requiring correction, it is proper in every instance to

observe the object in *reverse* positions of the declination-circle, that is, with graduation westward and graduation eastward; and one observation should be separated from the other by as small an interval of time as possible. As this method of observing in reverse positions tends to correct, by *mean* results, instrumental errors in various ways, it ought always to be adopted when the observer proposes to determine the places of objects by direct instrumental means.

283. In like manner formulæ might be obtained for calculating corrections of observed R.A. and P.D. for a given error of position of the axis of the declination-circle. But it is unnecessary to do this, because if this error of position be corrected mechanically in the manner described in Art. 275 (5), and if the rule of observing in reversed positions of the declination-circle be also attended to, a degree of accuracy is attained not inferior to that which would result from applying the above-mentioned corrections. So for a given error of collimation, obtained as indicated in Art. 275 (4), corrections might be calculated and applied just as in meridian-transits. But if the mean of the R.A. obtained, within a small interval, in reverse positions of the Telescope, be adopted, the effect of collimation-error will be very nearly eliminated. The only other instrumental corrections are those for reduction to the mean of the wires for omitted wires in taking the transits, and the corrections for Runs both in the readings off of the declination-microscopes and the hour-circle microscopes. The principles and the methods of applying the corrections of both kinds may be gathered from the explanations given in Arts. 116—126, and in Arts. 177—179, and need not be repeated here. When all the foregoing corrections have been carefully determined and applied, it is found that celestial places may be obtained independently by an Equatorial with a great degree of accuracy.

284. It remains to give account of the method of observing apparent R.A. and P.D. with an Equatorial by *differential* observations, that is, by comparisons with the R.A. and P.D. of a known star. This is the usual method of obtaining celestial places by an Equatorial, being, in fact, more accurate than the

independent method, inasmuch as it is more free from errors incident to the oblique position of the polar axis, and the place of the comparison-star may be ascertained at any time with as much accuracy as may be desired by means of meridian observations. The difference between the R.A. of the selected comparison-star, and the approximately known R.A. of the object compared, may be of any convenient amount; but the difference of their P.D. should be small, not, in general, exceeding two or three degrees. For observing the unknown object at or about a proposed mean, or sidereal, time, the instrument is set in the manner described in Art. 280, regard being had to the difference between its R.A. and that of the star. The hour-circle is then clamped to prevent movement about the polar axis, and the difference of the R.A. of the two objects, the difference of their P.D. being small, is considered to be measured with sufficient accuracy by the difference of the times of their transits across the mean of the wires, when bisected, one after the other, by the fixed wire.

285. The difference of their P.D. is obtained either by reading off the declination-circle and correcting for Runs, or by a measurement made by a micrometer-wire. In the former case the preceding object is bisected, as it crosses the mid-wire, by the fixed wire by means of the slow-movement in declination, the graduation-lines are then bisected by the cross-wires of the two microscopes (as in transit-circle observations), after which the declination-circle is moved for taking a like observation in R.A. and P.D. of the following object. The microscope-bisections are then read off, and new bisections are made and read off for the second observation. The observation is completed by recording the position of the graduation of the declination-circle as to eastward or westward, it being unnecessary to read off the hour-circle, because the apparent hour-angle, being the difference between the known R.A. of the comparison-star and the sidereal time of its transit across the mean of the wires, is always derivable from the observation. In the other case, that of the difference of P.D. being so small as to allow of its being measured by one of the micrometer-

wires, the preceding object is bisected, as before, by the fixed wire, and the following object by the micrometer-wire as it crosses the mid-wire. The difference between the micrometer-reading for the bisection, and that for coincidence with the fixed wire at the mid-wire, gives, after conversion into arc, the apparent difference of P.D. This method, as requiring no movement of the declination-circle, is likely to be more accurate than the other, and should be employed whenever the difference of P.D. is not too large for safe measurement with the micrometer. Even in cases of determining places of objects by differential observations, it is advisable to take observations in reverse positions of the instrument, as residual errors are thereby likely to be eliminated. The above described differential observations are much used for determining initial places of newly discovered planets and comets.

IV. FINAL CALCULATION OF RIGHT ASCENSIONS AND POLAR DISTANCES FROM EQUATORIAL OBSERVATIONS.

286. After getting rid, as accurately as possible, of all effects of instrumental errors by the processes described in Section III., it is required to apply in all instances corrections for *refraction* in R.A. and P.D., and in the case of moving bodies to correct also for *local parallax* in R.A. and P.D.

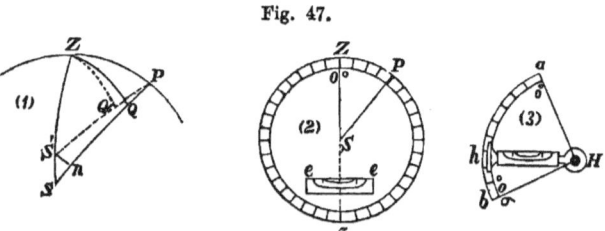

Fig. 47.

Formulæ for calculating the corrections for refraction in *independent* observations may be obtained as follows by reference to Figure 47 (1). Z is the zenith, P the pole of the heavens, S the place of object, PS its true polar distance, and ZQ is an

THE EQUATORIAL. 281

arc cutting PS at right angles. Let S' be the place of the object as altered by refraction, so that SS' is the vertical refraction. The observations, freed from instrumental errors, give PS', and the hour-angle ZPS'; and since PZ, the colatitude (l) is known, the apparent zenith distance ZS' may be calculated with these data from the triangle ZPS'. Let $SS' = r$, and $ZS' = z'$. Then if we put $r = A' \tan z'$, the factor A' may be calculated exactly by Bessel's Refraction Tables according to the example given in page LXIII. of the *Tabulæ Regiomontanæ*, or according to that given in page (iii) of Appendix I. to the *Greenwich Observations* of 1836. The vertical refraction, $A' \tan z'$, may then be obtained with as much precision as in the case of a meridian observation.

287. In Figure 47 (1), let $PS' = \delta'$, $PQ' = \Delta'$, ZQ' being perpendicular to PS', and let $\angle PS'Z = S'$. Then the refraction in P.D. $= Sn$, which without sensible error is equal to $SS' \cos S'$, or $A' \tan z' \cos S'$. But because $\cos S' = \tan Q'S' \cot z'$, refraction in R.A. $= A' \tan Q'S' = A' \tan (\delta' - \Delta')$, and Δ' is to be obtained from the equation $\tan \Delta' = \tan l \cos ZPQ'$. The refraction in R.A. $= \angle SPS' = S'n$ cosec $\delta' = SS'$ sin S' cosec $\delta' = A' \tan z' \sin S'$ × cosec δ'. But by the triangle $ZQ'S'$, $\sin Q'S' = \cot S' \tan ZQ'$, and $\cos S' = \tan Q'S' \cot z'$. Hence $\tan z' \sin S' = \tan ZQ' \sec Q'S'$. Consequently, refraction in R.A. $= A' \tan ZQ'$ cosec δ' sec $(\delta' - \Delta')$. To obtain A' from the Table, the argument z' is first to be found from the equation $\cos z' = \cos ZQ' \cos (\delta' - \Delta')$. It will be seen that the calculations might be shortened by means of tabulated values of Δ', $\tan ZQ'$, and $\cos ZQ'$, the argument being the hour-angle.

288. Before proceeding to obtain formulæ for the refraction-corrections in *differential* observations, it will be supposed that values of A', and also of $A' \tan z'$, have been tabulated for a series of apparent zenith distances separated by 5° or other convenient interval, with the view of deducing their values by interpolation for any intermediate zenith distance. In forming these tables it suffices to employ *mean* values of the readings of the barometer and thermometer as known from observations extending over one or more years, because as the difference of

the zenith distances of the two objects is small, the errors arising from employing mean instead of actual values may be assumed to be the same for the refraction-corrections of both, so as to disappear in the differences of R.A. and P.D. In the refraction-formulæ for differential observations the most convenient argument is the true zenith ZS (Figure 47 (1)), because, as the co-latitude PZ and the common hour-angle ZPS are known, this zenith distance can be calculated for the star from its known polar distance PS, and for the other object, with sufficient exactness, from its approximate polar distance inferred from the observed difference of P.D. uncorrected for refraction. Putting z for ZS, let us suppose that the vertical refraction is $A \tan z$. Then if its amount be r, we have $r = A \tan z = A' \tan z' = A' \tan (z - r)$. Hence it will be found, terms of the *second* order with respect to r being included, that

$$r = \frac{A'}{1 + A' \sec^2 z} \tan z, \text{ and } \therefore A = \frac{A'}{1 + A' \sec^2 z}.$$

If now r be obtained by interpolation from the above-mentioned table of values of $A' \tan z'$, the argument *pro hâc vice* being z (which is near enough), and then, with argument $z - r$, A' be interpolated from the other table, the value of A may be calculated from the foregoing equation, and that of r from the equation $r = A \tan z$. By the process indicated the vertical refraction may be obtained with the utmost precision; but generally it would be sufficiently accurate, after obtaining A' in the manner above stated, to take $A = A'$. On the foregoing principles a table of values of A, separated by suitable intervals of true zenith distance, might be calculated, and its value for any other true zenith distance be thence derived by interpolation. After thus obtaining $A \tan z$, the formulæ for the refractions in R.A. and P.D. become the same as those in Art. 287, after removing the dashes from A, δ, Δ, and Q. (See Fig. 47 (1).)

289. The following formulæ for calculating very small differences of refraction in R.A. and P.D., the hour-angle being given, are deduced by differentiation from the last-mentioned

formulæ for refractions in R.A. and P.D. Let a' be the apparent difference of P.D. of the two objects, as affected by refraction, and a be the true difference, for the same hour-angle. Then $a = a'(1 + A \sec^2(\delta - \Delta))$. Let β' be the observed difference of transits at the given hour-angle, and β the true difference. Then

$$\beta = \beta' - \frac{A a'}{15} \tan ZQ \sec^2(\delta - \Delta) \operatorname{cosec}^2 \delta \cos(2\delta - \Delta).$$

For greater accuracy the mean between the apparent and true P.D., viz. $\delta \pm \frac{a'}{2}$, should be substituted for δ.

290. It will be seen from the foregoing investigations that the calculations of the refractions in R.A. and P.D. would be facilitated by any ready means of obtaining with sufficient accuracy the zenith distance ZS' and the angle S'. This purpose is intended to be answered by *Ramsden's Refraction-piece* (already referred to in Art. 273), the construction and use of which I propose now to explain by means of Figure 47 (2) and (3). In (2) ZPz is a circular plate at the eye-end of the Telescope-tube, intersected by the axis of the tube perpendicularly at S, and graduated at its rim from 0° to 360°. The graduation is made to revolve about that axis by turning a milled-head-and-screw, and is read off by an adjustable index fixed to the tube. (The apparatus is, in fact, nearly the same as that of a *position-circle*.) A small spirit level ee is attached to the revolving plate, and may be placed, by turning the plate, so that the bubble is in mid-position when the Telescope is directed to any point of the heavens. By construction the plane ZSz passing through the zero of graduation is made to divide the level symmetrically in such manner that when the bubble is in mid-position the line SZ is, as nearly as may be, vertical. Accordingly the angle S' is obtained as follows. The Telescope being first pointed horizontally southward, the index at Z on the tube is adjusted to 0° exactly. Conceive, now, the Telescope to be directed to any projection of the point S on the heavens, and let PS be in the new plane of collima-

tion. Then to put the bubble again in mid-position the circular plate has to be turned through an arc equal to the change of position of the plane of collimation, which is exactly measured by the reading of the index at P on the tube. This determines the inclination of the plane of collimation to a vertical plane through the object, that is, the angle S'.

The construction for measuring the apparent zenith distance ZS' consists of a small graduated quadrant, represented by Fig. 47 (3), which is fixed by the mechanist with its plane perpendicular to the circular plate and intersecting it in a line parallel, as near as may be, to the fixed line ZSz, so that when the bubble of ee is in mid-position the quadrant is in a vertical plane. Attached to the quadrant is a small spirit-level Hh, movable about a pivot at H, and having an index at h for reading off. When the Telescope is pointed in the meridian horizontally southward, and the bubble of ee is in mid-position, the index at h should be adjusted to $90°$. Then for any other pointing, when the bubble is in mid-position, the reading of the index gives the zenith distance ZS'.

291. The reductions of equatorial observations of the R.A. and P.D. of Limbs to observations of Centres, by applying corrections for apparent semi-diameters, do not require explanations different from those given in treating of meridian observations. (See Arts. 113 and 209—213.)

292. Lastly, we have to obtain formulæ for calculating the corrections of equatorial observations of the R.A. and P.D. of bodies of the solar system for *Parallax*. In Figure 48, E is the centre of the Earth, A the place of observation, and σ the position in space of the observed body. The angular points of the spherical triangle $PZ'S'$ are situated on AP the direction of the Pole of the heavens, AZ' the prolongation of EA, and AS'' the direction in which the body at σ is seen from A. Draw AS parallel to $E\sigma$. Then the angle SAS'' is equal to the parallactic angle at σ, and AS is in the plane $Z'E\sigma$, which is the same as the plane $Z'AS'$. Hence AS cuts the arc $Z'S'$, and the parallax is measured by SS'. Now let $Z'Q'$ and Sn be drawn at right angles to PS'. Then taking EQ, the radius of

THE EQUATORIAL. 285

the Earth's Equator, to be unity, let $EA = \rho$, $\angle Z'Az$ the angle of the vertex $= \epsilon$, $\angle PAz$ the co-latitude $= l$, the arc $PZ' = l + \epsilon = l'$, and PS' the apparent P.D. of the object $= \delta'$.

Fig. 48.

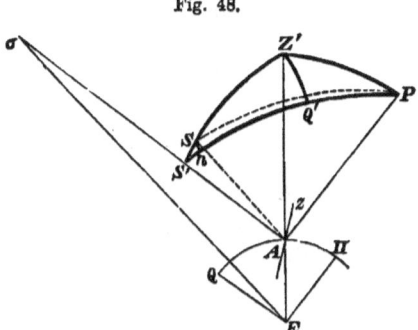

Also if P be the equatorial horizontal parallax, we have $\sin P = \dfrac{EQ}{E\sigma}$; and if P' be the horizontal parallax at A, $\sin P' = \dfrac{EA}{E\sigma} = \dfrac{EA}{EQ} \cdot \dfrac{EQ}{E\sigma} = \rho \sin P$. Putting p for the angle at σ, we have, by the rectilinear triangle $AE\sigma$, $\sin p = \dfrac{AE}{E\sigma} \sin Z'A\sigma$ $= \rho \sin P \sin Z'S'$. Then, the parallactic angle at σ being supposed small, the angle SPS', which is the Parallax in R.A., is very nearly $= Sn \operatorname{cosec} \delta = SS' \sin S' \operatorname{cosec} \delta' = \rho \sin P \sin Z'S'$ $\times \sin S' \operatorname{cosec} \delta' = \rho \sin P \sin Z'Q' \operatorname{cosec} \delta'$; and the Parallax in P.D. $= S'n = \rho \sin P \sin Z'S' \cos S' = \rho \sin P \sin Z'Q' \cot S'$ ($\because \sin S'$ $\times \sin ZS' = \sin Z'Q'$) $= \rho \sin P \cos Z'Q' \sin Q'S'$ ($\because \sin Q'S' = \cot S'$ $\times \tan Z'Q'$) $= \rho \sin P \cos Z'Q' \sin (\delta' - PQ')$. The calculations might be shortened by tabulating values of $\sin Z'Q'$, $\cos Z'Q'$, and PQ', the argument being the hour-angle. (For the Cambridge values of ρ and ϵ, see p. 227.)

293. The foregoing formulæ apply with sufficient accuracy to all the moving bodies except the Moon, the large parallax of which requires the investigation of more exact formulæ. Although it is not likely that an astronomer will have occasion

to employ an Equatorial for independent determinations of the Moon's place, I propose to obtain formulæ for the parallax-corrections in such observations, because the account of the determination of the places of all celestial bodies by an Equatorial is otherwise incomplete, and also because the same formulæ are available in calculations relating to occultations of bodies by the Moon, as will be subsequently shewn. In order to investigate the formulæ required for the present purpose, in addition to the designations by letters in the preceding Article of parts of Figure 48 it will be supposed that $ZS = z$, $ZS' = z'$, $SS' = Z' - Z = p$, $\angle ZPS = \theta$, $\angle Z'PS' = \theta'$, $PS = \delta$, and PS' (as before) $= \delta'$. Also, as in Art. 292, $\sin p = \rho \sin P \sin z'$. Now by the spherical triangles PSS' and $PZ'S$,

$$\frac{\sin(\theta' - \theta)}{\sin p} = \frac{\sin PSS'}{\sin \delta} = \frac{\sin \theta \sin l'}{\sin \delta' \sin z}.$$

Hence $\quad \sin(\theta' - \theta) = \rho \sin P \cdot \dfrac{\sin z' \sin l' \sin \theta}{\sin z \sin \delta'}.$

But since $\sin PZ'S = \dfrac{\sin \theta \sin \delta}{\sin z} = \dfrac{\sin \theta' \sin \delta'}{\sin z'}$, it follows that $\dfrac{\sin z'}{\sin z} = \dfrac{\sin \theta' \sin \delta'}{\sin \theta \sin \delta}$. Consequently, by substitution in the expression for $\sin(\theta' - \theta)$,

$$\sin(\theta' - \theta) = \rho \sin P \sin l' \operatorname{cosec} \delta \sin \theta' \ldots\ldots\ldots(1).$$

Again, $\quad \cos PZS = \dfrac{\cos \delta - \cos z \cos l'}{\sin z \sin l'} = \dfrac{\cos \delta' - \cos z' \cos l'}{\sin z' \sin l'};$

$$= \cos \delta' \cdot \frac{\sin \theta \sin \delta}{\sin \theta' \sin \delta'} + \rho \sin P \cos l'.$$

$$\therefore \cos \delta = \frac{\cos \delta' \sin z}{\sin z'} + \frac{\cos l' \sin(z' - z)}{\sin z'}$$

Hence $\quad \cot \delta - \cot \delta' = \cot \delta' \left(\dfrac{\sin \theta}{\sin \theta'} - 1\right) + \rho \dfrac{\sin P \cos l'}{\sin \delta}.$

But $\quad \cot \delta - \cot \delta' = \dfrac{\sin(\delta' - \delta)}{\sin \delta \sin \delta'}$

THE EQUATORIAL.

and
$$\frac{\sin\theta}{\sin\theta'} - 1 = \frac{\sin(\theta - \theta')\cos\frac{\theta+\theta'}{2}}{\sin\theta'\cos\frac{\theta-\theta'}{2}};$$

$$\therefore \sin(\delta' - \delta) = \frac{\sin\delta\sin\delta'\sin(\theta-\theta')\cos\frac{\theta+\theta'}{2}}{\sin\theta'\cos\frac{\theta-\theta'}{2}} + \rho\sin P\cos l'\sin\delta'.$$

Consequently, after substituting the value of $\sin(\theta' - \theta)$ from (1), it will be found that

$$\sin(\delta' - \delta) = \rho\sin P\left(-\cot\delta'\sin l'\cos\frac{\theta'+\theta}{2}\sec\frac{\theta'-\theta}{2} + \cos l'\right)\sin\delta'.$$

Hence, ϕ being a subsidiary angle calculated from the equation

$$\tan\phi = -\cot\delta'\cos\frac{\theta'+\theta}{2}\sec\frac{\theta'-\theta}{2}\ldots\ldots\ldots\ldots(2),$$

we shall have

$$\sin(\delta' - \delta) = \rho\sin P\sec\phi\cos(l' + \phi)\sin\delta'\ldots\ldots\ldots(3).$$

The equations (1), (2), (3) serve for calculating exactly $\theta' - \theta$, the parallax in R.A., and $\delta' - \delta$ the parallax in P.D. The results may be obtained conveniently by successive approximations, the known quantities θ' and δ' being put for θ and δ on the right-hand sides of the equations for a first approximation, and then with the consequent values of θ and δ proceeding to a second approximation; and so on[1]. It is evident that if the geocentric values θ and δ were given, the local values θ' and δ' might be calculated by an analogous use of the same formulæ.

294. If $S=$ the Moon's geocentric semi-diameter, and $S'=$ the apparent or measured semi-diameter, we have

$$\frac{S}{S'} = \frac{A\sigma}{E\sigma} = \frac{\sin z}{\sin z'} = \frac{\sin\theta\sin\delta}{\sin\theta'\sin\delta'}.$$

---

[1] The investigation here gone through shews that the formulæ in p. xxxiii. of Vol. XIII. of the *Cambridge Observations*, which are the same under another form as those in p. 23 of the Syllabus of my Astronomical Lectures, are not sufficiently accurate. The formulæ I have given (p. 298) in the Article which constitutes the Appendix to the Nautical Almanac for 1854 are correct.

After the foregoing calculation of θ and δ, the value of S may be calculated from this formula when S', θ', and δ' are given. Means of obtaining S' by equatorial observations have been indicated in Arts. 110 and 125. Accordingly, an Equatorial may be used for obtaining by means of observations of the Moon, when known to be full, values of her geocentric semi-diameter. The same formula will give S' when S, θ, and δ are known from Tables, and the local values of θ' and δ' have been calculated as above stated. [The local values θ', δ', and S' are required in the *Calculation of Occultations*.]

I have now gone through all that relates to the determination of celestial positions by an Equatorial. There are various other purposes for which this instrument may be employed, the consideration of which will be most conveniently entered upon after giving descriptions of it under other forms of construction and modes of mounting.

THE NORTHUMBERLAND EQUATORIAL OF THE CAMBRIDGE OBSERVATORY.

295. This powerful instrument, which at the time it was constructed was one of the largest of its kind, was presented to the Cambridge Observatory by the Duke of Northumberland, High Steward of the University 1834—1840. The erection, as regards both the Mounting and the Dome, was begun by Mr Airy in 1835, and completed under his superintendence after his appointment at the close of that year to the Greenwich Observatory. In 1844 he produced, as a separate publication, an account, illustrated by engravings, of the parts and appendages of the Instrument and Dome, which also forms an Appendix to Vol. xv. of the Cambridge Observations (1843). Any one desirous of acquiring full information respecting the details of this undertaking, which was, I believe, the first instance of the successful mounting of so large an Equatorial,

must have recourse to that "Account." For the purpose of the present work it will suffice to state the following particulars.

296. The object-glass is one of Cauchoix's: its effective aperture is $11\frac{1}{2}$ inches. All the brass-work, the graduations, and the eye-pieces, were executed by Simms. The focal length of the Telescope is $19\frac{3}{4}$ feet. The mounting is similar to that of the Five-feet Equatorial. The Polar Axis consists of six stout deal poles, the ends of which are fastened to two six-sided cast-iron frames, at the centers of which are the upper and lower pivots. The poles are braced across their middle by transverse iron bands, counter to which there are 24 deal spars, crossing each other two and two, and abutting near the middle of the poles so as to thrust them obliquely outwards, being made to act thus by means of screws which turn in shoulders fixed to their opposite ends and press against the iron frames. This apparatus answers the double purpose of giving stiffness to the Polar Axis, and adjusting the iron frames so as to be perpendicular to the axis of revolution of the instrument. The support of the upper pivot consists of two strong wooden beams connected by two cross iron bars, and surmounted by a triangular iron frame, at the apex of which is the Y for the pivot to rest in. The beams are inclined towards each other in a plane perpendicular to that of the meridian and deviating a little from the vertical towards the south: their lower extremities, which are armed with iron off-sets, are firmly embedded in massive brick-work, and their narrow faces are turned towards the middle point of the axis of revolution. By this construction the view of no part of the heavens is materially obstructed; Polaris and its companion can be well seen through the upper iron frame. The support of the lower pivot is a large stone slab resting on a deep mass of brick-work rising to a small elevation above the floor of the Dome. The pivot turns in a socket carried by a square mass of iron which is movable by adjusting screws to allow of bringing in the usual way (Art. 275 (2) and (3)) the axis of the instrument into coincidence with the axis of revolution of the heavens. This mounting is found to be remarkably steady.

297. The tube of the Telescope is made of well-seasoned deal. The pivots of the declination-axis, which turn in two hollow cylinders each formed of two brass pieces screwed together, are supported by two opposite poles of the Polar Frame in such manner that *one* of the pivots admits of being adjusted by screws so as to place the declination-axis perpendicular to the polar axis. (See Arts. 275 (5) and 276 (1)). The Telescope-tube just passes between the other four poles. Attached to one side of this tube is a flat brass bar nearly 6 feet in length, carrying at one end a small graduated arc perpendicular to its length, and turning at the other about a pin fixed to the Telescope-tube at the distance of $2\frac{1}{2}$ feet from the declination-axis. This is called the *Declination-Sector*. The graduation of the Sector is read off by a micrometer-microscope fixed to the Telescope-tube. For directing the Telescope to any required P.D., graduated brass rods of different lengths are provided, the graduation of which was effected by the intervention of an adjusted setting-circle attached to the tube, the Telescope being moved in the plane of the meridian, and marks being engraved on the rods (clamped as below stated) corresponding to different selected readings of the setting-circle. When each mark was drawn, the microscope pointed to a fixed reading about the middle of the sector-graduation. The marks were drawn at the edge of a cylindrical sliding-piece, capable of being clamped to the rod, which edge, consequently, afterwards served as index for setting according to this graduation. The rod proper to be used for a given P.D. was attached at its extremity, by means of a clamp, to a pin fixed in one of the poles of the Polar Axis, and by the above mentioned sliding-piece, previously clamped to the rod with its edge at the given P.D., it was similarly attached by a pin to the declination-sector. Near this pin, and carried by the sector, is a combination of one toothed wheel and two pinions turned by a milled head and acting on rack-work fastened to the Telescope-tube. As the sector is made stationary by the rod, the effect of turning the milled head is to cause the Telescope to move in declination. In

setting for an object the sector-reading has always to be brought by this movement to the above mentioned fixed reading; which is readily done, and with sufficient accuracy, by bringing a pointer fixed to the sector to the adjusted position of a zero attached to the tube. The rods also answer the additional purpose of steadying the Telescope when used for observing.

298. The construction above described is evidently adapted for measuring differences of P. D. which, although small (not exceeding for instance $1\tfrac{1}{2}°$), are too large for micrometer-measures. For this purpose I frequently employed it. The value in arc of the interval between the sector-divisions was obtained by measuring with the sector the difference of P. D. of two stars, and comparing the result with the difference as determined by reduced meridian observations, stars being selected whose difference of P. D. was not far short of the whole range of the sector. (See infra Art. 303.)

299. The Hour Circle is $5\tfrac{1}{2}$ feet in diameter. It is not permanently attached to the Polar Axis, but can be clamped to the lower iron frame at pleasure. There are two adjustible indices with verniers, pointing to the graduation of the hour-circle, one supported by the pier in or near the meridian on the south side of the polar axis, and the other fixed to the iron frame. The position of the latter may be adjusted by first setting the former to true sidereal time while a regulated clock is turning the detached hour-circle, and after pointing the Telescope to a known bright star so that the star is seen in the middle of the field, clamping the hour-circle to the instrument, after which process the hour-circle and Telescope are both carried by the clock (as explained below) according to the rate of sidereal time. The movable index should then point to the *right ascension* of the star. Accordingly, it has to be placed and adjusted so as to point exactly to that R. A. Or, supposing the movable index to be fixed *approximately* in position, the instrument detached from the hour-circle might be moved so that this index points to the R. A. of a known star, and then after clamping the hour-circle to the frame, pointing to the star,

19—2

292   PRACTICAL ASTRONOMY.

and setting the clock going, the fixed index might be adjusted to true sidereal time. After adjusting the indices, the pointing of the Telescope in R. A. for any star is conveniently effected by first setting the fixed index to sidereal time, and then setting the movable index to the R. A. of the star.

300. The graduation of the Circle has been performed with great care, and by the aid of the verniers can be read off with the naked eye to 1^s of time[1]. The outer rim of the Circle is cut with teeth to which an endless screw, connected by a brass rod with the clock, can be applied at pleasure, for the purpose of giving movement to the instrument about the polar axis. The clock is kept in motion by a heavy weight, the force of which is made to maintain the motion during the act of winding up. The clock's rate is under command, and can be regulated to sidereal time nearly. The clamping of the Hour-Circle to the iron frame is effected by a tangent-screw clamp fixed to the frame, by means of which, with the aid of a handle extending to the place of the observer, he can, when the endless screw is applied, give motion to the instrument, through a limited space, upon the hour-circle. The rate of motion given to the Circle by the clock is not affected by this movement. Hence supposing the Circle to be moved exactly according to sidereal time, small differences of R. A. can by this contrivance be measured by reading off the pointings of the movable index before and after the change of position. In any case correct measures may be obtained by allowing for the rate of motion of the Hour-Circle as inferred from the difference of hour-circle readings for two bisections of a star and the sidereal interval between. It will thus be seen that this Equatorial is adapted for measuring small differences both of R. A. and P. D. in differential observations. It is advisable, in every instance in which it can be done, to begin and end the series of measures with a bisection of the comparison star.

301. The foregoing descriptions will be made clearer by

[1] For greater precision in measuring differences of R.A., I added a graduated sector, analogous to the declination-sector, and arranged so that it could be clamped at pleasure to the hour-circle and be read off by a microscope.

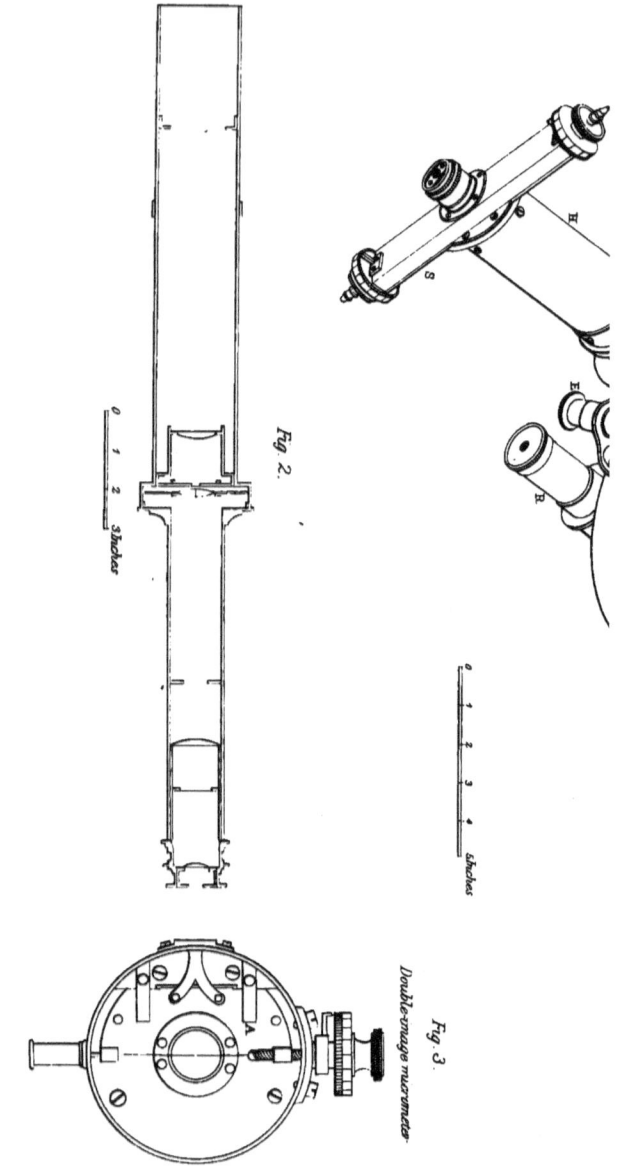

Fig. 2.
Fig. 3. Double-image micrometer

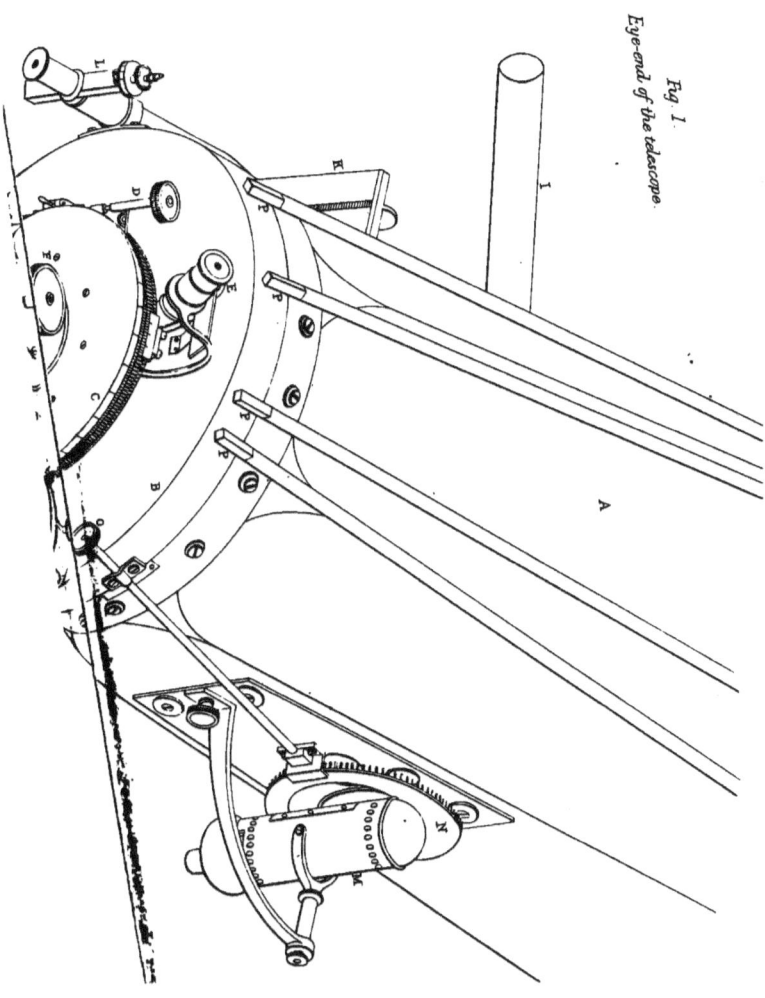

Fig. 1.
Eye-end of the telescope.

referring to the annexed Plates IV. and V. (which are the same as XVI. and XIX. in the "Account, &c."), and to the following explanations of the instrumental arrangements.

Plate IV., Figure 1, represents the eye-end of the Telescope with its double-wire micrometer mounted, as viewed from the side opposite to that to which the brass bar is attached. A is the wooden tube, and B a brass breech-piece. C is a position-circle, D its slow-motion-screw and clamp, and EE are microscopes for reading its divisions. F is the head of a pinion for adjustment to focal length, and G is the tube which is slid inwards and outwards by the action of F. H is an eye-piece inserted in G, carrying the double-wire micrometer S, and held by friction only. All the different eye-pieces are thus inserted in G. I is a declination-rod. K is the graduated arc of the brass bar, and L the micrometer-microscope for reading it. (See Art. 297.) M is the lamp for illuminating the field of view. N is a circular plate, turning on a screw at its center, and having a snail-shaped hole for limiting the aperture through which the light of M enters the side of the Telescope. The light is incident at an angle of 45° on a gilded face of an elliptical ring, through which the beam of light from the object-glass passes without interference. O is a milled head for turning a pinion whose teeth work in the teeth of the wheel N. $PPPP$ are the square ends of the rods by which the screws effecting the adjustments of the object-glass are turned. (See Arts. 17—19.) Small keys are provided which fit on these square ends; but as they are seldom required, they are not usually mounted. Q is a half-seconds chronometer fixed by screws in a cell; its winding-up-key may be seen projecting from one side. By the use of this chronometer there is no necessity for reference to a clock, which, with a Telescope of such a length, would be inconvenient; and consequently no clock is provided in the room[1]. R is the finder; its focal length is 28½ inches, and aperture 2¾ inches.

[1] In order to avoid exposing the eye to variations of light in observing very faint objects, I adopted the plan of having the seconds counted audibly from a chronometer by an assistant shut up in an adjoining closet.

294   PRACTICAL ASTRONOMY.

302. Plate V. is a north-and-south section of the building and instrument, intended to give a general view of the principal parts of every kind in combination, excepting the machinery of the shutters (for explanation of which see the Plate VII., and the descriptions in pp. 15—18, contained in the before mentioned "Account of the Equatorial and Dome"). It exhibits the interior appearance of the enclosing octagonal wall, of which WW and W are north and south vertical sections; the interior appearance of the Dome (MN); the fixed south steps (U), for observations of objects near the Pole northward; the clockwork box (m, m) to the east of the steps; the support of the upper pivot, consisting of two wooden beams, surmounted by an iron triangle (see Art. 296), one beam being represented by T, and one side of the triangle by S; the adjustible support (p) of the lower pivot, with a friction-wheel acting by a strong spring to diminish the wear of the pivot; the graduated hour-circle hh (traced by dotted lines); the general structure of the Polar Frame (see Art. 296); the bearing-piece for fixing the position of the declination-axis (Art. 297); the Telescope as mounted with its Finder; a declination-rod (a) in position, with its movable clamp attached (Art. 297); the brass bar with sector fixed at its end, together with the microscope for reading the sector's graduation, and the rack-work fastened to the tube, by means of which motion is given to the Telescope in declination by turning a milled-head (see Art. 297); the chair-frame and chair, movable in circular grooves by a windlass at the back of the frame, n, n being additional supports of the observer's head when the object is near the zenith and the chair cannot be used, and a separate seat being provided suitable for supporting the observer when he has to take a position between the poles of the polar frame. The Plate also exhibits the observer in the upper chair holding in his hand the long handle (b) by which slow motion in hour-angle is given to the polar frame relatively to the hour-circle (Art. 300), and having at his command the handle (e), by which motion is communicated to a wheel (f) whereby rotation is given to the chair-frame about a central pivot (o), as well as the handle (d),

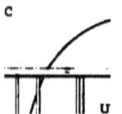

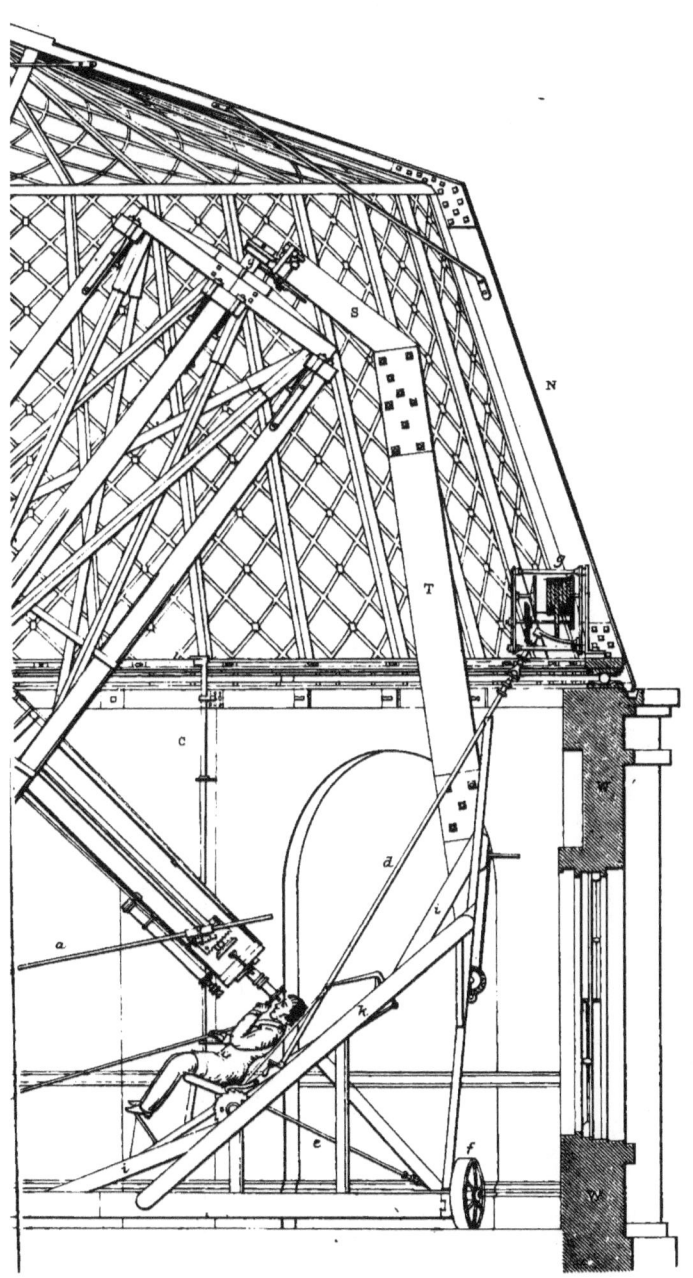

which by means of the apparatus at g gives motion to the Dome. C, C are two of six holdfasts by which the Dome is retained in any required position; at r are pulley-lines sustaining the weight, which is the moving power (Art. 300), hanging in a well below Q; y is a spring which being pressed by the key in winding up, moves a click whereby the action of the weight is continued during the process (Art. 300); w is a small weight, which by machinery connecting it with the endless screw acting on the hour-circle, was occasionally applied to supplement the moving power of the clock. The Dome rests on six cannon balls in a circular groove. For turning it through large spaces a lever is provided with fulcrum fixed to the wall, the upper part of which is made by force of hand to act on pins t, t, t fixed to the curb of the Dome.

It will thus be seen that all the movements during the continuation of a series of observations are under the control of the observer, without the necessity of his leaving his position in the observing chair. As, however, it is hardly possible for the observer by himself to perform with desirable ease and rapidity all the operations required in the manipulation of so large an instrument, I found it preferable to arrange that some one should always be present to assist in turning the Dome with the lever, rotating by hand the chair-frame about its pivot, shifting the observer's seat by the windlass, and recording, generally, the observations[1].

303. The value in arc of one revolution of the double-wire micrometer mentioned in Art. 301, (the value being assumed to be the same for both wires), was determined by means of transits of δ Ursæ Minoris and Polaris in the manner described in Art. 59 relatively to the micrometer of the Transit-instrument, the transits being taken at positions of each wire separated by a large number of integral revolutions. The adopted result was, $1^r = 16'',970$. The value in arc of the

[1] The large Equatorial of the Greenwich Observatory, the mounting of which is of the same kind as that of the Northumberland Equatorial, is described in full detail in an Appendix to the Volume of Greenwich Observations for 1868.

interval between consecutive divisions of the sector-arc (1^d) and the value of one revolution (1^r) of its microscope-micrometer (see Arts. 297 and 298) were obtained by measuring with the sector the difference of P.D. of two stars, λ Pegasi and μ Pegasi, eliminating the effect of refraction, and comparing the result with the difference of P.D. of the stars accurately known by meridian observations. (The difference of P.D., about 62', is considerably greater than any range of the arc ever employed.) The measure was obtained in integral division-intervals, together with integral micrometer revolutions and decimal parts of a revolution, and the value of one division-interval in micrometer revolutions was found by taking the mean of repeated measures of that interval by the micrometer. Thus the difference of P.D. of the stars could be expressed in integral division-intervals and a fraction of such interval, and be compared with the known difference of P.D. in seconds of arc. The means of numerous measures taken in this way gave $1^d = 204'',258$ and $1^{r'} = 10'',178$, which values are adopted in the reduction of differential equatorial observations taken by means of the sector. (The details of the measures for determining the values of 1^r, 1^d, and $1^{r'}$, are contained in pp. xxxi—xxxv of the Introduction to Vol. XII. (1839) of the Cambridge Observations.)

The Counterpoise Mounting of an Equatorial.

304. Figure 49 represents a wooden model, used in my Astronomical Lectures, which, although somewhat rude, may serve for describing the counterpoise mode of mounting an Equatorial. The main principle of this mounting consists in placing the Telescope on one side of the polar axis and counterbalancing it by a weight on the other side. The frame *f* represents the pier, or other kind of construction, on which the instrument rests. The position of the polar axis is indicated by the cylinder *a*, having an upper and a lower support, and capable of being turned about its axis of figure, whereby rotation is given to the Telescope and the Declination-axis. Near the lower end of the polar axis is the hour-circle *h*, which

is read by two opposite microscopes *nn*. The Declination-axis, the direction of which is indicated by *bb*, is fixed transversely to the polar axis, and carries at one end the Telescope *AE*, and towards the other the counterpoise *C*. The declination-circle

Fig. 49.

d is read off by the two opposite microscopes *mm*. The attachment of the Telescope to the declination-axis is nearer to *E*, the eye-end, than to *A*, the object-end, on which account two counterpoises *cc*, fixed to the ends of rods, are made to counterpoise the superior weight of the object-end.

305. This form of mounting has the advantage of giving to the eye-end of the Telescope a range of motion of less radius than that in the case of the Telescope being supported at its middle point, and on this account, as well as because it does not require the movement of a large polar frame for setting the Telescope, it is less laborious to the observer than the other mounting. At the same time it demands a Dome of larger capacity to secure that the object-glass shall be under cover in all positions of the Telescope, and, probably, a large telescope thus mounted is more liable to tremors than is the case with

the mounting of the large Equatorials of the Greenwich and Cambridge Observatories. Also with the counterpoise mounting it is not possible to have a large hour-circle without unduly weakening the support of the polar axis; and there is, besides, the disadvantage that for some movements of the Telescope across the meridian, especially when the pointing is between the Pole and the Zenith, the eye-end of the Telescope comes into contact with the support of the instrument. In that case, in order to point the Telescope continuously, it is necessary to reverse it so as to bring it to the opposite side of the polar axis. Means of adjusting by screws the positions of the axes of motion and the microscope readings are, of course, provided in this as in the other kind of mounting[1].

306. Figure 50 represents another mode of counterpoise mounting, the peculiarity of which is, that a cylindrical support PP, the axis of figure of which is directed to the pole of the heavens, is firmly attached to a vertical wooden support in such manner that the Telescope AE and its connected apparatus, together with the counterpoise C, may be made to revolve about the polar axis, as well as about the declination-axis, which is the axis of figure of the cylinder DD, without the possibility of encountering any obstacle, to whatever part of the heavens the Telescope be pointed. This construction is, therefore, especially adapted to *sweeping* for objects whose R.A. and P.D. are only very roughly known (these circumstances are alluded to by a remark in Art. 280, p. 276). A is the object-end, E the eye-end, and F the finder, of the Telescope; dd is the declination-circle, which, by means of a hook's joint and handle, may be fixed by the clamping-piece at l; hh is the hour-circle, which in like manner may be fixed by the clamping-piece at k; c is a leaden weight which, by means of the slit and screw at x,

---

[1] The counterpoise mounting of Fraunhofer's 14-feet refractor at Dorpat is represented in Plate VII. of Pearson's "Practical Astronomy," and the same plate is given in Coddington's Optics, Part II. The mounting of the Sheepshanks Equatorial of the Greenwich Observatory, which for the most part is the same as the counterpoise mounting, is described in the Introductions of the Greenwich Volumes of Observations, as, for instance, in pages xviii and xix of the volume for 1875.

THE EQUATORIAL.

Fig. 50.

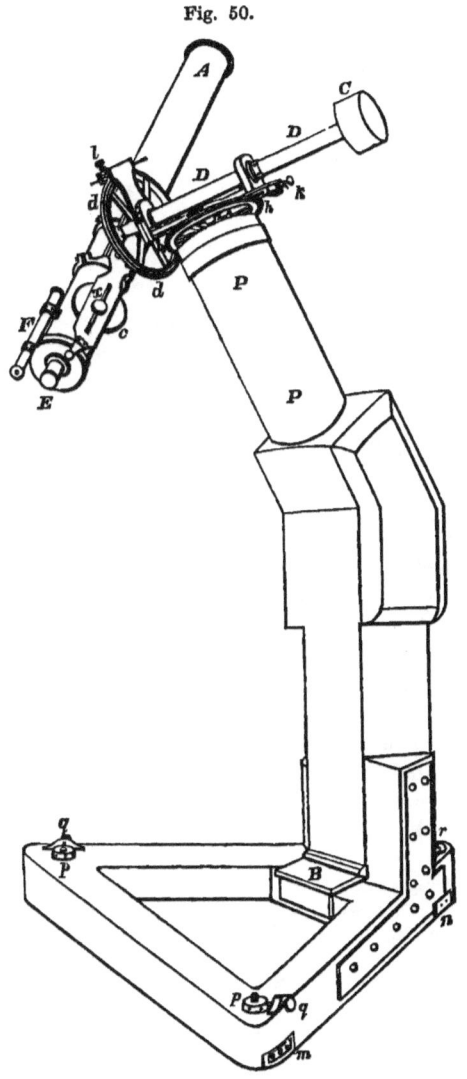

may be shifted and clamped so as to counterpoise the Telescope-tube. The upright portion of the support is attached at its base to a triangular frame, which is adjustible by vertical screws at p, p, r, and by horizontal screws at q, q. The vertical screws are turned by spikes inserted at m and a corresponding position (not seen in the figure), and at n. The screws q, q, acting against the capstan heads p, p, turn the frame about the vertical axis rn. After the adjustment of the polar axis is effected, the three capstan heads p, p, r are firmly screwed down. By having at command the vertical movements at p, p, r and the horizontal movements at q, q, the polar axis may be adjusted so as to be in the plane of the meridian and point to the Pole. The box B covers two small adjusted spirit-levels, one for levelling in the plane of the meridian, and the other for levelling in the transverse direction. The bubbles, supposed to be put in mid-positions after adjusting the polar axis, will afterwards indicate by change of their positions, whether a re-adjustment of the polar axis be required[1].

307. In making use of an Equatorial in sweeping for an object recognisable by its physical aspect, such as a comet, or a nebula, the Telescope should be clamped in R.A. at an arbitrary epoch so as to point considerably *westward* of the most probable hour-angle at that epoch of the object sought for, and be moved up and down on each side of the most probable circle of declination, till the object is detected in the field of view by its appearance. If the object be a small planet not distinguishable in appearance from a minute star, and a probable place be given, the sweeping may be effected by noting down the places of all likely objects in contiguous zones

---

[1] The stand represented in Fig. 50 was placed on a substantial brick wall, raised to a convenient height above a flat area at the top of the south-west angle of the Cambridge Museum Buildings. This wall, for the sake of steadiness, stands on a separate base, so as to be unconnected with the structural supports of the building. The mounting is so arranged that the Telescope (which is the 4·6-inch Dollond of the Cambridge Observatory), can, with its attachments, be readily dismounted when not required for use, and be deposited in an adjoining apparatus room, or, if left on its stand exposed to the open air, be protected by a suitable waterproof covering.

of declination of a certain length, the central point of the aggregate zone being the given place, and after a time sweeping over the same area, to see whether any star in it has changed position, or any new object has entered within its limits. To render this mode of sweeping sure and rapid, the instrument should be carried by clock-movement. On these principles I searched for the Planet Neptune in the year 1843 with the Northumberland Equatorial. (See Appendix No. 1 to Vol. XV. of the Cambridge Observations.)

Other Observations made with an Equatorial.

308. In addition to the use of an Equatorial for determining celestial positions, there are various kinds of observation for which this instrument is adapted, which we may now proceed to take into consideration. One kind, of special interest, consists of observations of the planet Mars in opposition, made for the purpose of determining his parallax, and thence inferring the parallax of the Sun. The determination of Mars's parallax is made by means of measures of small differences of declination between the planet and selected stars, corresponding measures being taken at places differing little in longitude, but considerably in latitude, such as Greenwich in the northern hemisphere and the Cape of Good Hope in the southern hemisphere. From a comparison of the differences of declinations for the same stars as observed at the two places at noted times, and reduced to the same epochs by taking account of the Earth's diurnal motion and the apparent motion of Mars, we may calculate the differences in P.D. of simultaneous directions of the two lines joining the North or South Limb of the planet and the places of observation, and hence infer, with sufficient approximation, the angle these lines make with each other at a given time. This angle is subtended by the line joining the two places, the length of which, together with its angles of inclination to the above two directions, may be inferred from the known dimensions of the earth, the known latitudes and longitudes of the places of observation, and

from the data furnished by the observations. The same angle is subtended by the perpendicular from one place of observation on the visual direction of the observed Limb from the other, the length of which is readily inferred from the previous results. Hence the angle subtended at Mars by the earth's equatorial radius, that is, his parallax at the time, may be at once calculated. It is desirable that as often as possible the differences of declination should be *micrometer* measures, and that the N. and S. Limbs of the planet should be taken alternately. A list of stars proper for use in this kind of observation is given in successive Nautical Almanacs. After thus obtaining Mars's parallax at a certain time, and allowing for the effects of the eccentricities of his and the earth's orbits, the Solar Parallax may be calculated from the known relation between the planetary mean distances and periods.

309. There is another method of taking observations of Mars in opposition for determining his parallax; namely, observing his displacement in R. A. when he is far east of the meridian, and far west of the meridian, as seen at a *single* Observatory. Conceive that by the diurnal and orbital motions of the earth an Observatory is carried from the position A to the position A', *in space*, in a certain time, and that the E. or W. Limb of the planet is compared with a suitably situated star in the first position, and with the *same* star in the other position. The observations give the means of determining the astronomical directions of the Limb from the Observatory in the two cases. The known effect of the apparent motion of Mars in producing the difference between these directions being eliminated, there remains the angle subtended by the straight line joining two positions the distance between which is wholly due to the earth's rotation; and since the length and position of this line can be calculated, the solar parallax may be inferred in a manner analogous to that stated in Art. 308[1].

[1] In an Article contained in the Monthly Notices of the Astronomical Society, Vol. xvii. No. 7, p. 220, the Astronomer Royal expresses the opinion that this is the best of all the methods of determining the solar parallax. The same Article contains a discussion of various modes of calculating this funda-

310. By observations with an Equatorial of the elongations of a satellite from its primary, the mass of the latter may, according to the principles of physical astronomy, be calculated. In this way, by observations with the Cambridge Five-feet Equatorial of the elongations of Jupiter's fourth Satellite, Mr Airy obtained a correction of the previously adopted mass of the planet. The details of the observations are given in Vols. V.—VIII. of the Cambridge Observations (for the years 1832—1835), and the calculation of the results in the Memoirs of the Astronomical Society, Vols. VI., VIII., IX., and X.

311. Measures of the Moon's Diameter by transits of the First and Second Limbs, when the Moon is exactly or nearly full, have already been under consideration in Arts. 110 and 125. The use of an Equatorial for this purpose has the advantage, generally, of giving the means of taking a series of such measures while the Moon is so nearly full that no calculations of corrections for defect of illumination are required.

312. An Equatorial is suitable for observing an *Eclipse* of a star, or a planet, or the Sun, by the Moon, these phenomena requiring, with very rare exceptions, observations out of the meridian. The observation of an Eclipse or Occultation of a *Star* consists in simply recording, with as much accuracy as the circumstances permit, the instant, as shewn by a time-piece, of the disappearance, or reappearance, of the star at the Moon's dark or bright Limb. The noted time has to be reduced to the true sidereal time of the place of observation; whence, on an assumed value of the Longitude of the place, the corresponding Greenwich Mean Time is inferred. The calculations required for obtaining results from observations of occultations of stars may be conducted as follows.

313. From data furnished by the Nautical Almanac, the Moon's Geocentric R.A. and P.D., the equatorial horizontal parallax, the geocentric semidiameter, and the star's apparent R.A. and P.D., may be obtained for the given Greenwich Mean

mental quantity, more especially that by means of observations of transits of Venus across the sun's disk.

Time. By means of these results the Moon's *apparent* R.A., P.D., and semidiameter have to be obtained by allowing for parallax according to rules which have been already investigated (see Arts. 292—294), and the apparent distance of the star from the Moon's centre has also to be calculated. *Each* of the quantities on which the calculations are based is supposed to require correction. Let the corrections of the Moon's tabular R.A. and P.D. be respectively x and y, and the Moon's true horizontal parallax at the place, and her true geocentric semidiameter, be equal to the calculated values multiplied respectively by $1 + 0{,}001m$ and $1 + 0{,}001n$. Also let e and f be the corrections of the star's R.A. and P.D., t a correction of the noted time of observation, τ the correction of the assumed longitude, and ν a correction which the assumed geocentric colatitude of the Observatory may require. Then if we call the Moon's calculated apparent R.A., P.D., semidiameter, and distance of the star from her centre respectively R, λ, S, and D, and their true values $R + \delta R$, $\lambda + \delta\lambda$, $S + \delta S$, $D + \delta D$, the values of δR, $\delta\lambda$, δS, δD admit of being expressed in linear functions involving the unknown corrections x, y, e, f, m, n, t, τ, and ν. A process for obtaining these expressions, and formulæ for calculating the numerical values of the coefficients of the corrections they severally involve, are given in the Appendix to the Nautical Almanac for 1854. By equating the true apparent distance of the star from the Moon's centre to the Moon's true apparent semidiameter (Art. 294), an equation of condition is obtained containing all the unknown quantities. Every observation of a disappearance, or reappearance, furnishes one such equation, and from the mean results of the calculation of a large number of observations, exact values of the corrections may be inferred. It may generally be assumed that x, y, e, and f are known from meridian observations. Supposing also m, n, and ν to have been sufficiently determined, and the recorded time to be so exact as to allow of assuming the correction t to be zero, the final equation gives the correction τ of the assumed longitude of the Observatory.

314. In observations of the eclipse of a *Planet* by the

Moon, four notes of time may generally be taken, namely, the times of external and internal contacts of limbs at disappearance, and those of internal and external contacts at reappearance. In observing the eclipse of either a star or a Planet, the observer should take the precaution of putting an equatorially adjusted wire on, or very near, the object before the disappearance, that he may have some guide as to where to look for it at reappearance, otherwise the instant of its first becoming visible might be missed. The phenomena at the Moon's dark limb generally admit of more accurate notes of time than those at the bright limb, the brightness of the limb causing some uncertainty. In forming the final equations derivable from the observations, e and f will represent the errors in R. A. and P. D. of the Planet's tabular place, and in addition to the terms containing m and n relating to the Moon, there will be terms containing m' and n' for corrections of the Planet's assumed horizontal parallax and semidiameter.

315. The observations and calculations applicable to an eclipse of the *Sun*, are analogous to those above mentioned relative to a planet, as far as regards the times of external and internal contacts; but the large disk of the Sun generally allows of taking in addition several series of differential transits and differential micrometer-measures. The differential transits may be taken between cusp and cusp, or between a cusp and the first or second limb of Sun or Moon, or between a limb of the Sun and a limb of the Moon. The micrometer-measures of differences of P. D. may be taken between cusp and cusp, or between a cusp and the N. or S. Limb of Sun or Moon, or between either the two N. limbs, or the two S. limbs, each micrometer-measure being accompanied by a note of the time. Each such differential quantity admits of being calculated from data furnished by the Nautical Almanac, and supposing that these data may all require certain unknown corrections, the true differential quantity may be expressed in a linear function involving these corrections, by equating which function to the observed measure, there results an equation of condition between the corrections. In this manner, when numerous differential measures have

306   PRACTICAL ASTRONOMY.

been observed, a large number of equations of condition may be obtained, which, being treated by the method of least squares, may serve to give accurate values of the unknown quantities[1].

316. Another class of observations, of much interest, namely, determinations of the relative positions of the components of *double, triple, and multiple stars*, are greatly facilitated by the use of an equatorially mounted telescope, carried by clock-movement. (The arrangements for this purpose pertaining to the Northumberland Equatorial may be gathered from Plates IV. and V., and the accompanying descriptions in Arts. 301 and 302.) These observations are made by means of a *position-circle* (C in Plate IV.), which is a small graduated circle fixed near the eye-end of the telescope, and capable of being turned by a tangent-screw, or by hand, about the telescope-axis. The graduation proceeds from $0°$ to $360°$ and can be read off to the accuracy of $1'$ by a single vernier, or, if great accuracy be desired, by two opposite verniers. The angular and linear measures may be taken either by means of a double-wire micrometer (Arts. 31 and 32), or a double-image micrometer (to be presently described), movable with the position-circle. The mode of using the former will be first stated.

317. In Figure 51 (1) and (2), AB and ab represent positions of the two parallel micrometer-wires. The line $BbBb$

Fig. 51.

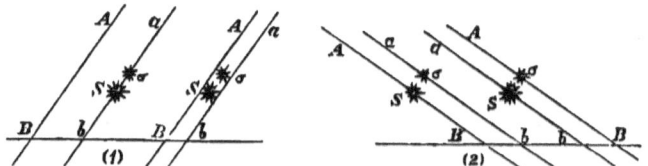

in (1), and $BbbB$ in (2), indicate the equatorial direction, which is determined by turning the position-circle till one of the component stars, S or σ, travels, the instrument being fixed,

[1] The Appendix to the *Naut. Alm.* for 1854 contains the numerical details of the calculation of an Occultation by the Moon in illustration of the method I have here proposed. For Bessel's processes of computing Eclipses of the Sun, and Occultations of Stars, with examples numerically worked out, see Loomis's *Practical Astronomy*, Sections IV. and V. of Chapter XI.

along one of the wires so as to be bisected both shortly after entering, and shortly before leaving, the field of view, and then reading off the position-circle. This gives the reading for the equatorial reference-direction for the angular measures. The *angle of position* may be defined to be the angle which the line drawn from the brighter to the smaller of the two stars makes with the circle of declination passing through the centre of the field and cutting the reference-direction at right angles. The telescope being supposed to invert images, and *pro hâc vice* to be pointed southward on the meridian, the lowest point of the field is called the north point (n), the extreme point to the right hand the following point (f), that uppermost the south point (s), and the extreme point to the left hand the preceding point (p). It has been agreed by modern astronomers to reckon position angles from $0°$ at n, through f, s, p in succession to $360°$ at n again. The circle is thus divided into four quadrants named in order nf, sf, sp, np, and the angles of position in these quadrants range respectively from $0°$ to $90°$, from $90°$ to $180°$, from $180°$ to $270°$, and from $270°$ to $360°$. In the case of two components the angle of position is measured by turning the position-circle so as either to bisect both stars with the wire used for recording the reference-position, or to place the two stars, according to the judgment of the eye, exactly intermediate to the two wires separated by a small interval, and then recording the vernier reading. Both processes are exhibited in Fig. 51 (1), S representing the larger and σ the smaller component. The second process is generally preferable for very close stars of unequal magnitudes, the eye judging with accuracy of the symmetry of the positions of the wires relative to the apparent disks of the two stars.

318. After thus obtaining the reading of the circle for angle of position, the position-circle is turned through $90°$ for the purpose of measuring *the distance* between the stars. This may be done by placing one of the wires, as AB, at a fixed micrometer-reading (See Fig. 51 (2)), bisecting the stars simultaneously by the two wires by means of the slow movement of the instrument combined with the micrometer movement of

the other wire ab, and then after passing this wire to the other side of AB, bisecting the two stars again by the same means. Let r_1 be the micrometer-reading for the first position of ab, and r_2 that for the second position, and let c represent the reading for coincidence of ab with AB. Then, supposing r_1 to be greater than r_2, we shall have

$$S\sigma = r_1 - c = c - r_2.$$

Hence
$$S\sigma = \frac{1}{2}(r_1 - r_2),$$

which value, since c is eliminated, gets rid of error incident to the determination of the coincidence reading. By means of the reference-direction and the two measures of angle of position and distance, the celestial position of one star relative to the other at the epoch of the observations is ascertained. (When the two stars are so nearly equal that it is hard to say which is the brighter, it may happen that the relative positions obtained on different occasions differ by 180°.) By repetition of the same processes the relative celestial positions of the components of a triple, or quadruple, star, or of any limited portion of a group of close stars, might be determined[1].

319. For the sake of illustration, two examples of actual observations of double stars, made with the double-wire micrometer of the Northumberland Equatorial, and of the calculations of the angles of position and distances, are here subjoined.

1842, Dec. 3, 7^h, after obtaining the reading 178° 13' of the position-circle for the equatorial direction, four readings were recorded for determining the angle of position of γ Andromedæ, the mean of which gave 151° 46'. By applying the reading for reference-direction

---

[1] The method of observing double stars here described is the same as that adopted by W. Struve for the composition of his great work, entitled *Stellarum Duplicium et Multiplicium Mensuræ Micrometricæ*. A large number of observations of double stars were made according to this method with the Northumberland Telescope in the years 1839-1843, principally by Mr John Glaisher. Subsequently the Equatorial was continually used in observing newly discovered planets and comets, and much time was expended in provisional calculations of their observed places, in consequence of which the observations of double stars were not resumed, and the reductions of those taken were deferred. Hitherto preparations have only been partially made for publishing the results of these observations.

(always *subtractive*), the result is $-26°\,27'$, to which is to be added $90°$ because the angle of position is measured from the north point. Hence the required angle is $63°\,33'$, shewing that the position of the small star relative to the other was north-following (*nf*). For measuring the distance, the values found for r_1 and r_2 were respectively $10''{,}596$ and $9''{,}375$, each the mean of two readings. Hence the distance $\frac{1}{2}(r_1 - r_2) = 0''{,}610 = 10''{,}35$, the value of 1^r being $16''{,}970$ (Art. 303). The following notes were made in the observer's memorandum book:—The star is $\Sigma\,205$ [i. e. No. 205 in Struve's Catalogue of Double Stars]; the small star followed [agreeing with the result of the measures]; the estimated magnitudes $2\cdot3$ and $5\cdot6$; the colour of the larger star, bright yellow, that of the other, a beautiful blue; the night was bad for definition, and the stars blurrish; the smaller star was not seen to be double [on other occasions it was observed as a double star]; the magnifying power, 275; the observer, G (Glaisher).

1842, Dec. 3, 8^h, the star was 55 Piscium ($=\Sigma\,46$); the reading of the position-circle for the equatorial direction, $178°\,14'$; the mean of four angular measures, $102°\,52'$. Hence the angle of position $= 102°\,52' - 178°\,14' + 90° = 14°\,38'$, according to which the smaller star followed. But the observer noted that the small star *preceded*; from which it appears that the opposite vernier was inadvertently read; and as the above angle is consequently to be increased by $180°$, the angle of position by the observation was $194°\,38'$, which accords with Struve's value. The value of r_1 was $10^r\,362$ and that of r_2 $9^r{,}627$, each by two micrometer-readings, consequently the distance $\frac{1}{2}(r_1 - r_2) = 0''{,}368 = 6''{,}25$. The observer's notes were:—Small star preceding; magnitudes $6\cdot7$ and 9; colours yellow and blue, very decided; sky misty; magnifying power, 275; observer, G.

320. I proceed now to explain in what manner the double-image micrometer is employed for observations of double stars. Eye-pieces of various powers furnished with double-image micrometers were executed by W. Simms, according to the Astronomer Royal's designs, and adapted for use with the Northumberland Equatorial. Their construction, and the mode of applying them for taking measures, may be understood from the following descriptions. In Plate IV., Fig. 2 represents the form of the eye-piece, and Fig. 3 the action of the double-image

micrometer. The eye-piece is the same as that called the erecting eye-piece in Art 272 (p. 261), where the disposition of the glasses, and the course of a pencil of rays through them, are shewn by Fig. 41. As the rays of each pencil fall *centrically* on the second lens reckoned from the object-end, and are limited by a small circular hole in a diaphragm close to the lens, advantage is taken of these circumstances to form two equal images of an object by dividing the lens through its centre into equal parts, each half furnishing an image. The separation of the images is effected either by a micrometer movement of one half lens relatively to the other fixed (as shewn in Fig. 3 of Plate IV.), or by equal movements of the two halves in opposite directions. The eye-glass, or that nearest the eye, is contained in a cylindrical tube prepared so that it carries a single wire crossing its axis at the focal distance of that glass, and therefore distinctly visible together with the image of any celestial object. This tube, usually kept in position by friction (as is also the case of the larger eye-piece tube), is furnished at its outer end with a milled rim, whereby it can be turned by the fingers about its axis, so as to be capable of adjusting the wire to any arbitrary angular position. The same thing might be done without moving the position-circle, by independently rotating the whole eye-piece by hand. Such being the mechanical arrangements, the observations of double stars are taken in the following manner.

321. After turning the micrometer-head for the purpose of separating the images of the binary star to an arbitrary extent, the images of one of the stars are placed by tentative movements of the position-circle and telescope so that the line joining them crosses the field near its centre in a direction judged to be nearly equatorial. Then by slow-movement of the telescope in R. A. and P. D. and by independent rotation of the wire, both images of that star can be bisected by the wire, the clock carrying the instrument. The clock being now stopped, if the two images do not leave the wire, both remaining well bisected, the required setting of the circle for the separation of the images in the equatorial direction is effected. Otherwise, by having independent rotations of the

wire and the position-circle at command, together with slow-movements of the telescope, it is possible by a few trials to satisfy the condition that the central points of both images be carried along the wire by the diurnal movement, as represented by Figure 52 (1). The reading of the vernier when this is the

Fig. 52.

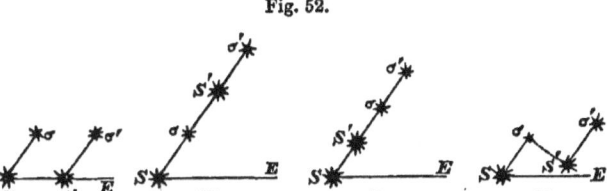

case, gives the equatorial reference-direction. The adopted reference-reading should be the mean of several recorded determinations. Next, the position-circle is turned till the four images are all bisected by the wire, as shewn in (2) and (3); and lastly, by turning the micrometer-head the three spaces intermediate to the images are made, according to the judgment of the eye, to be all equal. In (2) the separation of the images is double the distance between the stars, and in (3) it is half that distance. In the arrangement of (4) σS is judged to be equal and perpendicular to $S\sigma$, so that $S\sigma = \dfrac{SS'}{\sqrt{2}}$. The first arrangement, when practicable, is therefore to be preferred.

322. The reading of the position-circle for the new position of the wire (in (2) and (3)), and the micrometer-reading, may now be recorded. The former reading gives, by its excess above that for reference-direction, the angle $S'SE$, and the latter the distance between the stars multiplied, as the case may be, by 2, or $\frac{1}{2}$, or, $\sqrt{2}$, and subject to *index error*. In all cases these operations should be performed an equal number of times with the images of the stars separated in *opposite* directions, whereby probable errors incident to the determination of the angle $S'SE$ may be eliminated, and the necessity of obtaining the index error, or micrometer-reading for coincidence of the separated images, is avoided. The adopted angle of position is the mean of

several measures of the ∠ $S'SE$ increased by 90°, and the adopted distance is *half* the mean of several differences of pairs of micrometer-readings taken with opposite separations of the images. This measure is converted into arc by being multiplied by the value in arc of one micrometer-revolution obtained as follows. The images of a star of known P. D. are separated in the equatorial direction by an arbitrary large number of integral revolutions, and by the judgment of the eye the wire is placed at right angles to the line joining the images. The clock-movement being stopped, the interval between the transits of the images across the wire is recorded, and the operation is several times repeated. From the mean result in time, and the known P. D. of the star, the interval in arc of a great circle may be readily calculated, and from the given number of integral revolutions, the value of 1ʳ in arc be inferred.

323. We may next proceed to consider in what way the double-wire micrometer is used for *measuring the diameters of Planets*. For this purpose the two wires are placed simultaneously as tangents to the planet's disk at the extremities of the diameter proposed to be measured, whether it be its equatorial diameter, or polar diameter, or the major axis of the elliptical boundary of a gibbous or horned phase. For these positions of the wires let the reading of one micrometer (A) be r_1, and that of the other micrometer (B) be r_2, the readings increasing as usual towards the micrometer-heads. Then if the reference reading of A be chosen to be a, and the reading of B be $a+c$ when its wire coincides with that of A at the reference position, the distance of the limb touched by the wire of A from the reference point is $r_1 - a$ measured in the direction *towards* A, and that of the wire of B measured from the same point in the *opposite* direction is $r_2 - a - c$. (By construction it is provided that a may be an arbitrary number of integral revolutions when the wire of A crosses the hole of the comb, and that at the same time the coincidence reading $a+c$ of B shall differ very little from a, c being a small fraction of a revolution. Generally for the sake of convenience, a is assumed to be 10 revolutions.) Hence it follows that the measure of the

THE EQUATORIAL.

diameter is the positive value of $\pm (r_1 + r_2 - 2a - c)$. The coincidence reading $a + c$ might be obtained in the usual manner by calculating the mean of an even number of readings for contacts of the wire of B alternately on opposite sides of the wire of A at the fixed reading a (see Art. 51). But the following process gets rid of the necessity of taking coincidence readings. After recording the readings r_1 and r_2, let the wires be made to cross each other, and be transferred to the opposite ends of the diameter, and let r_1' and r_2' be respectively the new readings of A and B. Then the measure of the diameter is the positive value of $\mp (r_2' + r_1' - 2a - c)$, the sign being changed because as respects the former measure the direction of this measure is reversed. Hence

$$r_1 + r_2 - 2a - c = 2a + c - r_1' - r_2',$$

and therefore $\quad 2a + c = \tfrac{1}{2}(r_1 + r_2 + r_1' + r_2')$.

Consequently, by substituting for $2a + c$, the *plus* value of $\pm (r_1 + r_2 - 2a - c)$ is the same as that of $\pm(\tfrac{1}{2}(r_1 + r_2) - \tfrac{1}{2}(r_1' + r_2'))$, or $\tfrac{1}{2}((r_1 + r_2) - (r_1' + r_2'))$. If the number of sets of opposite measures be n, and S, S' be respectively the sums of the readings of A and B, it is easily seen from these results that the mean value of $2a + c$ is $\dfrac{1}{2n}(S + S')$, and the mean value of the measure of the diameter is $\dfrac{1}{2n}(S - S')$. (Data for numerical applications of these formulæ will be found in pp. 233 and 234 of Vol. XIII. of the Cambridge Observations, year 1840.)

324. Also the double-image micrometer may be conveniently employed for measuring the apparent diameters of Planets. In the instances of Jupiter, Mars in opposition, Saturn's Rings, and occasionally his ball, the micrometer is read for the external contacts of one image with the other at the opposite ends of the diameter. According as one image is fixed, or both move, the difference of the readings measures double the diameter, or simply the diameter. When the planet's phase is either gibbous or horned, the measures may be taken in the same manner, but the separation of the images has to be

made in the direction of the major axis of the boundary of illumination. (Examples of measurements of the diameters of Jupiter and Venus with the double-image micrometer are given in pages 234 and [142]—[143], of Vol. XIII. of the Camb. Obs.[1])

325. The optical principle of the divided object-glass micrometer, or *Heliometer*, is the same as that of the divided eyeglass micrometer, inasmuch as the pencils of rays from points of any celestial object fall centrically on the object-glass. The Heliometer, so called because on account of the large range of its micrometer-scale it is adapted for measurement of the Sun's diameter, was employed in a special manner by Bessel for determining the parallax of the double star 61 Cygni, this star being selected because from its large annual proper motion (5″) it might be supposed to have sensible parallax. The measures were taken between the middle point of the line joining the component stars and each of two very small stars presumed to have no sensible parallax, and distant from 61 Cygni about 12′ and 8′, approximately in rectangular directions. The measures were continued through $2\frac{1}{2}$ years (in 1837—1840) and from the variations of their values (the proper motion of 61 Cygni being taken into account) it was concluded that the star has an annual parallax of 0″,35, a result which has been confirmed by later observations[2].

326. I propose to conclude the account of the uses of the Equatorial by stating in what manner this instrument is employed for discovering comets and small planets, and by what means observations of these bodies are carried on after a discovery has been made. Whether it be the purpose of the observer to make an original discovery, or to find an object when only a rough estimation of its place can be gathered from an announcement of its having been discovered at a distant

---

[1] For the mode of employing Rochon's Double Image Micrometer, for measuring small angular spaces, see Godfray's *Treatise on Astronomy*, Arts. 113 and 114.

[2] See Grant's *History of Physical Astronomy*, p. 551, and Vol. IV. of the *Monthly Notices of the Astronomical Society*, pp. 152—160.

observatory, the search would be conducted by *sweeping* according to one or the other of the methods described in Art. 307. After detecting the object the observer immediately compares it by differential observations with a known star if one happens to be in the neighbourhood, or with an unknown star conveniently situated, the place of which is determined at the time by comparisons with a known star. The original discoverer, after thus obtaining two places, and deducing therefrom rates of motion of the body in R. A. and P. D., communicates with as much despatch as possible these data to other observers, for evidence of the originality of the discovery, and for the purpose of receiving aid, under possible contingencies of the weather, in securing additional observations. After *three places* have been satisfactorily ascertained, whether by observations made at the observatory of the discoverer, or by a combination of those made at several observatories, a first approximation to the orbit of the body can be deduced on the principles of the theory of gravitation, and from the elements so obtained a rough ephemeris of the apparent motion can be calculated. For this purpose the three places need not be far separated in time; they might be used even if deduced from observations made on three successive nights. This first ephemeris may facilitate carrying on the observations till three other places separated by larger intervals have been well ascertained, whereby a more accurate ephemeris can be calculated[1]. By

[1] The principles on which parabolic-orbits are to this day calculated from three observed places are fully stated and exemplified in Newton's *Principia*, Book III., Proposition XLI. In Proposition XLII. he enters upon considerations relative to elliptical orbits, particularly with reference to that of Halley's Comet. Various modes of applying Newton's principles, differing in the processes of approximation, have been adopted by modern geometers. For instance, Laplace's investigation is conducted by analytical expansion, and in that respect differs from the more geometrical process of Gauss. In the *Memoirs of the Royal Astronomical Society* (Vol. XVII. pp. 59-77) I have produced and exemplified a method, which, although it follows Laplace's in employing analytical expansion, differs from it in the details of the numerical application. Subsequently I found that slight modifications in the analytical reasoning of the same method gave rise to two other forms of expansion, one of which led to the well known equation III. in Gauss's *Theoria Motus Corporum Cælestium* (Lib. II.

continuing this process the elements of the orbit are obtained with increasing exactness, till finally an orbit is arrived at which satisfies as nearly as possible all available observations made at different observatories during this first apparition of the object, and which may serve to determine, in the case of a comet, whether the result is a parabolic orbit or one sensibly *elliptical*. In any instance of an elliptical orbit, whether a planet's or a comet's, the concluded elements are employed for calculating an ephemeris which may serve for detecting the body at the next apparition. Then by employing the results of the observations of the two apparitions, more correct elements might be obtained, and, it may be, of sufficient accuracy for calculating the amounts of perturbations produced by the large planets. By continuing this process, and combining the results as to the elements, given by the first two apparitions after taking account of the perturbations, with those similarly derived from subsequent apparitions, mean elements for a given epoch, which very approximately include the effects of planetary perturbations, might be arrived at[1].

OTHER FIXED INSTRUMENTS OF AN OBSERVATORY.

The instruments which will be treated of under this heading are, the Altitude and Azimuth Instrument, the Fixed Zenith Sector, and the Transit in the Prime Vertical.

Sect. 1., Art. 140), and of course to all the results deducible therefrom. As I had reason to conclude that of the three methods I have referred to, Gauss's was the most rapidly convergent, I have not thought it worth while to publish the additional analytical investigations.

[1] The Northumberland Equatorial was largely made use of in the years 1845-1858 for observations of newly discovered planets and comets, for the purpose of furnishing data for calculating their orbits by the processes above stated, but, excepting in the instance of the Planet Neptune, no attempts were made to obtain original places of these bodies. Provisionally reduced results of the major part of the observations have been published in the *Monthly Notices of the Astronomical Society*, and in the *Astronomische Nachrichten*.

THE ALTITUDE AND AZIMUTH INSTRUMENT.

327. The main principle of this instrument (named also for the sake of brevity an Altazimuth) may be indicated by means of Figure 53, which represents a wooden model used in

Fig. 53.

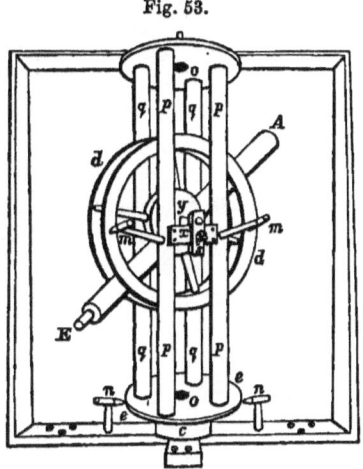

my Lectures. As respects the mode of supporting the telescope and two graduated circles, the arrangements resemble those of an Equatorial, but because the axis about which the whole instrument revolves is vertical instead of being inclined, in point of steadiness it is much superior to an Equatorial. The Figure represents that the instrument is supported in the vertical position by a frame, which practically might be constructed in any manner that ensures the condition of steadiness. The vertical axis of revolution, the direction of which coincides with a straight line joining the centres of the circular holes O, o, is the common axis of a lower and an upper pivot. The lower pivot turns in a perforation made in the fixed circular plate c, and the upper pivot is held in close contact with a horizontal Y, which at first should be so nearly adjusted, that the remaining errors of position of the axis may admit of being allowed for in the reduction of the observations (see next Art.). A is the object-end, and E the eye-end, of the telescope; dd is the altitude circle, and ee the azimuth circle, read off respectively by pairs of opposite microscopes, placed as indicated by m, m and n, n. The columns pp, pp and qq, qq, with their circular terminals, represent the material connection between the upper and lower pivots, which, whatever form be given to it, should especially be rigid. The axis of motion of the altitude circle, which is required to be in a direction at right angles to the axis of revolution of the instrument, should be mechanically adjustible to such position, unless the deviation be allowed for in the reduction of the observations (see next Art.). The Figure indicates, as a possible mode of adjusting the altitude-axis, that one of the pivots (y) rests in a support which is movable in the vertical direction relatively to a fixed transverse bar (x), and may be fastened to the bar by a nut and screw (s) in any required position.

328. A Fixed Altitude and Azimuth Instrument, the construction and mounting of which were planned by the present Astronomer Royal, has been made use of at the Greenwich Observatory since the year 1847, for the important purpose of increasing the number of exact observations of the Moon available for application in the Lunar Theory. I propose to give here some account of the methods of observing with this instrument, the special means employed for obtaining accurate results, and the processes of the calculation by which the observations are reduced[1]. It will, first, be proper to state that by means of a pier of brickwork forming the basis of a triangular arrangement of the parts of the frame which serve for fixing the position of the Y for the upper pivot, the condition of steadiness of the vertical axis is secured, and that for the sake of rigidity of the instrument, as many of the moving parts

[1] A full description of the Greenwich Altazimuth Instrument is given in the Introduction to the volume of Greenwich Observations for 1847, pp. iv.—xxvii., and in an appendix to that for 1867, illustrated by diagrams, and accompanied by a perspective view of the Instrument as mounted (previously published in the *Illustrated London News*, September, 1847). The methods of observing with it and reducing the observations are explained in successive volumes, for instance in pp. xiii.—xvii. and xci.—cxi. of the volume for 1875.

THE ALTAZIMUTH INSTRUMENT.   319

as possible are formed in one cast of metal, no small screws are used in the union of parts, and no power of adjustment is left in any part, it being intended that the observations shall be so arranged that every instrumental error shall be deduced from the ordinary observations, and that numerical corrections shall be applied in the reduction of the observations. Also the friction on the lower pivot is diminished by a weight acting by means of a lever from below in the vertical direction.

329. It is, in the first place, to be remarked that the methods of taking observations of azimuth and zenith distance are independent of each other, and may be described separately. For an *observation of azimuth*, the instrument, after being suitably set, as to time and pointing of the telescope, for catching the object, is clamped in azimuth, and in making the observation a moderately-slow-motion screw is used to cause the object to cross the middle of each vertical wire. The transit-times are recorded either by the chronographic method (Arts. 129—130), or by eye-and-ear observation. The mean of the times, corrected for personal equation of the observer, and for ascertained clock-error, gives the sidereal time of the observation, whether the transits be taken at all the wires, or omitted at some. The readings of the micrometer-microscopes (four in number) of the horizontal azimuth circle are then recorded. These are corrected for error of Runs on the principles stated in Art. 180 relatively to the Mural Circle. The formula used at Greenwich for this purpose is investigated as follows. Suppose that by the usual measurements for Runs an arc of $5'$ is found to be equivalent to $a-p$, $b-q$, $r-c$, $s-d$ micrometer revolutions of the respective microscopes, a, b, c, d being the measures to the nearest tenth of a revolution, and p, q, r, s very small fractions of a revolution. Then if w, x, y, z be the micrometer readings in any observation, the concluded circle-reading corrected for Runs is the Pointer Reading increased by

$$\frac{5'}{4}\left\{\frac{w}{a-p}+\frac{x}{b-q}+\frac{y}{c-r}+\frac{z}{d-s}\right\};$$

or, very nearly, on account of the smallness of p, q, r, s, and near equality of $\frac{w}{a}, \frac{x}{b}$, &c.,

$$\frac{5'}{4}\left\{\left(\frac{w}{a}+\frac{x}{b}+\frac{y}{c}+\frac{z}{d}\right)\left(1+\frac{p+q+r+s}{a+b+c+d}\right)\right\},$$

which form is convenient for calculation.

330. In the case of omitted wires another correction is required, the amount of which is equal to the angular distance between the mean of the wires observed and the mean of all the wires (six), multiplied by the cosecant of zenith distance to obtain the corresponding arc of the horizon. The calculation of this correction requires, therefore, a previous determination of the angular distances of the several wires from the mean of all, which is effected by the following process. From the recorded times of horizontal transit of a star across the vertical wires, taken as above stated in an observation for azimuth, the distance in time of each wire from the mean of all is inferred. Like inferences are drawn from the vertical transit of a star across the set of horizontal wires (to be presently mentioned).

Putting λ for the latitude of the Observatory, α for the azimuth of the star, calculated for the sidereal time of observation, (see Art. 329), and reckoned from the North or South point, and T for the interval in seconds of time for any wire as inferred from a vertical transit, it is easily shewn that the required vertical interval in arc $= 15\, T \cos\lambda \sin\alpha$. Hence, T' being similarly derived for a vertical wire from a horizontal transit of a star whose P. D. is δ across the vertical wires, the required horizontal interval in arc $= 15\, T' (\sin^2\delta - \cos^2\lambda \sin^2\alpha)^{\frac{1}{2}}$, the whole movement of the star in the time T' being $15T' \sin\delta$. The calculations indicated by the two formulæ having been gone through for each of a large number of transits, (say 50 vertical and 50 horizontal), the means of the several results give the intervals in arc of the respective wires from the mean of all for both sets of wires. After applying, in the case of an observation for *azimuth* the corrections calculated by the second formula for omitted wires, the concluded azimuth reading pertains in every instance to the mean of *all* the wires.

331. This concluded reading has next to be corrected for error of collimation, the amount and direction of which are obtained most readily by making use, in the usual way, of a mark on a level with the instrument. (At Greenwich another method is also employed, depending upon the concluded azimuth readings for a high star and a low star in reverse positions of the telescope. See vol. for 1875, p. cvi.) As the mark is necessarily bisected by a single wire, the immediate result of the operation is the horizontal error of collimation of the selected wire, from which that of the mean of the wires is derived by applying a correction for the already ascertained distance of the wire from the mean of all. The correction of the azimuth reading for error of collimation is evidently equal to the Horizontal Error (with the appropriate sign) multiplied by the cosecant of approximate Zenith Distance. A new determination of the constant of this correction is made at the end of each lunation from the aggregate of the values obtained in the course of the lunation.

332. The reading of the azimuth circle is also affected by any error of position of the altitude-axis, which is required to be exactly horizontal. For calculating such error, four levels, two upper and two lower, are attached to the instrument with their axes in directions parallel to the axis of the altitude-circle, and for each observation the two ends of each bubble are read off. From the eight data, regard being had to the order of the scale-graduations, and the values of the scale-intervals, a concluded Level-indication may be obtained by the calculation indicated in Arts. 68—70. The rule adopted for the Greenwich instrument is to add *all* the scale-readings, subtract 1000 from the sum, and divide the remainder by 8, the mean of the values of the scale-intervals being found to be as nearly as possible 1″. From this concluded level-indication, that for the horizontal position of the altitude-axis (obtained as shewn in the next Art.) has to be subtracted, and the remainder, which, with the proper sign, represents the angular elevation of one end of the axis, gives, after being multiplied by the cotangent of Z. D., the correction for level-error to be applied to the azimuth reading,

that factor being required on account of the displacement of the vertical axis in a plane at right angles to the plane of collimation of the altitude-circle. By subtracting from this finally corrected reading the zero, or index error, of the azimuth circle (see next Art.), the concluded azimuth of the object at the time of observation is obtained.

333. It only remains now to find the level-indication for the horizontal position of the axis of the altitude-circle, and the zero of azimuth. After applying corrections for collimation error, calculated with the mean value of the constant of the correction obtained as stated in Art. 331, let O_r and O_l be the readings of the azimuth circle for observations of a known star with graduated face of the altitude-circle to the right-hand and to the left-hand, C_r and C_l its azimuths *computed* for the times of observation, L_r and L_l the level-indications, D_r and D_l approximate computed zenith distances. Then putting y for the level-indication corresponding to the horizontal position of the altitude-axis, and z for the zero or index error of azimuth, we have, from what is shewn in Art. 332,

$$O_r - C_r = (y - L_r) \cot D_r + z, \quad O_l - C_l = -(y - L_l) \cot D_l + z,$$

from which equations it will be found that

$$y = \frac{O_r - C_r - (O_l - C_l) + L_r \cot D_r + L_l \cot D_l}{\cot D_r + \cot D_l},$$

$$z = \frac{(O_r - C_r) \cot D_l + (O_l - C_l) \cot D_r + (L_r - L_l) \cot D_r \cot D_l}{\cot D_r + \cot D_l}.$$

For the determination of y it suffices to use a single star, but a high star is preferable to a low one. The zero of azimuth adopted for any day's observations is either deduced from the star-observations of that day, or, if the zero appears to be steady, by combining with them the observations of other days.

334. Each observation of the Moon's First or Second Limb for azimuth, as often as it is possible, is made with the graduation of the altitude-circle Right and Left, and is accompanied by observations of azimuths of known stars, also made with graduation Right and Left. Measures for determining the coefficient of collimation by a mark or collimator are generally

grouped with the observations on any day on which the Moon is observed, and not unfrequently are taken on other days. The calculations of C_r and C_t for the given sidereal times are founded on places of the stars derived from a star-catalogue formed from observations made with the Greenwich Transit-Circle.

335. We may now proceed to consider the method of taking independently (see Art. 329) *observations of zenith distance*. In an observation for zenith distance vertical transits are observed across the set of horizontal wires by a process analogous to that for the horizontal transits (described in Art. 329), the Altitude-Circle being clamped, and the moderately-slow-motion of the azimuth-circle being used to make the transit of the object take place across the middle of each horizontal wire. It is, however, to be noticed that the mean of the times so taken does not give the time at which the object was at the mean of the wires unless its vertical motion were *uniform*, which is not strictly the case, especially for zenith distances observed near the meridian. Hence a correction of the mean of the times (generally a very small quantity) is required on this account, formulæ for the calculation of which are investigated in pp. xcv.—xcvii. of the Introduction to the Volume of Greenwich Observations of 1875. By applying this correction, together with corrections for personal equation and error of the clock, to the mean of the observed times, the sidereal time of the observation is obtained.

336. Next, the readings of the four microscopes of the Altitude-Circle are recorded. These readings are in micrometer revolutions, and the mean of their sum in arc, corrected for Runs, is calculated by a formula exactly analogous to that obtained in Art. 329 relatively to the azimuth circle. In case of the omission of any wires in observing the transits, the resulting sidereal time applies to a direction of collimation passing through the mean of the observed wires. Hence, as it is requisite that the direction of collimation should in every instance be made to pass through the mean of all the wires, a correction is required equal in amount to the interval in arc

between the mean of the observed wires and the mean of all. A formula has already been obtained in Art. 330 for deriving from vertical transits the interval for any single wire, viz. interval $= 15T \cos \lambda \sin \alpha$. Hence the required correction for the interval between the mean of the omitted wires and the mean of all may be readily inferred. When a mark, or collimator, is observed, the amount of correction is simply the interval in arc of the horizontal wire used for bisection from the mean of the wires. After calculating, as above stated, the corrected mean, in arc, of the microscope readings, the result added to the Pointer Reading gives the concluded reading of the altitude-circle.

337. There are two levels parallel to the plane of the altitude-circle, one vertically above the other, and in any observation the scale-readings of the two ends of each bubble are recorded. The directions of the graduations are so arranged that the mean of the four readings multiplied by $1'',075$, which is the mean value of one scale interval for both levels, gives a Level-indication inclusive of a certain constant depending on the assumed zeros of the scales (see Art. 332), which constant is, in fact, equal to the level-indication when the axis of the azimuthal rotation is vertical. The correction for refraction is calculated in the usual way (Art. 215) for an approximate zenith distance by employing an approximate zenith point. By applying the level-indication derived from the four scale-readings, and the correction for refraction, to the Pointer Reading, a *corrected* reading of the altitude-circle is obtained, which, however, involves the above-mentioned constant. The *difference* between such reading, and that for the zenith direction (named the Zenith Point), is the *concluded* zenith distance, supposing the Zenith Point to be obtained as follows. Let O_r and O_l be the corrected circle-readings in two successive observations of the same star in reverse positions of the graduation, C_r and C_l the corresponding zenith distances computed from an adopted place of the star, and Z the Zenith Point. Then assuming the computed values to be exact, we have

$$\pm (O_r - Z) = C_r \text{ and } \mp (O_l - Z) = C_l,$$

according as O_r or O_l is greater than Z, the quantities on both sides of the equations being necessarily positive. Hence it will be found that

$$Z = \frac{O_r + O_l}{2} \mp \frac{C_r - C_l}{2},$$

according as O_r is greater or less than O_l. If instead of a star, a mark, or collimator, be bisected, C_r and C_l will be equal, and the formula is simply

$$Z = \frac{O_r + O_l}{2}.$$

As either of these values of Z involves, in the same manner as the corrected altitude-reading, the constant depending on the level zeros, it will be seen that the concluded zenith distance, being the difference between that reading and Z, is free from the constant.

338. Thus the observations of the Moon consist of determinations, at ascertained sidereal times, of azimuths of the First or Second Limb, and of zenith distances of the Upper or Lower Limb, by the processes above described, the zero of azimuth being derived from accompanying star-observations of azimuth, and the zenith point from accompanying star-observations of zenith distance, excepting so far as instead of the latter a collimator is used. With this exception the lunar observations depend upon star-places obtained by means of observations on the meridian. (The star-places in the Greenwich reductions are deduced fundamentally from the "New Seven-year Catalogue of 2760 stars for 1864, 0, founded on observations with the Transit-Circle," and forming Appendix II. to the Volume of Observations for 1868.) Since it is possible to observe the Moon much nearer to her conjunction with the Sun with the Altazimuth than with the Transit-Circle, and in other respects the number of available opportunities as respects the state of the sky is greater, it is found practicable to obtain about twice as many places with the former instrument as with the other. In a Report of the Astronomer Royal to the Greenwich Board of Visitors dated June 3, 1876, the respective numbers per lunation are stated to be 180 and 88.

339. Having gone through all the processes of observation and calculation whereby places of the Moon sufficiently trustworthy for use in the Lunar Theory are obtainable with an Altazimuth, we may now consider by what processes of calculation such places are compared with theoretical places. As, however, this is a matter of such length and complexity as to be unsuitable for treatment in detail in a work like the present, I shall only undertake to give an abstract of the procedure adopted at Greenwich, and must refer for fuller information to the contents of Sections 9 and 10 (pp. c.—cxi.) of the Introduction to the Greenwich Observations of 1875. First, the Mean Solar Times of the observations both of azimuth and altitude are calculated in the usual way from the sidereal times obtained as shewn in Arts. 329 and 335. Next, the Moon's Tabular Geocentric R.A., P.D., Parallax, and Semidiameter, for the mean times of the observations of her Limbs, are interpolated to second differences from the data in the Nautical Almanac. (The Astronomer Royal has proposed a mode of interpolating to second differences, described in pp. ci.—ciii. of the above-named Introduction, whereby errors incident to changes of sign are avoided.) The Moon's place is then transferred from the Earth's centre as origin to the point of intersection of the Earth's axis with the normal to the elliptical meridian drawn at the position of the Observatory. A correction has to be applied on that account to the tabular P.D., the formula for which is obtained as follows. The semi-axis major and the semi-axis minor of the terrestrial ellipse being a and b, and the astronomical colatitude of the Observatory being l, it is known from Conic Sections that the distance of the foot of the normal from centre is $(a^2 - b^2)(b^2 + a^2 \tan^2 l)^{-\frac{1}{2}}$. Putting $a\epsilon$ for this quantity, we have, with sufficient approximation, $a\epsilon \times$ sin of Moon's P.D. for the distance of centre from the normal, and for the angle subtended at the Moon by this small line

$\epsilon \times$ seconds of Moon's Eq. Hor. Parallax $\times$ sin Moon's P.D.

This is the correction to be applied, always subtractively, to the Moon's tabular P.D. for transfer of origin. (For the latitude of Greenwich $\epsilon = [7,7174788]$, the ratio of the Earth's axes being taken to be that of 300 to 299.) From the three data, the Hour Angle (equal to the difference between the calculated R.A., whether of star or Moon's centre, and the sidereal time of observation), the calculated P.D. of the object, corrected for the Moon as just stated, and the colatitude of the Observatory ($= 38°. 31'. 21'',89$ for Greenwich), it is simply a problem in Spherical Trigonometry to infer the values of the arcs which may be termed the normal-centric azimuth and normal-centric Z.D. of the Star or the Moon's centre. (Formulæ for the calculations are given in p. civ. of the same Introduction.)

340. For comparing an observed azimuth of the Moon's first or second Limb with the corresponding tabular value, the calculated azimuth of centre is corrected for *semi-diameter* by the following process. To the interpolated semi-diameter (Hansen's, corrected by $+ 0'',49$ from results of Greenwich observations) a correction is applied for excess of the Moon's distance from the foot of the normal above her distance from the earth's centre, which excess is equal to $a\epsilon \times \cos$ ☽'s P. D. (see Art. 339). Hence if $S =$ the semi-diameter in seconds, computed as above stated, S' the required semi-diameter, and D the Moon's distance from the earth's centre, we have

$$\frac{S'}{S} = \frac{D}{D + a\epsilon \cos \text{☽'s P.D.}} = 1 - \frac{a\epsilon}{D} \cos \text{☽'s P. D., very nearly.}$$

Thus the correction to be applied to S to obtain S' is

$$- S \times \frac{a\epsilon}{D} \cos \text{☽'s P. D.}$$

Since $\frac{a}{D}$ is equal to the sine of the Moon's equatorial horizontal parallax, if we use the mean value of the parallax, and assume it to be 57', the correction becomes, with sufficient approximation, $- S \times [8,21958] \epsilon \times \cos$ ☽'s P. D.
(For the latitude of Greenwich $[8,21958] \epsilon = [5,93706]$, because (Art. 339) $\epsilon = [7,71748]$.) After correcting S by this quantity, the correction to be applied to the tabular normal-centric azimuth of centre for semi-diameter, is equal to

 ± corrected semi-diameter × cosec of normal-centric Z. D.

By applying this correction and comparing the result with the observed normal-centric azimuth of Limb (Art. 339), which is the same at the place of observation as at the foot of the normal, the apparent error of Tabular Azimuth is obtained.

341. The correction applied to the tabular normal-centric *zenith distance* (Art. 339) for *semi-diameter*, is the semi-diameter interpolated for the mean time of the observation of Z. D., corrected in this case by $+1'',01$ from results of Greenwich observations, and reduced, as before, to the foot of the normal. (The value of the last correction adopted at Greenwich is

$$-930'' \times [5{,}93708] \times \cos ☾\text{'s P. D.},$$

a mean value of the semi-diameter being used.)

342. The normal-centric Z. D. has also to be corrected for *Parallax*. For this purpose it is necessary to obtain the parallax depending on the normal-centric radius, or distance of the foot of the normal from the place of observation, on which account it is, first, required to multiply the equatorial horizontal parallax interpolated from the Nautical Almanac by the ratio of this distance to the semi-axis major. Using the same notation as in Art. 339, this ratio is known from Conic Sections to be

$$a \times (b^2 \cos^2 l + a^2 \sin^2 l)^{-\frac{1}{2}}.$$

(For Greenwich, the factor is $[0{,}0008851]$, a being to b in the ratio of 300 to 299.) Again, as the parallax must depend on the Moon's distance from the foot of the normal, another factor is required equal to the inverse of the ratio of this distance to the Moon's distance from centre, which, as we have seen (Art. 340), is equal to $1 - \dfrac{a\epsilon}{D} \cos ☾\text{'s P. D.}$ (For Greenwich, taking, as before, $a = D \sin 57'$, and $\epsilon = [7{,}71748]$, the factor becomes $1 - [5{,}93706] \cos ☾\text{'s P. D.}$) Supposing that $P =$ the result of multiplying the interpolated parallax by these two factors, $z =$ the normal-centric Z. D. (Art. 339), and $p =$ the required parallax in seconds, we have

$$\sin p = \sin (P + a) \sin z,$$

a being Airy's correction for parallax at Limb (see Arts. 124 and 232). The first approximation to the value of p being $P \sin z$,

if we substitute for $\sin p$ and $\sin P$ the respective approximations
$$p - \frac{p^3 \sin^2 z}{6R^2} \text{ and } P - \frac{P^3}{6R^2}$$
(R being the number of seconds in an arc equal to radius), and if we omit small quantities of an order that may be neglected, the above equation becomes

$$p = \left(P - \frac{P^3}{6R^2} \cos^2 z + \alpha\right) \sin z.$$

Hence, putting P' for the result of correcting P by the two small quantities within the brackets, the value of p is finally $P' \sin z$. After applying to the normal-centric Z. D. of centre the corrections (with their proper signs) for semi-diameter and parallax, the tabular apparent Z. D. of Limb is formed, the comparison of which with the observed Z. D. (Art. 337) gives the tabular error in Z. D. So by comparing the calculated azimuth of Limb (Art. 339) with the value deduced from observation (Arts. 332—334), the tabular error in azimuth is obtained. In order to make a direct comparison of the observed places with the Lunar Tables, the tabular errors in azimuth and zenith distance are converted into errors in R. A. and P. D., which is done by means of formulæ analytically investigated in pages cx. and cxi. of the before-named Introduction. In the application of these formulæ several observations made in the course of an evening may be grouped together, on the assumption that in the interval occupied, the Moon's P. D. did not vary much, and that the tabular errors in R. A. and P. D. were invariable. A Table of double entry is appended to the Introduction of the Volume for 1854, with the arguments Hour-Angle and Polar Distance, whereby the calculations made by means of the formulæ are much facilitated.

THE FIXED ZENITH SECTOR.

343. The particular purpose to which the Fixed Zenith Sector is applied is, the measurement with great accuracy of small angular distances from the zenith on the meridian. In

principle it is a Mural Circle adapted exclusively to these measures, and is generally reversible. Such an instrument, under different forms, has long been employed at Greenwich in observing zenith distances of the star γ Draconis, for determining the constants of aberration and nutation, that star passing the meridian very near the zenith. Bradley, who first made observations for this purpose, used a mounted Telescope of 12½ feet focal length, with plumb-line for the vertical direction, and a screw-micrometer with graduated head for measuring variations of Z. D.; but the telescope was not reversible. (This instrument, which was constructed by Graham (1727), is represented in Plate XXVII. of Pearson's Practical Astronomy, and a model of it is shewn in my Astronomical Lectures.) Pond substituted for Bradley's Zenith Sector, one which had a Telescope of 25 feet focal length and admitted of reversion, of which he gave an account in the Philosophical Transactions of 1834 and 1835. A description of "The Great Zenith Sector" is also given in successive Volumes of the Greenwich Observations to that for 1847, after which (in the spring of 1848) it was dismounted, the results of the observations made with it not appearing to justify the expenditure of time and trouble it demanded. The failure was considered to be due to its liability to tremors, and to difference of the effects of temperature at the upper and lower parts of an instrument of that great length. Accordingly the present Astronomer Royal has replaced Pond's Zenith Sector by one more compact as to the arrangement of essential parts, and less dependent on focal length. This instrument was constructed by Mr Simms in 1851, and was prepared for immediate use in the autumn of that year. A complete description of it under the name of "The Reflex Zenith Tube," with illustrations by Plates, forms Appendix I. of the Volume of Greenwich Observations for 1854. To this description recourse must be had for the details of the construction. For the present purpose it will suffice to state the following particulars.

344. The Zenith Tube is *fixed*, and the object-glass is carried by a cell which is supported by the Tube, and at the same time may be reversed in position by being turned about

THE FIXED ZENITH SECTOR.

the vertical through 180°. At a distance from the object-glass somewhat less than half its focal length a trough of mercury, independently supported, is placed so that the rays from a star after passage through the object-glass and reflection at the mercury, are brought to focus a little above the object-glass. (A special mode of suspension of the trough by the intervention of vulcanized caoutchouc is employed for diminishing tremors of the mercury.) The star selected for observation at Greenwich is γ Draconis, the mean north zenith distance of which on Jan. 1, 1875, was about 100″. This is small enough to allow of the axis of the zenith tube being fixed very approximately in a vertical position. The reflected image of the star is formed beyond the object-glass, and is looked at transversely by means of a *diagonal* eye-piece, constructed nearly in the same manner as that described in Art. 273. Naming the eye-glass of this eye-piece the first glass, and the field-glass the second glass, a prism just above the centre of the object-glass serves at the same time to turn the pencil of rays through 90°, and by the curvature of the vertical face to act as third glass, the image of the star and the wire for observation being as usual seen in coincidence through the fourth glass. Between the field-glass and the prism there is a vacant space, occupied only by a frame for carrying the eye-piece when moved by the micrometer-head, in order that the light from the star may be as little as possible intercepted. The eye-piece is furnished with a micrometer and micrometer-head, by which means the bisection of the star by a micrometer-wire may be read off in micrometer revolutions. The horizontal ordinate of the image depends on the star's zenith distance, and the distance of the image from a vertical line through the optical centre of the object-glass will be exactly that corresponding to the star's zenith distance with radius equal to the focal length of the object-glass. It is, therefore, required to measure the horizontal distance of the image from the vertical line through the centre of the object-glass.

345. This is done by attaching a micrometer to the cell of the object-glass at the distance at which the star's image is formed, and by placing the diagonal eye-piece so as to receive

the pencils which have been reflected from the mercury and obtain a distinct view of the micrometer wire. This apparatus might be used to measure the *changes* of the star's zenith distance from day to day: but it cannot be used to measure the star's absolute zenith distance, because the vertical line through the centre of the object-glass is not discoverable by direct observation. The arrangement by which the Astronomer Royal effects the absolute measurement is described in page {iv} of the above-mentioned "Description" in the following terms. "Let the micrometer be attached to and carried by the cell of the object-glass, and let the cell be reversible in azimuth, by being turned horizontally round a vertical line parallel to that which passes through the centre of the object-glass, that is, any vertical line, and let the micrometer-wire be placed upon the image of the star, first, when the object-glass cell is in a certain original position, secondly, when the cell is reversed in azimuth; then it is evident that wherever the centre of the object-glass may be, the star's image is in the two observations formed at equal distances from the vertical passing through that point, and in the same direction from the vertical as referred to the earth, or in opposite directions from it as referred to the reversible object-glass-cell. And therefore the distance between the positions of these two images upon the micrometer-frame is double the distance of either from the vertical line, or double the distance which gives the means of computing the angular zenith distance. Thus the practical problem of measuring the star's zenith distance is, in its most important and fundamental point, completely solved." It is, however, necessary to satisfy two conditions; first, parts of the micrometer, such as the *appui* of the micrometer-screw, must be so firmly connected with the object-glass-cell as to retain strictly the same relative positions when the object-glass-cell is reversed, which is insured by the ordinary connections of instrument-maker's work. Secondly, the object-glass must rotate about a vertical line so as to preserve the same inclination to it in the reversed positions; or, if it does not preserve the same inclination, we must have the means of computing the effect of the difference of inclinations. The Astronomer Royal then shews how each micrometer-reading may be corrected for inclination of the object-glass-cell by the indications of a spirit-level attached to the cell.

346. The mean of the two micrometer-readings in the reversed positions of the cell gives what may be called "the zenith point," that is, the reading for the vertical direction through the centre of the object-glass. Instead of using the zenith point given by a single set of observations in reverse positions of the cell, it is preferable to adopt for zenith point the mean of a considerable number of such determinations, and then the measure of the star's zenith distance as given by a particular bisection is the difference between the adopted zenith point and the micrometer-reading for the bisection. This measure is in micrometer revolutions. To convert it into celestial arc, the diagonal eye-piece is replaced *pro hâc vice* by one which might be represented by the figure in Art. 273 turned upside down. This new eye-piece is furnished with cross-wires which by turning the micrometer may be made to bisect the series of wires by which the observations are made. The intervals between the wires are thus determined in micrometer revolutions. To obtain them in celestial arc the cell is turned through 90°, and transits of γ Draconis across all the wires are taken with the new eye-piece (without the cross-wires), and hence the intervals in arc may be computed, and by comparison of the resulting values with those expressed in terms of micrometer revolutions, the value in arc of one division of the micrometer is found.

By continuing such measures of the zenith distance of γ Draconis through a whole year, and through a succession of years, the constants of aberration and nutation are obtainable with great precision by the method of computation first practically applied by Bradley.

THE TRANSIT IN THE PRIME VERTICAL.

347. The foregoing method of ascertaining the constants of aberration and nutation is plainly of a local character de-

pending on the small difference between the polar distance of γ Draconis and the co-latitude of the Greenwich Observatory. It will, therefore, be required to describe another of more general application. With this view I have selected the method adopted by Professor F. G. W. Struve at the Imperial Observatory of Poulkova, which depends on observations made with a transit-instrument in the Prime Vertical. A description in detail of the large instrument constructed by Repsold for this purpose is given in the *Bulletin* of the Imperial Academy of St Petersbourg, Tome x. No. 14—16, and also in the *Astronomische Nachrichten*, Nos. 468, &c. Respecting the process of using this fine instrument for effecting the required determinations it will suffice for my purpose to make the following statements.

348. The Telescope is attached transversely at its middle to one end of a horizontal axis of revolution, and is counterpoised by a weight at the other end. The position of the axis, as respects its deviation from horizontality, is ascertained by means of a spirit-level of great sensibility, (see the note to Art. 67, p. 57), which is permanently fixed to the axis. The plane of collimation of the Telescope is adjusted, so as to be as nearly as possible in the plane of the prime vertical, and consequently the wires at which the transits are taken, are perpendicular to the plane of the meridian when the Telescope is pointed to the Zenith. The polar distance of any star used for taking the transits exceeds by only a small arc the colatitude of the Observatory, so that its path in the field of the Telescope is very oblique relatively to the wires, and the interval between its transit across any one wire and the next is considerable (in one of Struve's instances, 44'). These arrangements being understood, we may proceed to the explanation of the process of observation by which the instrument effects its intended purpose.

349. In Figure 54 let P be the North Pole of the heavens, Z the zenith of the place of observation, PZ the colatitude, $S\sigma S'$ a portion of the diurnal path of the star, cutting the prime vertical in S and S'. Let S be the place of the *first* transit of the star across the prime vertical, where, consequently, it

THE TRANSIT IN THE PRIME VERTICAL.   335

passes from the north side to the south side. In the field of view there are seven wires parallel to the plane of collimation, and so placed (regard being had to inversion by the Telescope) that the image

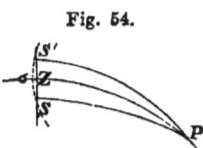

Fig. 54.

of the star crosses them in the order from I. to VII., *before* it crosses the plane of collimation; the seventh being at a convenient small distance from that plane. By shifting the Telescope about the axis of revolution the star is made to cross the wires at their middle points. Quickly after the transit across the seventh wire the instrument is lifted from its bearings and the position of the Telescope is reversed, say, from southward to northward. By this operation the set of wires is transferred to the opposite side of the plane of collimation, and the star, pursuing its course, crosses them *again*, but in the order from VII. to I. The Telescope retaining its position, the *second* transit across the prime vertical (that at S') is observed in the same manner as the first, the order in which the star's image crosses the wires before crossing the plane of collimation being from I. to VII., because the direction of crossing is now from the south to the north side of the prime vertical. After recording the transits at the seven wires, the instrument is again quickly reversed, and the transits are taken in the order from VII. to I. The spirit-level is read before and after each reversion for calculating corrections for displacement of the axis from the horizontal direction. It thus appears that transits of the star are taken at *each* wire in both sets of transits, both before and after passing the prime vertical. The apparatus devised by Repsold for lifting the instrument from its supports and reversing the axis end for end, gives the means of performing the operation within the considerable interval allowed by the oblique movement of the star[1].

350. The mean of the two times of transit of a *given* wire in the first set of transits, is the time of its transit across the

---

[1] In Struve's *Description of the Central Astronomical Observatory of Poulkova* (St Petersbourg, 1845), in page 178, an example is given in detail of the record of an observation such as that here explained.

prime vertical at S, and similarly the mean of the two transit-times of the *same* wire in the second set, is the time of its transit at S'. The excess of the latter time above the other gives the angle SPS'. The transits at each of the other wires in like manner give a value of the same angle, and by the mean of all the values the angle is determined with great precision. The result is not sensibly affected by refraction, and it is considered that errors arising from want of exact adjustment of the axis of revolution, and other incidental causes, are very approximately eliminated by the process of observing in reverse positions of the Telescope. By the spherical triangle SPZ,

$$\tan PZ = \tan PS \cos \frac{SPS'}{2}.$$

Hence since PZ is constant,

$$\delta . PS = \tfrac{1}{4} \sin 2PS \tan \frac{SPS'}{2} \times \delta . SPS'.$$

By means of this formula the variation $\delta . PS$ of the star's P.D. due to aberration and nutation may be derived from the observed values of $\delta . SPS'$, and consequently the required constants be inferred from continuous series of such observations, as already indicated (Art. 346). The observational means above described are those specially employed by W. Struve for determining the constant of aberration, $20'',4451$, cited in Art. 143 (3), p. 135.

351. As a corollary to the preceding results it may be stated that the sum of *all* the times of transit of the star divided by 14, half their number, gives with great exactness the time of its transit across the meridian. Hence the difference between this time and that of transit at S obtained as above shewn, is the angle SPZ, from which, and PS, the P.D. of the star, supposed to be known, PZ the colatitude of the place of observation may be computed from the right-angled triangle SPZ. Similarly PZ may be obtained from the triangle $S'PZ$, and the mean of the two results may be regarded as the most trustworthy determination of the colatitude derivable from one set of observations made with the given star. By employing in

like manner repeated observations of several known stars, a final result of great certainty might be arrived at. This is in principle Bessel's method of determining the latitude of an observatory by means of a transit-instrument set up in the prime vertical.

I have now gone through, with, I think, as much completeness as was required by the character of this publication, the descriptions of the fixed instruments of a fixed observatory, and the processes of using them for astronomical determinations. What is effected by such means may be generally stated to be *fundamental* with respect to the purposes to which Practical Astronomy is applied in the use of transportable instruments, such as those I am now about to treat of. Having already several times adverted to the science as being essentially *logical* with respect to the relation of its different parts (see, for instance, Arts. 10, 102, and 137), I take occasion to remark here that wherever in the subsequent portion of the work, places of stars, or the places and apparent magnitudes at given times, of the Sun, Moon, and planets, are cited as *known*, it is to be understood that the data are inferred, either directly or by computation, from observations made according to methods described in the antecedent portion.

OBSERVATIONS WITH TRANSPORTABLE INSTRUMENTS.

352. *Ramsden's Zenith Sector.* This instrument, constructed for use in the Trigonometrical Survey of England and Wales, was applied from the beginning of the Survey to 1836, in measuring by means of a plumb-line the meridian zenith distances of stars culminating near the zenith of each place at which it was temporarily set up, the purpose being to compare the difference of the astronomical latitudes of two places on a given meridian with the difference as inferred from actual measurement on the earth's surface, and from theoretical computation based on an assumed figure of the earth. Data were

thus obtained for testing, or correcting, the constants adopted in the theoretical calculation. It was also used for determining the latitudes of certain stations fixed upon in the course of the triangulation[1].

353. *Airy's Zenith Sector.* This name is given to an instrument the design, and superintendence of the construction of which, were undertaken by the Astronomer Royal in consequence of an application from Col. Colby, superintendent of the Ordnance Survey of Great Britain and Ireland, who desired to replace Ramsden's Zenith Sector by a new and more perfect instrument. In the "Account of the Ordnance Survey," published in 1858 by Leiut.-Col. H. James, the principles involved in the construction of "Airy's Zenith Sector" are thus stated in page 57: "The first principle was the arrangement for making successive observations in two positions of the instrument, face east and face west, at the same transit. The second principle was the substitution of a level or system of levels for the usual plumb-line. The third principle was the casting in one piece, as far as practicable, of each of the different parts of the instrument, in order to avoid the great number of screws and fastenings with which most instruments are hampered, and to secure, if possible, perfect rigidity." The inferences of latitude from the observations of zenith distance are all made by the intervention of *known* stars[2].

(Examples of the Transportable Instruments that remain to be described are exhibited in the Lecture-Room, with the exception of the Altazimuth, the principle of which is included in the construction of the Theodolite.)

354. *The portable Altazimuth Instrument.* Like the fixed

[1] An account of Ramsden's Zenith Sector and the mode of using it is given in pp. 533—546 of Pearson's *Practical Astronomy*, illustrated by Figures in Plates XXVI. and XXVII., exhibiting its component parts, and the instrument as a whole.

[2] In his work entitled *On the Theory of Errors of Observations and the Combination of Observations* (Macmillan and Co., 1875), Mr Airy has laid down principles for calculating the probable errors of the results obtained in Trigonometrical Surveys by observations made with the Theodolite and the Zenith Sector. See particularly instances in Sections 12 and 15.

Altazimuth, this instrument is furnished with graduated altitude and azimuth circles, but it suffices for the purpose to which it is applied to read off the graduation by *Verniers*. It is supported by a tripod stand, and the height of each foot is adjustible by a screw. The principal adjustment is that of the axis of revolution to verticality, which by means of two spirit-levels fixed to the supporting frame in rectangular directions and having marks for the mid-positions of the bubbles, is effected in the following manner. The adjusting screws being supposed to rest on a firm basis, the instrument is turned about its axis till the direction of the axis of figure of one of the levels is parallel, as nearly as may be estimated by sight, to the line joining two of the feet. The bubble of that level is then adjusted to mid-position by turning either a screw of one of the two feet or the screws of both. The screw of the third foot is then turned till the bubble of the other level is in mid-position. This completes the adjustment. A first trial, ought, however, to be tested by turning the instrument quite round to see whether the bubbles retain their positions; for if they do not a fresh adjustment has to be made. The axis of the altitude-circle may be supposed to be sufficiently adjusted by the instrument maker so as to be perpendicular to the axis of revolution. The field of the Telescope has vertical wires and a transverse wire which, after the adjustment of the vertical axis, is readily adjusted to horizontality by a fixed land-mark. The error of collimation of the intersection of the middle vertical wire and the horizontal wire may be corrected in the usual way by a mark, with reversion by means of the azimuth movement. Instead of vertical wires, there might be two wires crossing each other and the horizontal wire at the middle of the field.

355. After these arrangements the use of a portable altazimuth has the particular advantage of requiring, for finding the time of day and latitude of a place in an unknown territory, no other instrumental adjustment than that of the axis of revolution to verticality in the manner above stated. This adjustment being made, suppose a known star, or the N. or

S. limb of the Sun, to be bisected before culmination, by the horizontal wire where it intersects the mid-wire or the cross of wires, the Telescope being previously clamped to the altitude-circle, and the bisection being effected by the azimuth movement. The chronometer-time of bisection having been recorded, and the Telescope remaining clamped, let the time of bisection of the same object after culmination be similarly noted. When the object is a *star* the mean of the two times is the chronometer-time of its transit across the *meridian*. Also if the azimuth-circle be read for the two positions, the mean of the two readings is the azimuth-reading for meridian. Hence, supposing the observer to be provided with a sidereal chronometer and the Nautical Almanac for the year, a portable altazimuth might be employed, in the same manner as a transit-instrument in the meridian, for rating the chronometer by means of stars, and for determining local sidereal time and local mean time. Again, after correcting the index error of the altitude-circle by means, for instance, of its readings for bisections of a land-mark in reverse azimuths, if the instrument be then set to the meridian according to a previous determination of the meridian azimuth-reading, the meridian altitude of some known star might be observed, and thence, after correcting for refraction, the *latitude* of the place of observation be inferred. Lastly, if the observer is able to ascertain, by any of the processes which will be explained in the next *Division*, the Greenwich mean time contemporaneous with any local mean time obtained by the instrument, the difference of the two times gives the *longitude* of the place. Thus the portable altazimuth is a very serviceable instrument for enabling a voyager after landing on any coast, or an inland explorer, to make approximate geographical determinations of unknown localities.

356. After finding the azimuth-reading for meridian by observing, as above stated, equal altitudes of a star, or rather after obtaining the mean of several such determinations, and thereby placing the plane of collimation in the plane of the meridian, the chronometer-time of *apparent noon* may be inferred by observing a transit of the Sun. This determination is, how-

ever, often made by noting the times of bisections of the Sun's N. or S. limb at equal altitudes before and after culmination, in which case a correction is required for change of the Sun's P.D. in the interval between the bisections. The correction may be calculated as follows, supposing the azimuth readings to be taken at the two bisections, and the latitude to be known by the process explained in Art. 355. Supposing P to be the North Pole of the heavens, Z the zenith of the place of observation, and S the place of the bisected limb, let $PZ = \lambda$, $ZS = z$, $PS = \delta$, and the hour angle ZPS *west* of the meridian $= h$. Then we have to find the change Δh of hour angle corresponding to the change $\Delta \delta$ of the P.D. of the bisected point in the interval from the first to the second bisection. Putting S for the angle PSZ, it may be readily seen from the small variations of h and δ, z and λ being constant, that

$$\Delta h = - \Delta \delta \cot S \frac{\operatorname{cosec} \delta}{15},$$

in which expression $\Delta \delta$ may be supposed to be obtainable from the Nautical Almanac; but S and δ have to be determined. Now if $A =$ the angle PZS, or the difference between the azimuth-reading for meridian and that for the second bisection, we have by the spherical triangle PZS,

$$\operatorname{cosec} \delta = \frac{\sin h}{\sin z \sin A}, \qquad \sin S = \frac{\sin h \sin \lambda}{\sin z},$$

in which equations, without sensible error, h may be taken equal to the excess of the noted time of the second bisection above the mean of the two times, and A equal to the difference between the second azimuth and the mean of the two azimuths. Thus $\operatorname{cosec} \delta$ and S may be calculated from observed quantities, and the value of Δh be obtained from the foregoing formula. The correction to be applied to the time of the second bisection is $-\Delta h$, and the mean between the corrected value and the time of the first bisection is the chronometer-time of *apparent noon*. (The correction for difference of the refractions at the two positions is too small to be worth taking into account.)

The correction $-\Delta h$ has the same sign as $\Delta\delta$, if S be less than $90°$, and consequently is positive or negative according as the Sun's P.D. is increasing or decreasing. The opposite signs would apply if S should be greater than $90°$.

357. *The portable Transit-instrument*[1]. The parts, mode of mounting, and adjustments, of this instrument, being like those of the fixed Transit, need not be described in detail. It will suffice to mention the following particulars respecting an example (exhibited) of Troughton's construction. The pivots rest in Ys supported by two standards firmly attached to a brass circular ring, which has three feet adjustible as to height by screws, and requiring to be placed in contact with a solid horizontal basis. One of the screws is vertically below one of the Ys, and is used for the level-adjustment, the azimuth-adjustment being performed by a horizontal screw-movement of the other Y. The other two screws are so placed that a line joining the feet is approximately in a direction parallel to the telescope's plane of collimation. The setting-circle, which is graduated for indication of altitude, is fixed to one end of the axis of revolution, with which, consequently, it revolves and is reversed, the support of a lamp at the other end for illumination of the field of view being also reversible. The indications of the setting-circle are read by a double vernier, having an attached tongue whereby it may be adjusted and clamped, the tongue being acted upon by one of two sets of opposing screws fixed to the standards. Index error may be got rid of by pointing the Telescope to a star of known altitude, and then adjusting the vernier-reading to this altitude as altered by refraction. After this correction the bubble of a spirit-level attached to the vernier should be put in mid-position, in order to give the means of adjusting the vernier by the tongue after a reversion, so as to be again free from index error.

358. The adjustments of the plane of collimation of a portable Transit, required to be set up in an unknown locality,

---

[1] A full description of this instrument as constructed by Troughton (illustrated by a Figure), and of its adjustments and uses, is given in Simms's *Treatise on Mathematical Instruments, &c.*," 8th edition, pp. 68—92.

may be performed as follows. The error of collimation of the middle wire might be corrected by bisections of any suitable fixed point (such as a bright point seen by reflection of sunlight from a convex glass surface), and by reversing the instrument. This correction is usually made by the instrument-maker. Before correcting the errors of level and azimuth, the plane of collimation has to be placed approximately in the plane of the meridian. This could be done by employing a portable altazimuth for drawing a meridian line by observations of equal altitudes of a star (Art. 355), or, in default of such means, the process described in Art. 10 might be adopted. The axis of motion may then be adjusted to horizontality by applying the feet of a corrected striding spirit-level to the pivots, and turning the foot-screw for level-adjustment (mentioned in Art. 357), to put the bubble in mid-position. The adjustment to the meridian may then be effected by transits of a circumpolar star in the manner indicated in Arts. 88 and 89. Supposing that, after making one such adjustment, it should be required to place the instrument repeatedly in the same position, this may be conveniently done by means of the following expedient. The screw for level-adjustment, terminating in a rounded surface, is applied to a horizontal plane surface, while the other two screws have pointed feet, which fall into an angular groove cut in the direction of the line joining their extremities after the initial adjustment has been completed. Thus the instrument may, at least approximately, be placed in the adjusted position. A more exact adjustment, if thought good, might then be made by the usual process.

359. The portable transit is principally used for rating chronometers by means of transits of known stars, and for calculating local sidereal time and local mean time. In many cases, determinations of these quantities such as may be obtained by a portable transit having a telescope of twenty inches focal length, are sufficiently approximate. But certain observations, for instance some that are available for calculating solar parallax from a Transit of Venus, require to be accompanied by a determination of local time with the precision that can only

be attained by stable mounting of transported transit-instruments of considerable size.

360. *The Theodolite.* Figure 55 represents a Theodolite made for me by Mr J. Simms for exhibition to the class in my lecture-room. Excepting in being smaller, it differs very little from the instruments furnished by Troughton and Simms for the Great Trigonometrical Survey of India. In principle it is a portable Altazimuth and a portable Transit combined, and accordingly what has been said in Arts. 354—359 respecting the parts and adjustments of those two instruments applies in great measure to the Theodolite. As shewn in the Figure, A is the object-end and E the eye-end of the Telescope. At the focus of the object-glass there is a system of wires consisting of two vertical wires, one on each side of the plane of collimation, two wires equally inclined to it and intersecting at an acute angle at the centre of the field, and a horizontal wire passing through the point of intersection. For fixing the position of the wire-frame, and correcting error of collimation, two sets of antagonist screws with capstan heads are placed in the position indicated by i, one for vertical, and the other for horizontal, movement. By turning a milled head projecting from the telescope-tube near the object-end (not seen in the figure), the distance of the object-glass from the wire frame may be regulated so that the image of a land-object can be seen distinctly with the eye-piece at the same time as the wires. The mounting of the Telescope on the supporting frame is the same as that of a transit-instrument, one of the pivots resting in its Y being shewn at e. The two screws $g\ g$ hold down a plate, which is moveable about the one with capstan-head on the left hand, when both are unscrewed. After unscrewing them the plate can be turned aside, and a like operation having been performed with respect to the other pivot, the instrument can be lifted from the Ys for reversing it. The capstan-headed screw between the two screws g has a piece of cork at its lower end, and is made to press on the pivot for keeping it in position. The vertical circle dd is graduated for reading off altitude by means of a double vernier vv, the position of which relative to the

Fig. 55.

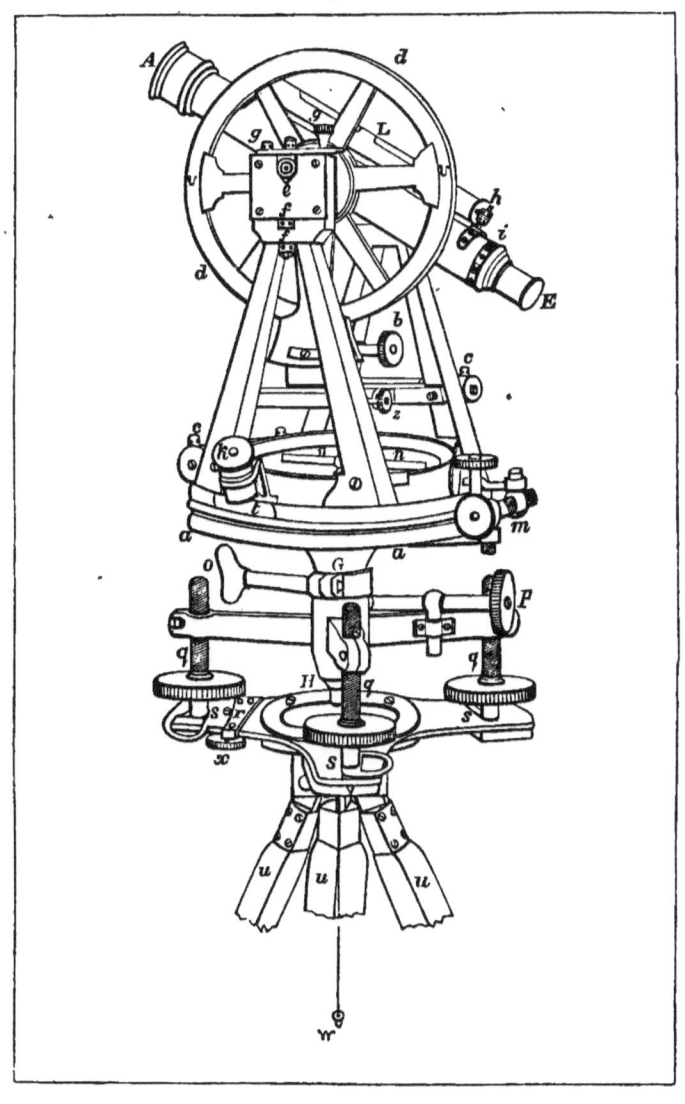

graduation is adjusted and fixed by the action of antagonist screws (one of which is seen at z) on a projection from the frame, the screws moving a vertical tongue rigidly connected with the vernier and tapped for the action of the screws. The graduation is from 0° to 90° in the same direction in each of the four quadrants, both indices pointing to 0° when the telescope is horizontal. Slow motion is produced by the tangent-screw and clamp at b. The azimuth circular plate aa is graduated from 0° to 360° on a chamfered edge. Above this is another circular plate opposite portions of which directly below the pivots, as shewn at t, have chamfered edges, and are graduated to serve for indices and verniers, the graduation of the lower plate being read off by the microscope k and an opposite one. The vertical and azimuthal circles are both graduated to a third of a degree, and by the microscopes and verniers can be read off to 20″. The two above-mentioned plates have conical axes one within the other, the outer one represented by G, being a hollow cone fixed to the azimuth plate and the inner one fixed to the upper plate, turning smoothly within the other. The two axles are held together by a screw and circular plate H fixed to the lower end of the interior one. The vernier plate and all the parts of the instrument above the azimuth plate turn with that inner axle, and may either be moved by hand through large angles, or partake of slow motion by means of the tangent-screw and clamp at m. Also after turning the handle o, so as to tighten the collar about the axle G, slow motion of rotation may be given to all parts of the instrument in common by means of the screw p, which works in a tapped projection from the collar (not seen in the Figure).

361. Such being the description of the parts of a Theodolite, it is next required to give an account of its *adjustments*. It will, at first, be supposed that its three feet s, s, s rest on a firm plane approximately horizontal, the usual means of support by a tripartite staff being considered afterwards. The first adjustment is to set the common axis of revolution so as to be exactly vertical; which, after unclamping m and o, is effected by means of two corrected levels with marks, attached in

rectangular directions to the supporting frame, in positions indicated by the letters c, c. The adjusting screws are q, q, q, by which, and by the three legs when used, the azimuth plate is first placed horizontally as nearly as the eye can judge. The exact adjustment is then made by placing the axis of figure of one of the levels in a position judged to be parallel to the line joining two of the feet of the screws, putting the bubble of that level in mid-position by means of one or both these screws, and finally putting the bubble of the other level in mid-position by the third screw. The accuracy of the adjustment should always be tested by turning the instrument quite round to see whether the bubbles retain their positions with so much exactness as to shew that a fresh adjustment is not required. (See what is said respecting this adjustment in Art. 354.) After the foregoing process, the axle G should be clamped by the handle o, and it is then advisable to repeat the operation by turning independently the parts connected with the vernier plate. Next, the error of collimation of the cross of wires is to be corrected in the usual way by a fixed horizontal mark, and reversing the horizontal axis end for end. The resulting plane of collimation being supposed, *at present*, to be exactly vertical, let a fixed horizontal mark be bisected by the horizontal wire at the intersection of the cross-wires, and the altitude-circle be read off by the verniers v, v. Then if the same process be gone through after reversing the telescope by the azimuth movement, the mean between the two readings gives the reading for the direction of the vertical axis; and if it be desired that this direction should be indicated by $90°$, an index error has to be corrected equal to the excess of that mean reading above $90°$. This correction is effected by means of the screw z, which regulates the position of the verniers on the limb of the circle. After correcting the index error, if the vernier readings be set to $0°$, the line of collimation of the telescope will point horizontally. In that case the large level L, which has its axis of figure parallel to the axis of the telescope, is to be adjusted by putting and fixing the bubble in mid-position by means of the two capstan-headed screws at h. Afterwards by means of this level the telescope can be pointed horizontally in any azimuth.

362. It is here to be remarked that the foregoing adjustments will not be affected if the plane of collimation, instead of being vertical as above supposed, be inclined by a small angle from the vertical direction in consequence of a small deviation of the axis of motion of the telescope from horizontality. It is, however, necessary that that axis should be exactly horizontal, and the deviation has consequently to be corrected; which may be done as follows. Bisect with the cross of wires some well-defined distant point near the horizon, and record the azimuth readings of the two verniers. Then by horizontal rotation of the instrument set the vernier readings at exactly 180° from the former positions. If the transverse axis be exactly horizontal, the same distant point will be bisected by the cross of wires after this reversion. But if it deviates from horizontality by the small angle θ, its new position will make an angle 2θ with the first, and the plane of collimation will cut the horizon at some point different from that originally bisected. *Half* the distance between the two points is to be corrected by means of the two capstan-headed screws ff, whereby the end e of the axis is moved and fixed. The axis will then be adjusted so as to be, at least approximately, horizontal. For greater certainty the operation should be repeated; after which the index error of the azimuth circle may be corrected by the following process.

363. Below the telescope there is a compass-needle $n\, n$, which may be clamped by turning a small exterior handle, and, after adjusting the vertical axis, be unclamped and allowed free suspension, as occasion requires. The compass-card, which rotates with the telescope, is so placed that the plane of collimation passes nearly through its N. and S. points. Hence at a locality where the magnetic declination is known, the plane of collimation can be at once approximately adjusted to the meridian. For instance at Cambridge, where the magnetic declination is about 19° westward, the telescope may be brought nearly to the meridian by being turned, in common with the

compass-card, till the northward end of the needle points to 19° eastward, as marked on the card, from the N. point. If it be desired to fix the azimuth reading at 0° when the telescope points northward in the meridian, the vernier-index is first to be set to 0° by turning the vernier-plate whilst the azimuth-plate is clamped by the handle o, and then after connecting the two plates by the clamp at m, and unclamping o, the plates are to be moved together till the plane of collimation is shewn by the compass to be nearly in the meridian, the telescope pointing northward. The required adjustment of the zero of azimuth may in this way be made, at least approximately.

364. For an exact adjustment recourse must be had to astronomical observation, which, after the foregoing approximation, may be such as follows. With the instrument in its actual state observe the altitude and azimuth of a known star at a recorded time, and compare therewith the altitude (inclusive of refraction) and azimuth of the star calculated for the given time by the usual formulæ from its known R.A. and P.D. The excesses of the calculated above the observed values are the corrections to be applied for index errors of the altitude and azimuth circles. The result in the case of the altitude circle will serve to test or rectify the correction already applied (Art. 361); and the mechanical correction of the residual index error of the azimuth circle may be effected by the following process. Suppose the azimuth reading for bisection of a well-defined terrestrial point to be A reckoned from the apparent N. direction, and, the correction for index error having been found by the foregoing method to be $+a$, let the azimuth reading be altered to $A + a$ by the tangent-screw and clamp at m. The cross of wires will then no longer bisect the same point, but may be made to bisect it again by the slow-motion screw p, the axle G being clamped. By this means the index error is corrected so that the reading of the azimuth-circle will be 0° for the N. direction. It should be added that generally it would be preferable to find the index errors of both circles by repeated observations of stars, and apply the resulting mean corrections in the reductions of the observations; in which case, although

OBSERVATIONS WITH TRANSPORTABLE INSTRUMENTS.

it is necessary to be provided with the means of executing these corrections mechanically, it would only be occasionally required to make use of them.

365. It remains to describe the mounting of the Theodolite on a three-legged stand. In the Figure, uuu are upper portions of the legs, connected in common at their upper ends by joints with a large screw, on which a tapped three-armed plate is screwed for support and adjustment of the instrument by the three screws qqq. These screws have conical feet with flanges, and the plate consists of two parts in contact, the upper of which, after lifting the spring r, can be moved for bringing into position three openings through which the conical feet have to pass in order to rest in prepared angular grooves. The moved plate is then restored to its first position so as to be in close contact with the flanges, the spring r is let down, and the two plates are screwed together by the screw x, by which means the positions of the screws qqq are fixed, and the instrument is firmly connected with the three-legged support. w is a small plummet, suspended from a hook vertically under the centre of the azimuth circle, and nearly reaching the ground, so that it indicates the spot at which the instrument may be considered to be set up. When it is required to move the instrument to a new position, the parts uuu being turned on their hinges are put together so as to form a single tapering staff, round which three brass rings, temporarily held on by friction, keep the parts together. In the foregoing description of the adjustments, mention was several times made of two related screws, one for adjusting, and the other for fixing the adjustment. In consequence of this precaution the Theodolite can be borne from place to place with the staff resting on the observer's shoulder without disarranging the connection of the parts[1].

366. The account of the Theodolite may be appropriately concluded by describing, as follows, an example of its use which may serve to shew in some degree how it is employed in large

---

[1] The adjustments of a Theodolite differing in several particulars from that represented by Figure 55, are given in detail in Simms's *Treatise on Mathematical Instruments*, pp. 17—22.

trigonometrical surveys. In order to determine the difference of longitude between the transit-instrument of the Cambridge Observatory and the place of a clock in the Museums Building, for the purpose of inferring true local mean time from comparisons of this clock with the transit-clock Hardy, the following process was adopted. Two stations A and B were fixed upon in the neighbourhood of the Observatory, from both of which the N.E. pinnacle of King's College Chapel could be well seen. The former station was nearly opposite the middle of the steps of the Portico of the Observatory, and the differences between its latitude and longitude, and those of a point T vertically under the central point of the transit-instrument, were obtainable simply by measures parallel and perpendicular to the meridional direction through T. The Theodolite being set up at A, a vertical adjustment of the axis of azimuthal motion was carefully made by the three adjusting screws and the two attached levels (see Art. 361), the adjustment was next tested by turning the large level L, supposing it to be already corrected for indicating the horizontal pointing of the telescope, after which the index error of the altitude-circle was corrected. The index error of the azimuth circle was left of arbitrary amount, the adopted method of observing being such as not to require its correction. At B, a point nearly due south of A, and close to the road to St Neots, a staff was set up vertically by a plumb-line, and being furnished with an adjustable mark which could be placed at the same height from the ground as the central point of the telescope, the positions of that point and the mark could be interchanged by means of the plummet w (in the Fig.). The highest point P of the above-mentioned pinnacle being visible both from A and B, the process of taking the observations was such as follows.

367. With the Theodolite at A, the instrumental altitudes and azimuths of the point P, and of M, the position of the mark, were recorded. The Theodolite was then transferred to B and the staff to A, and the altitudes and azimuths of P and M were observed from B. Without changing the position of the Theodolite, an altitude of a limb of the Sun, and a transit of

both vertical limbs (see Art. 329), were taken at times indicated by a sidereal chronometer whose error on true sidereal time was ascertained. The distance between A and B, which corresponds to the base-line of a trigonometrical survey, was measured by a Gunter's chain, the whole length of which, and the mean length of the links, were obtained by means of a four-feet measure (lent for the occasion by Professor Adams), which had been compared with a standard of length. This distance was measured along the ground, which declined regularly from A to B. Supposing l to be the measured length, and α the angular depression, as inferred from the altitude of A observed from B, the horizontal interval between A and B is $l \cos \alpha$. Thus there were given the difference of the azimuths both of the horizontal projections of AP and AB, and of the horizontal projections of BP and BA, and the length of the projection of AB. With these data the *lengths* of the horizontal projections of AP and BP were calculated, two angles and the included side of a plane triangle being given. Next, the altitude of the bisected N. or S. limb of the Sun was obtained, for the given time of observation, by calculating, according to known rules, from the R.A., P.D. and semi-diameter derived from the Nautical Almanac, and the result compared with the instrumental altitude, in order that the index error of the altitude-circle might be tested, and, if of sensible amount, be corrected. From analogous data derived from the Nautical Almanac, the Sun's azimuth at the given time of the observation of azimuth was calculated, and compared with the instrumental azimuth, for the purpose of ascertaining the amount of correction to be applied to the latter for index error. After correcting thereby the instrumental azimuth of the projection of BP, the *astronomical azimuth* of this projection became known, and, by consequence, that of the projection of AP, the two projections making with each other the angle of the above-mentioned plane triangle subtended by the projection of AB. Thus both the length and the azimuthal bearing of the projection of AP were determined, and consequently the differences between the R.A. and P.D. of the points A and P, as measured on the earth's surface, were readily inferred. Also the relative positions in R.A. and P.D. of the points A and T having been found by direct measurement (see Art. 366), the differences in R.A. and P.D. between the positions of the pinnacle P and the Transit T were at once obtained.

368. Again, two positions C and D were chosen in the South Court of the Museums Building, the former on the east side and the other on the west side, and from both the top of the N.E. pinnacle of the Chapel could be seen. But neither of them was suitable for measurement of its distances in R.A. and P.D. from the position, say K, of the clock, or for taking observations of the Sun; on which accounts a position Q on the roof of the S.W. angle of the building was made use of as an intermediate station. First of all, the azimuths of P, and of the mark M adjusted after placing the staff at D, were taken from C, and then, the Theodolite retaining its position, the staff was transferred to Q, and, the adjustment of M remaining the same, its azimuth in the new position was observed from C. The length CD was measured by the Gunter's chain, and, as it happened, no correction was required for inclination of the ground. Lastly, the Theodolite being transferred to D and the staff to C, the azimuths of P, and of the mark M first at C and then at Q, were taken from the position D, in the same manner as the analogous azimuths were taken before from C. It will thus be seen that data were obtained for calculating (as shewn in Art. 367) the *lengths* of the projections of CP and CQ, but not their *directions*. To obtain these the Theodolite was transferred to the position Q, the staff was placed at C, the heights of the Telescope and the mark M were adjusted to the values they had before their positions were interchanged, and the azimuths of M and of the Sun at a noted time, were observed from Q. The observed azimuth of the Sun was compared with its value at the given time as computed from the data of the Nautical Almanac, and the excess of the latter above the former was applied as correction of index error of the observed azimuth of the projection of QC. Hence the *astronomical azimuth* of that projection was determined, and that of the projection of CP was at once inferred from the antecedent

observations of the azimuths of CP and CQ from C. Thus both the lengths and the directions of the projections of CP and CQ were ascertained, and consequently the differences of R.A. and P.D. between C and P, and between C and Q, and, by inference, between Q and P, admitted of ready calculation (see Art. 367). Hence, after obtaining the relative positions in R.A. and P.D. of the point Q and the place K of the clock, which could be done by direct measurement, the relative positions of K and P became known. The relative positions of T and P having been already found (Art. 367), the differences of R.A. and P.D. between T and K, that is, between the Transit at the Observatory and the Museums' clock, were immediately inferred.

369. The observations to which the foregoing statements refer were taken at various times between October 27 and November 5, 1866, and between October 29 and November 4, 1867, partly in presence of my astronomical class. The measured length of the base-line AB was 6584,3 inches, and that of CD 3259,4 inches. The positions A and B were westward, and those of C and D eastward, from the point P. The immediate purpose of the observations with the Theodolite and the auxiliary measures was to determine the differences of R.A. and P.D. between T, the place of the Observatory Transit, and T' that of a small portable Transit, which was occasionally set up, in the manner described in Art. 358, on the wall erected for supporting out-door instruments. (See Note to Art. 306.) This wall was raised a few feet above the level of the roof mentioned in Art. 368, so that the Transit could be conveniently placed and adjusted for approximately obtaining the local sidereal time. By measurement A was 2941 inches south and 535 inches east of T. By calculation from the observations at A and B, it was found that the horizontal projection of AP was 70482 inches, and, taking the above measures into account, that T was north of P by 42191 inches, and west of P by 59076 inches. Again, the distances in R.A. and P.D. between T' and Q having first been obtained by auxiliary measurement, by taking these into account and calculating from the observations made with the Theodolite in the South Court and on the Roof, it appeared that T' was

south of P by 7457 inches and east of P by 6788 inches. Hence it follows that T was north of T' by 49648 inches, and west of T' by 65864 inches. By adopting these values, which are measures on a surface cutting the direction of gravity at right angles, and applying small corrections due to the figure of the earth, it resulted that the N. latitude of the Museums' Transit was less than that of the Observatory Transit by $40'',8$, and its longitude eastward greater by $5^s,875$. Hence from the known latitude and longitude of the Cambridge Transit it results, that the latitude of the Museums' Transit is $52°.12'.10'',8$ north, and its longitude $28^s,62$ east of Greenwich. It may be noticed that the previously known geographical place of the Cambridge Transit here answers the same purpose as determining by astronomical observations the geographical places of certain stations in Trigonometrical Surveys.

By measures taken on March 15, 1867, it was ascertained that the latitude of K, the place of the Museums' clock, was greater by 132 inches than that of T', and its east longitude greater than that of T' by 70 inches. Hence, finally, the longitude of the clock is $28^s,63$ east of Greenwich, and its latitude $52°.13'.32'',3$ N.

370. *The Sextant.* Figure 56 was taken from a drawing representing one of Ramsden's sextants, which differed little in construction from those afterwards made by Troughton[1]. By reference to the figure, the different parts and their uses may be described as follows. The rays from an object are first incident on the plane reflector I, called the index-glass, and after reflection are incident on a second glass, called the horizon-glass, the lower half of which, i, is silvered for producing a second reflection. After this reflection, the rays are received by the Telescope AF and reach the eye at F, at the same time that a second object is viewed directly with the Telescope by rays that have passed through o, the unsilvered part of the same glass. The celestial arc between the two objects subtends,

---

[1] A sectant of Troughton's construction is figured in page 50 of Simms's *Treatise*, and the parts, adjustments, and modes of using it are described in pages 50—56 of that work.

as is known from Optics, an angle equal to double the mutual inclination of the planes of the reflectors. The arc BC is so graduated that this double angle is immediately read off by means of the vernier V, which is connected by a bar with the index-glass I, and turns with it about an axis passing through the centre of the graduation. The vernier is detained by the elastic plate s, and after being clamped to the limb by the screw m, is movable by the slow-motion screw n. The graduation proceeds from B towards C, extending 137° from zero in

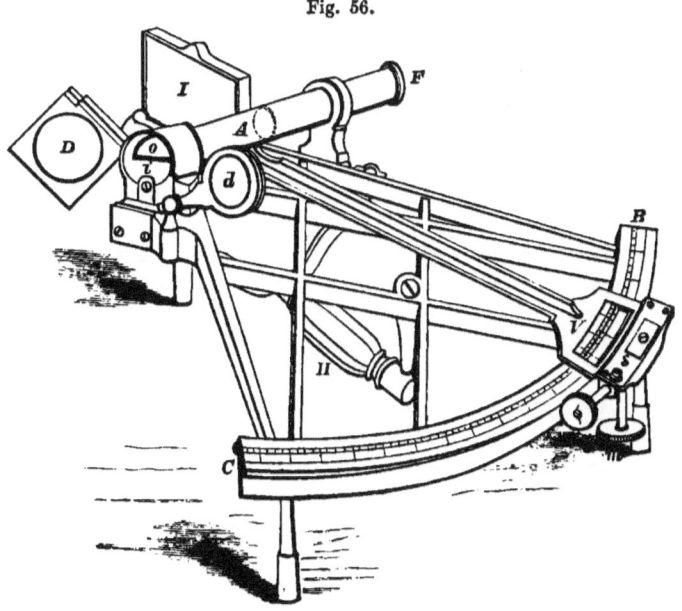

Fig. 56.

that direction, and 2° from zero in the opposite direction, and each graduation-interval is 20'. As 39 graduation-intervals are equal to 40 vernier-intervals, it follows that the arc can be read off to 30", and by estimation to a fraction of that quantity. [In Troughton's sextants, the graduation-interval is 10', which by the vernier is subdivided into intervals of 10″, so that the graduation can be read off to an estimated fraction of 10″.] Two sets of dark glasses D and d are so placed that by turning the former the sun-light reflected at I may be intercepted, and by turning the other the sunlight entering at o may be intercepted, whereby the light in each case can be sufficiently moderated before it enters the eye. In certain circumstances afterwards mentioned, it is preferable to place darkening glasses at the eye-end F of the Telescope.

371. The *adjustments* of a Sextant are executed by the instrument-maker, and generally so as not to require, or admit of, alteration by the observer. It is right, however, that the observer should know how to test their accuracy, the methods of doing which will accordingly here be given. The requisite adjustments are these: (1) the index-glass is required to be perpendicular to the plane of the instrument; (2) the horizontal glass should be perpendicular to the same plane; (3) after setting the index of the vernier to 0° of the arc, the horizontal-glass must be parallel to the index-glass: (4) the optical axis of the Telescope should be parallel to the plane of the instrument. To test the first adjustment, hold the instrument horizontally with the index-glass towards the eye, and look at the graduated arc partly directly and partly by oblique reflection at the glass. If the parts thus viewed appear at their junction to form a continuous arc, the index-glass is adjusted. In fact, by this means the plane of the instrument is practically defined. The second adjustment is tested by sweeping with the index-bar to see whether the reflected image of a star, or other object, passes exactly over its position as seen directly with the Telescope. If not, the error can be corrected by means of a capstan-headed screw (not seen in the figure), which acts transversely on the lower part of the frame that supports the horizontal glass. The third adjustment is verified, after setting the vernier-index at 0°, by finding that the images of an object, for instance the Sun, seen directly and by reflection coincide, or appear as one. If this be not the case, the deviation is due to index error, the amount of which, as no instrumental means are provided for correcting it,

is allowed for in the reduction of each observation, after being ascertained by the following process. The vernier-index being first set and clamped to a reading of about 30′ on the side of zero towards C, on looking towards the Sun two images will be seen nearly in contact, or a little overlapping, and may be brought to exact contact by turning the tangent-screw n. This is, in fact, the process for measuring the Sun's diameter by the Sextant, the images in contact being those of opposite limbs. Generally it is best to hold the Sextant horizontally for this purpose, the horizontal measure not being affected by refraction. The measure thus obtained is to be compared with the Sun's diameter as deduced for the given time from the Nautical Almanac, and the algebraic excess of the latter above the measured diameter is the required correction for index error. This correction is also obtainable as follows by instrumental observation without having recourse to the Nautical Almanac.

372. The graduation being extended, as stated in Art. 370, $2°$ on the side of zero towards B, the index may be set about 30′ from zero on that side, and then an exact contact of the two images be effected as before. The interval between zero and the place of the index for this contact may then be derived from the graduation by means of the vernier. Hence if D, D' be the first and second measured intervals and $D+e$ be the interval between the *true* place of zero and the vernier index on the side towards C, $D'-e$ will be the corresponding interval on the opposite side. Hence as each interval measures the Sun's diameter, $D+e = D'-e$. Consequently the correction e of D is equal to $\tfrac{1}{2}(D'-D)$, and is positive or negative according as the right-hand interval D' is greater or less than the left-hand interval D. The adopted value of e should be the mean of several determinations. In taking the above-mentioned observations of the Sun for determining the index error, a dark glass should be placed before the eye-piece at F instead of using the shades D and d, because some error might be induced by refraction through the sets of glasses of which these are composed. (See Art. 370.)

373. For testing the accuracy of the fourth adjustment two parallel bars are fixed in the field of view of the Telescope

very nearly at equal distances from its centre. These bars may be placed so as to be very approximately parallel to the plane of the instrument by causing the reflected image of a slow-moving star, as Polaris, or of a fixed terrestrial point, to move along one of the bars by combining a rotation of the Telescope by hand with a change of inclination of the reflector I by the bar. This arrangement being made, let us suppose that the plane of collimation of the Telescope exactly bisects the interval between the bars. Then selecting two objects, as the Sun and Moon, or two stars distant 90° at least from each other, bring their images into contact at the bar nearest the plane of the sextant, clamp the index, and after bringing by change of position of the instrument the images upon the other bar, notice whether they are still in contact. Assuming the axis of the Telescope to be parallel to the plane of the instrument, these two measures of the distance between the objects would be taken in planes equally inclined to the telescope's plane of collimation on opposite sides, and therefore ought to be equal. If a separation of the images in the second measure shews that this condition is not fulfilled, the requisite adjustment of the axis is to be effected tentatively by shifting two screws which fix the place of an up-and-down piece, whereby the object-end of the Telescope is raised or lowered. This adjustment, as originally performed by the maker of the instrument, rarely requires alteration.

374. The Sextant is an instrument indispensable to the navigator, being capable of effecting on ship-board several of the purposes for which, as already explained, a portable altazimuth is employed on land. Thus the altitude of the Sun, or other heavenly body, is obtainable by taking the sea-horizon for the second object, the Sextant measuring the vertical arc between any object and the horizon. (When used for such purposes the instrument is held with the left-hand by the handle H (Fig. 56), while the vernier-bar is moved for completing an observation with the right-hand by turning the slow-motion screw n.) The local time of day may be deduced from the chronometer times of taking equal altitudes of a known star, or of the Sun, before and after its meridian

passage (see Art. 355). In the case of the Sun a correction has to be applied for change of its declination in the interval between the observations (see Art. 356), and, whether the Sun or a star is observed, a small correction is required for the *ship's way* in the same interval, the rate of sailing (exclusive of the effect of currents) being ascertained by casting the log, and the course being indicated by the compass[1]. The practical method of finding the latitude of the place of the ship is to watch a little before noon the changes of altitude of the Sun's limb, as shewn by the readings of the vernier for measures of altitude till the maximum height is reached, and then to verify this value by seeing whether the reflected image of the limb just grazes the horizon-line when the instrument is moved to and fro to right and left. The Sun's altitude at the time of apparent noon of the place having been thus observed, and corrections for refraction and parallax applied, the *Latitude* of the place is readily inferred from the Sun's geocentric declination at the time, as calculated from the data of the Nautical Almanac. This calculation, however, implies that the longitude of the ship's place is at least approximately known by comparing a mean time chronometer regulated to shew Greenwich mean time with the chronometer used for determining the local mean time. The foregoing method of finding the ship's latitude is suitable for a voyage, and is much practised by seamen; but evidently when the error of a sidereal chronometer on local sidereal time is found, and consequently the time at which a *known* star crosses the meridian may be determined, the latitude might be more simply deduced from an observation of the meridian altitude of such star.

375. The Sextant is also employed in a very important manner for measures of *Lunar Distances*, that is, distances in celestial arc between the Moon's bright limb and the Sun's limb, or the limb of a Planet, or a bright known star. In taking this observation the Sextant has to be held so that the eye of the observer and the two objects are in the plane

[1] A formula for calculating this correction is investigated in Godfray's *Treatise on Astronomy*, Art. 146.

of the instrument as defined in Art. 371. Each measure requires to be accompanied by a note of time, and also by *simultaneous*[1] observations of the altitudes of the two points between which the measure is taken, for the purposes of calculating the difference of the azimuths, and clearing the observed distance of refraction and parallax. The corrections of the observed altitudes for refraction may be calculated from the usual Tables, the argument being Apparent Zenith Distance. The correction (p) for parallax in the case of the Moon may be calculated with sufficient exactness by the formula $\sin p = \sin P \sin z$, P being the horizontal parallax, and in the case of the Sun, or a Planet, by the approximate formula $p = P \sin z$. After applying these corrections the geocentric distance is calculated from the corrected zenith distances and the difference of azimuths, and from the result, as will be shewn in the next *Division*, the *Longitude* of the ship's place is inferred.

376. The altitude of a celestial object is also sometimes observed with a Sextant on land, in which case it is necessary to provide an *Artificial Horizon*. The one exhibited for illustration in the Lectures consists of a small oblong iron trough containing mercury for a reflecting surface, and placed under a covering having two slanting sides of plate glass for protection against disturbance by the wind. The plates are equally inclined from a vertical plane in opposite directions, so that a ray in a transverse plane, after passing through one plate, and undergoing reflection at the mercury, passes through the other. The Sextant measures the angle between the actual direction of the object and that of its reflected image, and consequently gives the double of its altitude above the horizon. In practising this method, observations should be taken an equal number of times with reverse positions of the protecting covering, and the mean result be adopted, in order to eliminate any error that may arise from unequal refractive action of the two plates on the ray. For a

---

[1] A rule is given in Simms's *Treatise*, p. 56, according to which the three arcs can be virtually measured at the same time by a *single* observer, the required altitudes being obtained by interpolation from altitudes taken before and after the measure of distance.

like reason the eye-glass at F should be armed with a single dark glass in place of using the shades D and d (see Arts. 370 and 372).

377. Troughton's *Reflecting Circle*. The construction and uses of this instrument are for the most part the same as those of the Sextant, from which it differs chiefly with respect to the following particulars. It can be mounted on a tripod stand having three adjusting foot-screws, and being at the same time counterpoised, and capable of movements about a vertical and a horizontal axis, it admits of adjustments whereby the plane of collimation of the telescope is made to pass through the two objects the angular distance between which has to be measured. By means of a plumb line, or by the application of a small spirit-level to the arms bearing the foot-screws, the position of the vertical axis of motion may by the screws be approximately adjusted, whence that of the horizontal axis follows by reason of the mechanical construction. The Reflecting Circle thus mounted is convenient for use on land in taking measures of celestial arcs. Again, it is furnished with handles by which when dismounted it can be so held in the hand as to allow of measures being taken in reverse positions of the graduated circle. This circle is graduated all round, and there are three vernier-indices for reading its indications, separated by intervals of 120°, the minutes and seconds being read from each, and the degrees being taken from that to which the tangent-screw is attached. It was considered by Troughton that the mean of the three readings got rid of errors that might arise from eccentricity of the centre of the graduated arc relative to the axis of motion of the index-glass, and that by the mean of the two sets of readings in reverse positions the index error was corrected, and errors due to refraction through the dark glasses were counteracted. In other respects the adjustments are the same as those already explained relatively to the Sextant (Art. 371), and need not be stated here. By the processes above mentioned the indications of the Reflecting Circle can be read with a precision hardly required by the observations for which it is adapted, which do not generally admit of a high degree of

accuracy. For seamen the Sextant is a more serviceable instrument[1].

378. Borda's *Repeating Circle*. This instrument resembles in some respects a Troughton's Reflecting Circle mounted on a tripod stand, being like the latter counterpoised, and capable of having the plane of collimation directed through two objects by means of rotations about a vertical axis and a horizontal axis. It differs, however, essentially from the Reflecting Circle in the respect that instead of measuring celestial arcs by the intervention of reflections, it employs for this purpose *two* telescopes, which can be turned about a common axis passing through the centre of a circle graduated on one face for taking the measures. These telescopes and the circle (which is between them) may be moved about that axis independently of each other, and each telescope admits of being clamped to the circle, and adjusted in position by slow-motion screws. One of the telescopes has four arms radiating from it in rectangular directions, and carrying four verniers by which the graduation is read off. One of the arms has at its end a tangent-screw-and-clamp apparatus by which the telescope is attached to the graduated face of the circle and adjusted, the clamping being effected by turning with the fingers the teeth of a small racked circle. The other telescope is clamped to the circle and adjusted by like apparatus. To this telescope a large spirit-level is fixed, which has to be adjusted so that its bubble is in mid-position when the axis of the telescope is pointed horizontally (see Arts. 361 and 366). For adjusting the axis of horizontal motion vertically, the feet of the tripod-stand are adjustible by screws, which rest in three brass sockets the heights of which are also regulated by screws, by which means and by the adjusted spirit-level, the required vertical adjustment may be effected and verified. This being done, the adjustment of the horizontal axis of motion follows from the mechanical construction. After clamping a collar and connected

[1] Of Troughton's Reflecting Circle, an example of which is exhibited in the Lectures, a particular description, accompanied by a Figure, is given in Simms's *Treatise*, pp. 57—60.

tongue to the vertical column supporting the instrument, the whole may be turned by means of a slow-motion screw about the vertical axis. By like means, when the telescopes are clamped to the circle, slow motion about the axis of revolution passing through the centre of the graduation is given in common to the circle and the telescopes. After the plane of collimation is adjusted for an observation, motion about the horizontal axis is prevented both by the counterpoise and by a clamp-and-screw acting upon a small fixed arc concentric with that axis. The mode of using this instrument for measuring celestial arcs is such as follows.

379. Let S_1 and S_2 be the two objects in the plane of collimation, the direction from S_1 to S_2 being that in which the graduation proceeds. Let the telescope which carries the four verniers, and is consequently contiguous to the graduated face of the circle, be signified by A, and the telescope contiguous to the opposite face of the circle by B, and let O be the position of the centre of the graduation. By the motions about the vertical and horizontal axes, and by the apparatus for clamping the telescopes to the circle, first point B towards S_2 and A towards S_1, and by the slow-movements make the pointings as exact as possible. Next, turn the circle and clamped telescopes till B points to the object S_1, and A is consequently made to point farther in the direction opposite to that of the graduation by the angle $S_1 O S_2$. Hence if A be now unclamped, and by independent movement, with adjustment by clamp and tangent-screw, be made to point exactly towards S_2, it will be moved in the direction of the graduation through an angle equal to *twice* $\angle S_1 O S_2$. Hence half the excess of the second reading for the pointing of A above the first gives the arc between the two objects. A *second* measure may be obtained by proceeding as follows. Turn the circle, with the telescopes clamped to it, so as to make A point again towards S_1, by which operation B will be brought to point farther in the direction from S_2 towards S_1 than its original pointing by twice the angle $S_1 O S_2$. Then bring B by its independent motion to point to S_2, and adjust the pointing of A to S_1 solely by the slow

motions of the circle about the vertical and horizontal axes (because the circle-reading for the pointing of A must not be altered), and that of B to S_2 by combining the same means with its proper tangent-screw adjustment relative to the circle. Supposing the telescopes thus to be correctly pointed, A to S_1 and B to S_2, all circumstances will be the same that they were at first, and the measure may be *repeated* by proceeding exactly in the same manner as before. Accordingly the telescope A will be still further advanced in the direction of the graduation by twice the angle S_1OS_2. If n such measures be taken, the required angle will evidently be the excess of the last reading for the pointing of A above the first, divided by $2n$. In the same manner a succession of altitudes of Polaris, or other star, might be obtained, the horizontal pointing of the telescope which carries the spirit-level supplying the place of a second object. It was formerly thought that by the repeating principle errors of reading off and of graduation might in great measure be eliminated. But the methods of graduating and subdividing the intervals are now brought to such perfection that there appears to be no reason for having recourse to that principle, of which, moreover, it is to be said that it leaves uncorrected any error that is common to all the measures.

380. The foregoing explanations apply to a Borda's Repeating Circle (exhibited in the Lectures) which was made by Troughton according to the construction that he adopted at first. Subsequently he produced the instrument in the modified form represented by Fig. 1 in Plate XXIII. of Pearson's *Practical Astronomy*, and described in pp. 498—509 of that work. The same form is figured and described in Loomis's *Introduction to Practical Astronomy*, pp. 103—106. The principal difference between this and the previous instrument is, that in the new form a graduated horizontal circle surrounds, and is fixed to, the supporting column at its base, and has a vernier for reading the graduation, with clamping and slow-motion apparatus. As the circle moves with the column about the vertical axis, the repeating circle may by this addition be used as an altazimuth instrument.

MISCELLANEOUS ADDITIONAL SUBJECTS.

381. *Different methods of determining Terrestrial Longitude.* The methods all depend on the principle of finding the mean solar, or sidereal, time at a certain reference meridian (which I shall suppose to be that of Greenwich), contemporaneous with a given instant of the mean solar, or sidereal, time at the meridian of the place whose longitude is required, the longitude being equal to the difference, less than 12^h, between the two times, and named East or West Longitude according as the time of day at the reference meridian is earlier or later than that at the other. The various means that have been employed for determining differences of longitude may be described as follows:—

(1) By the transfer of chronometers. This method requires arrangements to be made for setting up a transit-instrument and a sidereal clock at each of the two stations whose difference of longitude has to be found, for the purpose of ascertaining by transit-observations of known stars the errors of the two clocks on the true local sidereal times. The transit-instruments should be well adjusted to the respective meridians, the chronometers employed should each be transferred several times alternately from one station to the other, and be compared at the beginning and end of each transfer with the respective clocks at the places of starting and arrival. If sidereal chronometers be used, the comparisons are to be made by the intervention of a solar chronometer (see Art. 264). Supposing that the rates of the chronometers, well ascertained at the same time, are allowed for, it will be seen that by these means the true sidereal time at one of the places is transferred to the other, and that as the difference of simultaneous local sidereal times is thus known, the difference of the longitudes is determined. (See what has been said on this method in Art. 262.)

By observations conducted on the principles above stated, and occupying the interval from the end of June to the end of

September, 1844, the Astronomer Royal obtained the Longitude West from Greenwich to Liverpool, from Liverpool to Kingstown, from Greenwich to Kingstown directly, and from Kingstown to a station in the Island of Valentia at the western extreme of Ireland. For taking the transit observations of stars at Greenwich and Liverpool, the transit-instruments of the Observatories at those localities were used, and for obtaining like observations at Kingstown and Valentia, transit-instruments and sidereal clocks were temporarily provided, and special attention was given to steadiness of mounting, and to the usual exact methods of correcting transit-observations for instrumental errors. According to the final results the stations at Liverpool, Kingstown, and Valentia were respectively $11^m.59^s,86$, $24^m.31^s,13$, and $41^m.23^s,12$ west of Greenwich, with an average probable error of $0^s,04$. (All particulars relating to these observations are given in Vol. XVI. of the *Memoirs of the Royal Astronomical Society*, pp. 55—276.)

(2) By Galvanic Signals. Supposing that the chronographic method of taking transit-observations has been arranged at two localities (see Arts. 129 and 130), and a galvanic connection by conducting wire is established between them, the difference of their longitudes may be obtained as follows. The transit of a given star, taken and recorded according to that method by means of the transit-clock and galvanic apparatus at the transmitting Station, is recorded also at the receiving Station by noting the times, by the transit-clock of the latter, of the jerks of a galvanic needle set in motion by the contacts made at the instants of the passage of the star across the transit-wires. Assuming that the errors of the clocks on the true local sidereal times have been accurately determined, and that the transmission of the signals between the stations is instantaneous, each difference between corresponding recorded times corrected for the clock-errors is equal to the difference between simultaneous local sidereal times, and is therefore a determination of the difference of the longitudes. The concluded result should be the mean of all such determinations derived from repeated series of observations. The result may,

however, be affected by difference of the personal equations of the observers, to obviate the influence of which the series of observations should be repeated after an interchange of the places of the observers. Also it is desirable to transmit several sets of signals in equal numbers in opposite directions, to ascertain whether the transmission occupies a sensible interval of time, which would be the case if the difference of longitude be found to be different for the two directions of transmission. (See Arts. 263 and 264.) For the convenience of recording at the receiving station each set should be divided into batches of about six transmissions, commencing at times previously agreed upon.

A determination of the Longitude of the Cambridge Observatory from Greenwich was made according to the foregoing method on May 17 and May 18, 1853, being, I believe, the first accomplished in this country by such means. All the above-mentioned precautions for ensuring accuracy were attended to; but as the galvanic connection with Greenwich did not extend beyond the Cambridge Telegraph Station of the Great Eastern Railway, the Cambridge times of sending and receiving signals were recorded at that Station by chronometers, which before leaving and after returning to the Observatory were compared with the Transit-clock. The clock-errors were obtained by two independent methods. In one, the errors were deduced in the usual way from the apparent R.A. of *fundamental* stars included in the respective lists that were in use at the two Observatories, excepting that the adopted mean R.A. 1853,0 of the stars observed at Cambridge were all the same as their mean R.A. in the Greenwich list. In the other method the clock-errors were inferred from transit-observations of *identical* stars taken by the two observers on the nights of May 17 and 18, both before and after recording the signal times, the observers changing places after the first night. In calculating the difference of the simultaneous sidereal times, the *same* apparent R.A., derived from the Greenwich Twelve-year Catalogue, were used both for the Greenwich and the Cambridge transits, in consequence of which this process, while it increased the number of available stars, was not affected by small errors in their assumed apparent R.A. The mean results of the two methods differed by only $0^s,001$. The final result of these calculations gave for the Longitude of the Cambridge transit-instrument $23^s,027$ East. Subsequently (in 1854), I applied a correction derived from a computation of the effects of the forms of the pivots according to the formulæ obtained and exemplified in Arts. 93—102, whereby the Longitude East was altered to $22^s,753$, which is less by $0^s,79$ than the value previously obtained by transfer of chronometers[1].

The difference of Longitude between the Greenwich and Paris Observatories was determined by operations carried on from May 26 to June 4, and from June 12 to June 24, 1854, which closely resembled those described above as regards both the galvanic signalling and the transit-observations, excepting that there was direct galvanic communication between the two Observatories. The observers were M. Faye and Mr Dunkin, and it resulted from their observations that the Longitude of the Paris Observatory is $9^m. 20^s,63$ East from Greenwich, which is about $0^s,9$ less than the value which had been previously accepted. (The particulars of this determination are given in Vol. XXXIX. of the *Comptes Rendus*, 1854, pp. 552—566.)

(3) By observations of Lunar Distances. This method of finding the Longitude is practicable on ship-board, and is specially important in long voyages. It is mainly applied in determining from time to time the error of a chronometer which should shew the navigator Greenwich mean time. We have already considered in what way the Sextant is employed for measuring the great-circle arc between the Moon's bright Limb and either a bright star, or the Limb of the Sun or a Planet, and how by applying corrections for refraction, semi-diameters, and parallaxes, the geocentric, or *reduced* distance between the two objects is obtained. It remains to state

---

[1] This account of the process employed for obtaining the Longitude of the Cambridge Observatory by galvanic signals is intended to be supplementary to the statements made in Arts. 263 and 264. For full details see Vol. IX., pp. 487—514, of the *Camb. Phil. Transactions*.

the process by which the longitude of the spot where the observation was taken is thence inferred. This is done by means of the Tables of *Lunar Distances* from certain bright stars, and from the centres of the Sun and bright Planets, given in the Nautical Almanac for every third hour of Greenwich mean time each day of every month, the calculation of the distances being made according to the Lunar Theory, and the places of the objects selected for such calculation being well known, and convenient for observation of the distances. The G. M. T. corresponding to the reduced lunar distance may then be inferred from the calculated value nearest that distance by a simple proportion, supposing the Moon's motion to be uniform during an interval of 3^h. If, however, the results obtained by first differences be not sufficiently accurate, the calculations may be carried to second differences by the following formula adapted to this case from the known general formula of interpolation. Let D be the lunar distance reduced from the observations, D_1, D_2, D_3 three consecutive tabular distances, of which D_2 is the nearest to D, and let $t_2 - 3^h$, t_2, $t_2 + 3^h$ be the corresponding Greenwich Mean Times. Then if t be the required G. M. T. the interpolated value is given by the equation

$$t = t_2 + \left\{ \frac{(D-D_1)(D-D_2)}{(D_3-D_1)(D_3-D_2)} + \frac{(D-D_2)(D-D_3)}{(D_1-D_2)(D_3-D_1)} \right\} 3^h,$$

as may be readily verified by substituting separately D_1 and D_3 for D. The longitude results from the difference between t and the local mean time of the measure.

EXAMPLE. Let it be required to find the G.M.T. at which the reduced distance between the Moon's centre and a Aquilæ was $39°\ 50'\ 31''$ on March 15, 1879.

By data in the Nautical Almanac, t_2 is midnight of March 15, $D_1 = 40°\ 25'\ 44''$, $D_2 = 39°\ 21'\ 6''$, $D_3 = 38°\ 18'\ 49''$. Hence as $D = 39°\ 50'\ 31''$, we get $D - D_1 = -2113''$, $D - D_2 = 1765''$, $D - D_3 = 5502''$, $D_3 - D_1 = -7615''$, $D_3 - D_2 = -3737''$, $D_1 - D_2 = 3878''$. The remainder of the calculation may be conducted by five-figure logarithms as follows:

$L. (D_2 - D_1) = 3,88167\, n$ $L. (D_1 - D_2) = 3,58861$
$L. (D_3 - D_2) = 3,57252\, n$ $L. (D_3 - D_1) = 3,88167\, n$
$L.(D_3 - D_1)(D_3 - D_2) = \overline{7,45419}$ $L.(D_1 - D_2)(D_3 - D_1) = \overline{7,47028}\, n$
$A.C.\quad ,,\qquad ,,\quad = 2,54581$ $A.C.\quad ,,\qquad ,,\quad = 2,52972\, n$
$L. (D - D_1) = 3,32490\, n$ $L. (D - D_2) = 3,24674$
$L. (D - D_2) = 3,24674$ $L. (D - D_3) = 3,74052$
$L.\ 10800^s = 4,03342$ $L.\ 10800^s = 4,03342$
Sum $(= L. - 1415^s,4) = \overline{3,15087}\, n$ Sum $(= L. - 3551^s,4) = \overline{3,55040}\, n$

$\therefore$ G. M. T. $=$ March 15, $12^h - 1415^s,4 - 3551^s,4$,
$\qquad\qquad =$ March 15, $10^h\ 37^m\ 13^s$.

This example is calculated in p. 509 of the Naut. Alm. for 1879, and the resulting G. M. T. is March 15, $10^h.\ 37^m.\ 19^s.$, which, if not as accurate as that obtained by the preceding calculation, has the advantage of having been computed according to rules which dispense with using $+$ and $-$ signs.

(4) By observations of occultations of stars and planets by the Moon. These phenomena have been under consideration in Arts. 312—314, where it is shewn that by calculations founded on a noted time of the disappearance or reappearance of a star or planet at the Moon's Limb, an equation of condition can be formed involving, as unknown quantities, corrections of the adopted R. A. and P. D. of the bodies and of assumed semidiameters and parallaxes, together with corrections of the assumed latitude and longitude of the place of observation. For the use of such an equation in determining terrestrial longitude, it is necessary to deduce subsequently, from the results of Meridian Observations (as those of Greenwich), and from other independent means, *all* the corrections except that of the assumed longitude, in order that the equation may give this correction and thereby the required. longitude. It is, however, to be remarked that the calculation of the equation of condition implies that an approximate value of the longitude is already ascertained, as by the observation of a lunar distance at the same place, or other available means. Supposing the local sidereal time of the occultation to be well ascertained, and one or more observations of lunar distances to be taken about the same time, the observer may proceed to calculate an approxi-

mate longitude by the foregoing method (3). The observations may be made either on land or ship-board; but in the latter case there must be at least one observation of lunar distance not long before the occultation, and one of the *same* star not long after, and from each the geocentric lunar distance is to be calculated in the usual way, in order to deduce by interpolation the geocentric lunar distance at the time of the occultation. After obtaining by such means a value of the longitude sufficiently approximate for calculating the equation of condition, and correcting this approximation by means of that equation as shewn above, the resulting longitude is likely to be much more exact than one obtained directly by a measured lunar distance, inasmuch as in the case of the occultation the lunar distance, being the Moon's semi-diameter, is accurately known. The method, as applied at sea, gives the longitude of the spot at which the time of the occultation was noted, independently of the change of the Moon's place and the ship's way, if both motions be uniform in the interval between the observations of the lunar distances. (See for details of calculation the Appendix of the Naut. Alm. of 1854.)

(5) By transit-observations of the Moon and Moon-culminating Stars. Stars are designated as moon-culminating, when they are near the Moon's parallel of Declination, and have Right Ascensions not much differing from the Moon's. From transits of stars so situated, taken together with a transit of the Moon's bright limb, across the meridians of two places on the same day, the difference of the Longitudes of the two places may be calculated on the following principles. On account of the variation of the Moon's R. A., the interval between the transit of a given star and the Moon's Limb will be different at the two meridians, and the difference will be equal to the change of R. A. of the bright limb in the interval between *its* transits. Now that change is obtained by comparing the intervals between the transits of star and limb at the two places; and since the Variation in Right Ascension of the Moon's *bright limb* in the interval between transits across two meridians equidistant from Greenwich, and one hour distant

from each other, is given in the Naut. Alm. for the Upper and Lower Culminations (being derived from data furnished by the Lunar Theory), it follows that the difference of the Longitudes may be inferred from a simple proportion. If, however, the difference so found be large, the calculation should be repeated with the value of the variation of the R. A. of the Limb in one hour of Longitude corresponding to the middle of the interval between the observations, which value, since the epoch would be sufficiently known by the result of the first approximation, might be obtained by interpolation to the second order from values given in the Naut. Alm.

This method, as being differential, is independent of small errors in the places of the Moon and Stars. It is also applicable when the two localities are far apart. For instance, moon-culminations observed at the Cambridge Observatory have been supplied for determining by means of corresponding observations the Longitudes of places in Chili.

(6) By observations of *Eclipses* of Jupiter's Satellites. The Naut. Alm. gives, as results of theoretical calculation, the configuration of the Satellites every day on which they are visible from the earth, the G. M. T., to the nearest minute, of Disappearance and Reappearance in occultations, of Ingress and Egress in transits of Satellites and shadows of Satellites across the Planet's disk, and the G. M. T., to the nearest *second*, of the Disappearance and Reappearance of a Satellite in Eclipses. Although the theoretically calculated times of Disappearance and Reappearance, and especially the observed times, cannot be supposed to be very accurate in eclipses, such observations afford the navigator a ready means of at least approximating to his longitude, requiring only a comparison of the local mean times of the phenomena with the times computed in the Naut. Alm. Observations of these eclipses should therefore receive attention from voyagers and explorers. As the phenomena of the eclipses, as seen from different places on the earth's surface, occur sensibly at the same instants, differences of longitude may be inferred from the times of the *same* phenomenon as recorded at two or more places; but in

that case the circumstances of the observations, as regards the powers of the telescopes and atmospheric conditions, should be, as nearly as may be, the same. At the Greenwich Observatory the eclipses are constantly observed, and the mean times of Disappearance and Reappearance strictly calculated; whence by comparisons with the mean times of identical phenomena at other localities, the longitudes of these from Greenwich may be inferred with greater certainty than by comparison with the theoretical times. The method, however, of comparison of observed times is not, like the other, available on the spot, an interval necessarily intervening before the comparison can be made. The other phenomena above mentioned can scarcely in any instance be observed with the optical means on ship-board accurately enough for determinations of longitude; on which account the data in the Naut. Alm. only serve to indicate approximately the times of their occurrence for observations on land. At Greenwich these observations are made and reduced in the same manner as those of the eclipses, because the reduced times may be employed for correcting the theoretical calculation of the motions of the Satellites.

(7) By Trigonometrical Surveys. The principles of this method have already been exemplified by the process, explained in Arts. 366—369, for determining by means of a Theodolite and Gunter's chain the difference of longitude between the transit-instrument of the Cambridge Observatory and the Museums' Clock. For complete information respecting the triangulations employed in large surveys for ascertaining relative geographical positions, and the use made of the Transportable Zenith Sector for that purpose (Arts. 352 and 353), the published accounts of the Trigonometrical Surveys of Great Britain and Ireland, and of the Great Trigonometrical Survey of India, ought to be consulted. I shall here only add the remark, that at the request of the Astronomer Royal, Col. Colby ascertained, by connecting a secondary triangulation with the principal one, the geodetic value of the difference of longitude between Greenwich and the Island of Valentia, for comparison with the value obtained by transfer of chronometers, and that

the chronometric value was found to exceed the geodetic value by not more than 0˚,16, the earth's polar radius being taken to be 20,853,810 feet and the equatorial radius 20,923,713 feet. It was concluded by Mr Airy from the results of comparing the measures by the two methods of the arcs mentioned in Art. 381 (1), that no improvement can be made in the assumed elements of the earth's figure, and that consequently the length, 101·6499 feet, of an arc of 1″ perpendicular to the meridian in latitude 51°. 40′ deduced from those elements, may be accepted as a very certain geodetical *datum*. (See *Memoirs of the R. Ast. Soc.*, Vol. XVI., pp. 277—289.)

382. *Solar Parallax*. The Sun's Horizontal Parallax at any place is the ratio, in seconds of arc, of the distance of the place from the earth's centre to the mean distance of the earth from the Sun. If the place be at the Equator, the ratio is Equatorial Horizontal Parallax. Hence to calculate the horizontal parallax for any given place, it is first of all necessary to know the earth's figure and dimensions. These are obtainable by such operations and calculations as those referred to in Art. 381 (7), which rest on the principle of combining results of geodetical and astronomical observations with information derived from theoretical calculation on the hypothesis of gravitation. But to deduce therefrom with exactness the parallactic angle, and thence the value, in miles, of the earth's mean distance from the Sun, (which is the dimensional unit of the Solar System), is a problem of much difficulty on account of the smallness of that angle, and various methods of solving it have been proposed. As I have already in Arts. 308 and 309 treated of the determination of solar parallax by means of observations of the planet Mars in opposition, I shall here only take account of the methods of employing observations of *Transits of Venus across the Sun's disk* for the same purpose.

383. This method, which is the one that has been chiefly relied upon for calculating the Sun's parallax, is only available at epochs separated by irregular intervals. According to the results of calculations in Delambre's *Astronomy* (Vol. II., Chap.

XXVII.), the intervals are 8 years, 121½ years, and 105½ years, the longer intervals alternating with the short interval, and occurring alternately. The observations at any place consist essentially of recording as exactly as possible, in local times, the instants of external contact of the limbs of the Sun and Planet and complete immersion of the Planet, and the instants of internal contact and complete emersion. The notes of time are rendered somewhat uncertain by the formation of what is called *the black ligament*, or *drop*, which is, apparently, a diffraction phenomenon, and in greater or less degree according to Telescopic and atmospheric (perhaps, also, physiological) conditions interferes with discerning at what instants the two limbs have actually a common tangent. (The phenomenon can be produced by artificial representation of the disks and relative motion of the Sun and Planet, by which means an approximation might possibly be made to the correction which an observer should apply on this account to his noted time of observation in the case of the natural phenomenon[1].) As respects the principles on which the calculation of the parallax from data furnished by any observations of a Transit of Venus should be conducted, and the circumstances determining the selection of the most favourable places of observations in the instances of the Transits of 1874 and 1882, I cannot do better than refer the student to the Article by the Astronomer Royal in the *Monthly Notices* of the R.A.S. (Vol. XVII., No. 7, p. 208), which has been already cited in a note to Art. 309 (p. 302) relatively to observations of Mars for determining solar parallax. I shall only add here an account of the process of obtaining analytical formulæ of calculation appropriate to any instance of an observed Transit of Venus, adopting for that purpose, and in certain respects correcting, the investigation given in Woodhouse's *Astronomy*, Vol. I., Part II., Chap. XXXVIII[2].

[1] See an article on this subject by Prof. Simon Newcomb in the *M. N.* of the R. A. S., Vol. XXXVII., No. 5, p. 237.

[2] Transits of Mercury occur much more frequently than those of Venus, but on account of the small difference between his parallax at conjunction and that of the Sun, they are comparatively useless for determining the solar parallax. (See, on the Transits of both planets, Godfray's *Astronomy*, Chap. XXII., Arts 349—351).

384. Let T be the Greenwich Mean Time of the conjunction in longitude of Venus and the Sun, as deduced from the theoretical Tables of their motions, or from data contained in the Nautical Almanac, and let l_0 be their common geocentric longitude, and λ_0 the geocentric latitude of Venus, at that time. Suppose also the horary variations of the longitudes of Venus and the Sun to be m and m', and the horary variation of the latitude of Venus to be n, as gathered for the same time from the same Tables. Then if S and s be the contemporaneous semi-diameters of the Sun and Planet, and $T+t$ be the time of any contact of limbs as seen from the earth's centre, we shall have very nearly

$$(mt - m't)^2 + (\lambda_0 + nt)^2 = (S \pm s)^2,$$

according as the contact is external or internal. From this quadratic equation, by taking account of the double sign, four values of t may be obtained, two applying to the external contacts, and two to the internal. If t be now taken to represent one of these values, and if $l = l_0 + mt$, $l' = l_0 + m't$, and $\lambda = \lambda_0 + nt$, it follows that

$$(l - l')^2 + \lambda^2 = (S \pm s)^2 \ldots \ldots (a)$$

according as t applies to an external or internal contact. The values of l, l', and λ, which suppose the spectator to be at the earth's centre, may either be inferred from the above equalities for any of the given values of t, or, as is preferable, be directly computed from the Tables for any of the known epochs $T+t$. To obtain an equation applicable to the case of an observation made at any given place on the earth's surface, the parallaxes of the Planet and Sun in longitude and latitude have to be taken into account. Let a, a' be their respective parallaxes in longitude, and δ, δ' the parallaxes in latitude, and putting T' for $T+t$, assume $T'+t'$ to be the Greenwich M.T. of observation. Then the apparent distance between the centres at that time is the hypothenuse of the triangle the sides of which are

$$l - l' + (m - m')\, t' - (a - a') \text{ and } \lambda + nt' - (\delta - \delta').$$

Hence, rejecting the squares and products of the small quantities $a - a'$, $\delta - \delta'$, and t', and equating the sum of the squares to $(S \pm s)^2$, it will be found, after taking account of the equation (a), that

$$t' = \frac{(l - l')(a - a') + \lambda(\delta - \delta')}{(l - l')(m - m') + \lambda n}.$$

This equation may be further simplified as follows. If P and P' be the equatorial horizontal parallaxes of Venus and the Sun at the time of the observation, and if $a = aP$, $a' = a'P'$, we shall have $a - a' = aP - a'P' = a(P - P') - (a' - a)P'$, which is very nearly equal to $a(P - P')$, because a and a' must be nearly equal on account of the apparent proximity of the two bodies. So if $\delta = bP$ and $\delta' = b'P'$, $\delta - \delta' = b(P - P')$ very nearly. Consequently

$$t' = \frac{(l - l')a + \lambda b}{(l - l')(m - m') + \lambda n} \times (P - P') = f(P - P').$$

Suppose the observation to give H for the local time of contact. Then if L be the Longitude East of the place of observation, the Greenwich mean time of the observation is $H - L$. Hence

$$H - L = T' + t' = T' + f(P - P')$$
$$\therefore f(P - P') = H - L - T'.$$

Such an equation as this is furnished by each of the two notes of time both at ingress and at egress. Hence if f_1, H_1, T_1' and f_2, H_2, T_2' represent respectively the means of the two values at ingress and egress, we shall have

$$f_1(P - P') = H_1 - L - T_1'$$
and
$$f_2(P - P') = H_2 - L - T_2'.$$

From either of these equations $P - P'$ may be calculated when L is known. Hence as the ratio of P to P' is known from the Tables, the value of P', the Sun's parallax at the time of observation, may be inferred, and thence, by the Solar Theory, the parallax at the earth's mean distance from the Sun be deduced. This is *Delisle's method*, which is applicable when only the ingress, or only the egress, is observed at any place, but requires, as we see, an exact knowledge of the longitude of the place. A good result, however, may be looked for from the mean of a large number of such determinations. When both the ingress and the egress are observed at the same place, we get, by subtracting one equation from the other,

$$(f_1 - f_2)(P - P') = H_1 - H_2 - (T_1' - T_2'),$$

from which L is eliminated. The deduction of the solar parallax by means of this equation may be made in the manner just shewn. This is *Halley's method*, which is preferable to the other in not requiring an exact determination of the longitude of the place of observation, but is generally available on the occasion of any Transit in only a few localities. The formula involves in an important manner $H_1 - H_2$, the *duration* of the Transit.

385. The following particulars are mentioned for the purpose of giving information respecting the actual state of the determination of the Sun's parallax. From the Transits of Venus observed in 1761 and 1769, especially the latter, Professor Encke obtained $8'',5776$ for the mean parallax, giving for the Earth's mean distance from the Sun about 95 millions of miles. This value was adopted in successive Nautical Almanacs, till in that for 1870 it was changed to $8'',95$, which is the value indicated by Leverrier in p. 114 of his Solar Tables published in 1858 in Vol. IV. of the Annals of the Imperial Observatory of Paris. This, again, is changed to $8'',848$ in the Nautical Almanac for 1882, from the result of calculations by Professor Newcomb contained in the Vol. of the Washington Observations for 1865, Appendix II., p. 29. A Report to Parliament of the results as to the Solar Parallax derived from the Telescopic observations made in the British Expedition for observing the Transit of Venus on Dec. 8, 1874, was prepared by the Astronomer Royal in 1877 and printed the same year. From the contents of this Report, and from supplementary considerations published in the *Monthly Notices* of the R. A. S., Vol. XXXVIII., No. 1, pp. 11—16, he concludes that the most complete result obtainable from these observations is a mean parallax of $8'',754$ corresponding to a mean distance from the Sun's centre equal to 93,375,000 miles. In p. 199 of the same Vol., Captain Tupman, who was at the head of the British Expedition, gives $8'',76$, subject to future correction, for the general result of the comparison of *similar* phases observed at different localities. Mr Stone, who has given particular attention to the influence which the phenomenon of the black-drop might have had, on the observations of 1769, arrived at $8'',897 \pm 0'',020$ as the mean of the several results given by observations of Mars in 1862, the Earth's parallactic inequality, the Moon's parallactic inequality, and a rediscussion of the Transit of Venus observations of 1769. Finally he deduces from a comparison of Contacts in the observations of 1874 the result $8'',88 \pm 0'',02$, which he considers to be the value entitled at present to the greatest weight. (See *M. N.* of the R. A. S., Vol. XXXVIII., pp. 279—295 and pp. 341—346.) From

the observations of Mars in 1862 Dr Winnecke obtains 8″,96, and Professor Newcomb 8″,855. Hansen's deduction from the Moon's parallactic inequality is 8″,916. The experimental determination of the velocity of light by M. Cornu leads to 8″,86. As a parallax of 8″,89 corresponds to a mean distance equal to 91,940,000 miles, it will be seen that all the modern determinations give a smaller mean distance than that obtained by Encke. It would be premature to calculate from them a definitive value because the photographic observations have not yet been fully scrutinized, and it is desirable to include the information that may be gained from the Transit of Venus on Dec. 6, 1882, which, it is expected, will be more favourable for good results than that of 1874.

386. *Reticles.* A Telescope mounted on a stand, without apparatus for moving it about a fixed axis, may be used for observing differences of R. A. and P. D. between a known star and any other object, if at its focal distance, it has either a reticle formed of straight wires enclosed in a circular boundary, or a *Ring Micrometer.* Figure 57 represents various arrangements for this purpose. In every case the observations are made by recording the times by a clock or chronometer of transits at straight lines or circular edges.

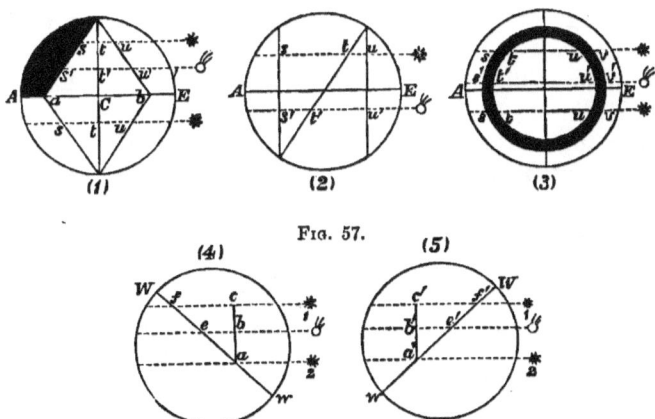

Fig. 57.

387. Fig. 57 (1) shews the form of the reticles employed by

Lacaille at the Cape of Good Hope for obtaining the places of the stars in his *Cœlum Australe Stelliferum*[1]. The reticles had the shape of a rhombus, three sides of which were fine wires, and the fourth was the straight edge of a brass plate. The opposite angles were joined by wires, and the longer diagonal was equal to the diameter of the circular boundary of the field, while the shorter was of different lengths in different reticles, according to the distances of the star-zones from the Pole. In the majority of Lacaille's observations the shorter diagonal ab was equal to the radius of the circle. By the construction the two diagonals were at right angles to each other and C the point of intersection was at the centre of the field. AE is a diametral wire formed by the prolongation of ab both ways. The reticle could be turned by hand about the axis of the Telescope, and the only adjustment preliminary to the observations was to place the diameter AE in coincidence with the path of a star through the field, in which case the other diagonal coincides with the circle of declination through C. Suppose now repeated transits of stars to be taken at the points A, a, C, b, E, for the purpose of determining exactly the ratios of CA to Ca, and of CE to Cb, and let the mean of the ratios so found be m. This being premised, let the observations give the times of transit of the known comparison-star at u, t, s, and those of the compared object at u', t', s'. Then if the mean of the former times be T, and the mean of the latter be T', the algebraic excess of the R.A. of the compared object above that of the star is $T'-T$, and as the latter R.A. is known the former is immediately inferred. Again, let τ be the mean of the intervals occupied by the passage of the comparison-star along ut and ts, and τ' the mean of the intervals occupied by the passage of the other object along $u't'$ and $t's'$, the intervals in both cases being derivable from the transit-observations. Now the ratio of Ct to $Cb-ut$ is, by similar triangles, equal to the ratio of the radius of the circle to Cb, that is, the ratio m. Hence $Ct = m(Cb-ut)$. So $Ct' = m(Cb-u't')$. Hence $tt'(= Ct - Ct') = m(u't'-ut)$. But ut and $u't'$ in arcs of a great

[1] A reduction to 1750,0 of the star-places of this Catalogue, in number 9766, was undertaken by a Committee of the British Association, and published in 1847.

circle, are respectively equal to $15\tau\cos\delta$ and $15\tau'\cos\delta'$, δ and δ' being the declinations of the two objects. Consequently tt' the difference of the declinations is

$$15m\,(\tau'\cos\delta' - \tau\cos\delta) = 15m\,(\tau' - \tau)\cos\delta_1$$

very nearly, if δ_1 be the mean between δ and the value of δ' derived from a first approximation to the value of $\delta - \delta'$. It has here been supposed that the two objects pass in the upper half of the field, and that Ct is greater than Ct'. If they pass in different halves the difference of the declinations is $15m\,(\tau' + \tau)\cos\delta_1$. We may hence derive the following general rule:—The difference of the P.D. is $15m\,(\tau' \sim \tau)\cos\delta_1$, or $15m\,(\tau' + \tau)\cos\delta_1$, according as the objects cross in the same or different halves of the field, and in all cases the P.D. of the compared object is equal to that of the other $\pm$ the difference, according as it is the higher or lower in crossing the field of an inverting Telescope.

388. Fig. 57 (2) gives the form of Valtz's reticle. Two parallel wires, which are chords of the circular field of view, are separated by arcs of $60°$, so that the distance between them is equal to the radius of the circle. Two contrary ends of the chords are joined by a diametral wire, which, consequently, makes an angle of $30°$ with each chord; and a second diametral wire AE, passing through their middle points, cuts both at right angles. The latter wire is used for adjusting the reticle, just as in the case of Lacaille's reticle (Art. 387) by rotating AE so as to place it in coincidence with the path of a star, in order that the chords may be parallel to the circle of declination through the centre of the field. The observations are then made by recording the times of transit of the two objects across the three wires; as, for instance, the transits of the star at u, t, s, and those of the comet, or other compared object, at u', t', s'. Let t_1, t_2, t_3, be the former recorded times, and t'_1, t'_2, t'_3, the latter. Then if $t_1 + t_3 = 2\tau$ and $t'_1 + t'_3 = 2\tau'$, the algebraic excess of the transit-time of the compared object across the circle of declination above the transit-time of the star is $\tau' - \tau$. Hence the required R.A. of the object is obtained by adding

$\tau' - \tau$ to the known R.A. of the star. Again, the difference of the P.D., namely ss' or uu', is readily seen to be, in arc of a great circle, very nearly equal to

$$15 (\tau - \tau' - t_2 + t'_2) \cos \delta_1 \cot 30°,$$

δ_1 being put for the mean of the declinations of the two objects. As this expression remains the same, whether the objects cross the field on the same side, or on different sides, of AE, Valtz's reticle has the advantage of not requiring the circumstance in either case to be noted. If q be the value of that expression, which is necessarily *positive*, the P.D. of the compared object is obtained by adding to that of the star $\pm q$, according as its path across the field of an inverting Telescope is higher or lower than the star's.

389. The method of observing differences of R.A. and P.D. by means of a Ring Micrometer may be explained by referring to Fig. 57 (3). A brass ring of uniform breadth is fastened to a transparent glass plate, and placed at the focal distance of the Telescope, so that an object, as it crosses the field, can be well seen through the plate before it reaches and after it leaves the ring, the space enclosed by the ring being vacant. Also as the object and the ring are both seen distinctly with the eye-piece, transits of a known star and a compared object (for instance a comet) may be taken, as indicated by the figure, at the outer and inner edges of the ring, the times of the disappearances behind the ring and of the reappearances being both noted. Let t_1 be the mean of the transit-times at v and u, and t_2 the mean of those at t and s, for the comparison-star, and t'_1, t'_2 the analogous times for the comet. Then if $t_1 + t_2 = 2\tau$ and $t'_1 + t'_2 = 2\tau'$, $\tau' - \tau$ will be the algebraic excess of the R.A. of the comet above that of the star. Hence from the known R.A. of the star, by adding $\tau' - \tau$ the required R.A. of the comet, or other compared object, is obtained. Again, let r be the radius (expressed in arc) of the circle which is intermediate to the circular boundaries of the ring, and let δ and δ' be the respective declinations of the star and comet. Then if

$$\sin \theta = \frac{15}{2r} (t_2 - t_1) \cos \delta, \text{ and } \sin \theta' = \frac{15}{2r} (t'_2 - t'_1) \cos \delta',$$

the difference of P.D. is the sum or difference of $r\cos\theta$ and $r\cos\theta'$, according as the transits were taken in different or the same halves of the field (as shewn by the Fig.). This difference of P.D. is to be added to, or subtracted from, the P.D. of the star according as the comet or star was the higher in the field, and the result is the required P.D. of the comet. It is to be remarked that in order to obtain trustworthy results the transits of the objects should be at a distance from the centre of the ring, and not very near the points of greatest and least P.D. This differential method clearly does not require any preliminary adjustment. It is necessary, however, to determine the value in arc of the radius r; which may be done as follows. Take a transit of the Sun's disk across the field, noting the times of exterior and interior contacts at the circular boundaries of the ring. Then if T be the interval between the two exterior contacts with the circle of mean radius r, as inferred from the observed times, and d be the Sun's semi-diameter, it is readily seen that T will be proportional to $2d + 2r$. Similarly it appears that T', the interval between the interior contacts, is in the same proportion to $2d - 2r$, d being supposed greater than r. Hence as T is to T' as $d+r$ to $d-r$, it follows that r is to d as $T-T'$ to $T+T'$, from which proportion, since d the Sun's semi-diameter in arc at the time of the observation is known, the value of r in arc may be calculated. The adopted value should be the mean of several such determinations.

390. The following method of taking differential observations, which, like the last, does not require preliminary adjustment, was proposed by M. Boguslawski, the astronomer at Breslau. A single wire passing through the middle of the field, and movable about the Telescope-axis, is put first in one *arbitrary* angular position, and then in a second, as represented by Fig. 57 (4) (5), and transits of *two* known stars and an unknown object (a comet) are taken at the wire in each position. Let t_1, T, t_2 be the observed times of transit of first star, comet, and second star at the points f, e, a, in the first position, and t_1', T', t_2' the analogous times for the second position; and conceive abc and $a'b'c'$ to be drawn parallel to the

circle of declination through the centre of the field. Then if a_1, a_2 and δ_1, δ_2 be respectively the R.A. and P.D. of the stars, and the R.A. and P.D. of the comet at the time of its transit be A and D, we shall have $\delta_2 - D$ to $\delta_2 - \delta_1$ in the proportion of ab to ac. Also the transit of a star along fc occupies the interval $(a_2 - a_1) - (t_2 - t_1)$, and that along eb the interval $a_2 - A - (t_2 - T)$. Hence by similar triangles

$$\frac{\delta_2 - D}{\delta_2 - \delta_1} = \frac{a_2 - A - (t_2 - T)}{a_2 - a_1 - (t_2 - t_1)} \quad \ldots\ldots\ldots(1)$$

similarly, if $A + \alpha$ and $D + \epsilon$ be the R.A. and P.D. of the comet at the time of its second transit,

$$\frac{\delta_2 - (D + \epsilon)}{\delta_2 - \delta_1} = \frac{t_2' - T' - (a_2 - A - \alpha)}{t_2' - t_1' - (a_2 - a_1)} \quad \ldots\ldots\ldots(2).$$

The two equations (1) and (2) suffice to determine A and D, the changes α and ϵ being supposed to be known from an ephemeris of the comet, or from nearly contemporaneous determinations of its place by observation.

391. The *Vernier*, so called from the name of the inventor, is used for subdividing the intervals of a graduation in cases which do not require the degree of precision attainable by a microscope-micrometer. As I have not hitherto explained the

Fig. 58.

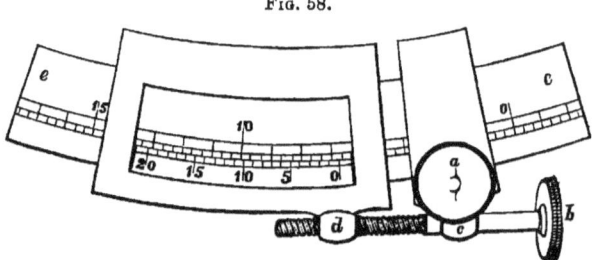

principle of this contrivance and the mode of reading its indications, I propose to supply the omission here, referring for this purpose to Fig. 58. If D be the interval between the graduations of the arc ee to which the vernier is applied, and D' the interval of the vernier-graduations, in all cases $(n + 1)D' = nD$,

n being an integer. Also the vernier-intervals are reckoned in the *same* direction as the arc-intervals. From the above equality we have $n = \dfrac{D'}{D - D'}$; and if, as is often the case in small setting circles, the degree be divided into three equal parts so that $D = 20'$, and it be proposed to read to the accuracy of $1'$, we shall also have $D - D' = 1'$. Hence $D' = 19'$ and $n = 19$, and accordingly 19 arc-intervals have to be equal to 20 vernier-intervals, as is the case in the example shewn by Fig. 58. To read the vernier-indication it is first to be noticed that the zero of the vernier is opposite to $6°$ and something more than two intervals, or $6°\,40'\,+$. The additional quantity is found by observing, on looking along the vernier, that the seventh mark from zero coincides with a mark of the arc-graduation. Hence the required reading is $6°\,47'$. If the reading be to the accuracy of half a minute, we should have $D - D' = 0',5$, $D' = 20' - 0',5 = 19',5$, and $n = 39$. Supposing the degree to be divided into six equal parts, and the reading to be accurate to $10''$, we shall have $D = 10'$, $D - D' = 10'' = \dfrac{1'}{6}$, $D' = 10' - \dfrac{1'}{6} = \dfrac{59'}{6}$, and consequently $n = 59$ (see Art. 370). After the foregoing explanation it will be easily seen that the zero of the vernier can be set to any proposed reading by means of apparatus such as that exhibited in Fig. 58, where a is a clamping-screw for attaching to the arc the nut c which supports a slow-motion-screw, and d is a tapped nut joined to the vernier, whereby the same screw, being turned by the milled head b, works so as to give to the vernier the slow-motion required for the setting.

392. The *Transit-reducer*. For the time-reduction of any transit-observation to the meridian it is requisite to calculate the quantity

$$(a + b \cos z + c \sin z) \frac{\operatorname{cosec} \delta}{15},$$

a, b, c being respectively the corrections of the collimation, level, and azimuth errors, and z, δ the Z.D. and P.D. of the object observed (see Art. 92). To facilitate the calculation tables are formed of the particular values of the factors of a, b, c, with argument P.D., for the clock-stars, and of such other values as may be suitable for obtaining by interpolation the factors for any given P.D. (The Introductions of successive Volumes of the Cambridge Observations contain tables proper for these purposes.) The results of multiplying a, b, c by the respective factors are obtained at Greenwich by means of sliding scales, one for each product. In place of this method I began in 1849 to make use of a machine, constructed according to my directions by Mr Simms, whereby the total value of the foregoing formula could be calculated by a single operation. The construction of this machine may be gathered from the following argument. In Fig. 59 the line $OAQA$ is drawn in a fixed direction, OM

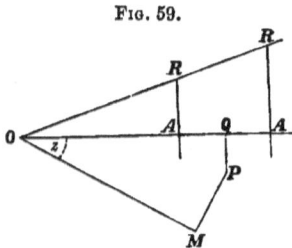

Fig. 59.

makes with this line an angle z equal to the zenith distance of an object, the angle MOR is equal to the latitude of the Observatory, and consequently the $\angle AOR$ is the declination of the object, or $90° - \delta$. Set off $OM = b$, make MP, drawn at right angles to OM, equal to c, then letting fall the perpendicular PQ or OA, make $AQ = a$, and draw AR at right angles to OA. Then it follows that $OA = \pm a + b \cos z + c \sin z$. If now OR be measured on a scale the unit of which is fifteen times that of the scale of the measurements of OM, MP, and AQ, we shall have

$$OR \text{ (in time)} = \frac{OA}{15} \sec AOR = (\pm a + b \cos z + c \sin z) \frac{\operatorname{cosec} \delta}{15}.$$

For the mechanical calculation of this quantity, a space six inches square, on a circular plate, was ruled all over with lines

MISCELLANEOUS ADDITIONAL SUBJECTS. 389

parallel to the sides of the square, and one-twentieth of an inch apart, which interval was assumed to be the measure of $0'',1$. Hence the side of the square measured $12''$. Accordingly from the centre of the square, which coincided with the centre of the circle, and corresponded to the point O in the Fig., distances less than $6''$, equal to the level and azimuth corrections, could be set off as rectangular coordinates, the positive and negative directions being indicated by $+$ and $-$ signs along the sides of the square. In this way the position of the point P is marked. The position of the line OA is determined by its passing through the centre of the circle, and by its being *parallel* to a fixed fiducial edge constituting an essential part of the machine. A plate, contained in a rectangular trough, and held with its plane face against one side by a spring, is movable along that edge, and carries with it two fine parallel wires perpendicular in direction to the face of the plate. One of the wires is adjustable so that it can be placed, by means of a fixed scale, and an index moving with the wire, at a distance from the other equal to the collimation-correction irrespective of its sign, the unit of the scale being $1''$. Two brass frames carrying the wires, are marked with the signs $+$ and $-$, to indicate the wires to be used respectively for positive and negative collimation-corrections. These arrangements being understood, it remains to shew how to set the circle, and to read off the indication of the machine, for a particular astronomical observation.

393. The circle can be turned smoothly by a handle about an axle at its centre, and is graduated close to the circumference from zero in an arbitrary position to $130°$ in the direction reckoned positive, and to $60°$ in the opposite direction, to meet the cases of P.D. above and below the Pole. In order to determine the place of an index and vernier for reading the graduation of the circle, the line OR, which with its graduations is engraved, requires to be put, by turning the circle, into coincidence with one of the parallel wires, which is, in fact, to make the angle ROA equal to $90°$. But this angle, as we have seen, is equal to the declination of the object. Hence in that

case the zero of the graduation is in the position corresponding to 0° of P.D. This result determines the position of an index intended for setting in degrees and minutes, by means of an appropriate vernier, for the given P.D. of a proposed object. The mechanical calculation then simply consists in first setting the circle to the given P.D. by the index and vernier, and, after bisecting the point P by the + or − wire according as the collimation-correction is positive or negative, reading the part OR cut off by the other wire. This reading, by reason of the scale of the graduation, gives the reduction to the meridian *in time*.

394. In the instance of the machine above described the diameter of the circle was 15 inches, the dimensions of the scales were such that the greatest amount of correction that could be read was $\pm 1^s$, and on account of the small angles made by the wire with the time-scale it could not be safely used for P.D. less than 30°. But for the more usual P.D. it was capable of calculating the correction to the nearest hundredth of a second of time; whereas by the use of the sliding-scales in the same cases, requiring three multiplications and three additions, the error might amount to $\pm 0^s,02$. To serve for the cases of small P.D., an additional graduation, the unit of which was 1″, was engraved along and contiguous to the fiducial edge, and reckoned in opposite directions from a zero, the position of which was defined by its being bisected by either wire made to cross the centre of the circle. Then, after determining the position of P, and using the parallel wires, just as before, the part OQ of the scale cut off by the second wire is

$$\pm a + b \cos z + c \sin z.$$

The calculation of this quantity by the machine for Polaris and δ Ursæ Minoris, or any stars within 30° from the Pole, is abundantly accurate, and might, apart from other applications, save much time[1].

---

[1] A fuller account of this instrument is given in the *Monthly Notices* of the Astronomical Society, Vol. x., p. 182. The one made for use in the Cambridge Observatory was shewn in the Great Exhibition of 1851, and received the award of a Bronze Medal.

MISCELLANEOUS ADDITIONAL SUBJECTS. 391

395. The *Huyghenian Eye-piece*. In addition to what is said on this eye-piece in Art. 21, I have only to direct attention to Fig. 60, in order that by a comparison with Fig. 2 in p. 23, the difference between the courses of the rays in Huyghens's and Ramsden's eye-pieces might be readily perceived. In the former the distance between the two lenses is usually half the

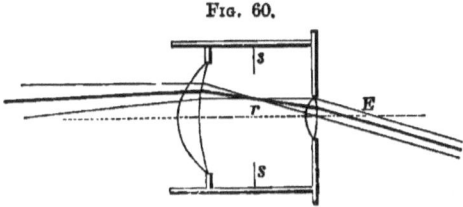

Fig. 60.

sum of their focal lengths, and the image is formed at the position of a diaphragm ss midway between the lenses. The four-glass erecting eye-piece may be made, like Huyghens's, free from injurious colour, but on account of the loss of light by passage through four glasses it is not to be preferred to the latter for observing the *physical phenomena* of celestial bodies, the details of which, especially in the cases of comets and nebulæ, are often very faint and difficult to recognise.

396. *Magnifying powers*. The following process, called "the method of double vision," is suitable for finding the magnifying power of a small hand-telescope. A white staff is set up, graduated by black transverse marks at intervals, say, of one inch, and one of the intervals is blackened. An observer, on looking at this interval through the Telescope, and opening the unemployed eye, may perceive an enlarged image of the black space thrown upon the staff. By counting the number of intervals which this image covers, the magnifying power of the combination of the lenses of the Telescope is obtained.

397. The power of *any* optical instrument composed of lenses, or of lenses and mirrors, and adapted for magnifying, is best ascertained by means of a small double-image micrometer, called a *Dynameter*, invented for this purpose by Ramsden and afterwards improved by Dollond. Figures 61 and 62 represent

the parts of a Dollond's Dynameter, the construction of which is very nearly the same as that of a Dynameter made by

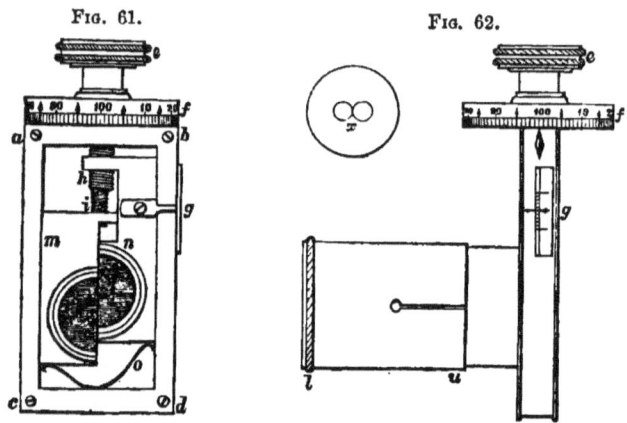

Fig. 61. Fig. 62.

Troughton and Simms, which I made use of for finding the magnifying powers of the various eye-pieces pertaining to the Telescopes of the Cambridge Observatory, and afterwards exhibited to my astronomical class. In the two Figures f is the circular micrometer-head divided on its rim into 100 equal parts, and e is a milled button for turning the micrometer. The screw h (Fig. 61), which is *right-handed*, and is fixed to the micrometer-head, acts in a tapped projection from the plate n, and thereby *draws* the plate together with an attached half-lens l. At the same time another screw i, which is *left-handed*, turns within h, the two having axes in a common direction, and as the thread-intervals are the same in both, it follows that the screw i *pushes* the plate m and the attached half-lens k just as much as the screw h draws the plate n and the half-lens l. The relative motion of the two half-lenses is therefore double the motion of each. The foregoing apparatus is contained in a box $acdb$. The external scale g is moved with the plate n. The spring o keeps the two screws in bearing. It is next to be remarked that when the eye is situated at some distance from the eye-glass of any optical instrument, and looks in the same direction as when it is applied to the eye-glass for observing an object; there is seen in front of, and near, this glass a small bright circle, which is, in fact, the image of the object-glass or -mirror, formed by pencils of rays converging to this position, after emergence from the eye-glass. The purpose of the Dynameter is to measure the diameter of this circle, because, as is known from Optics, the ratio of the diameter of the object-glass or -mirror to that of the small circle is the magnifying power. (I have added in an Appendix the proof of this Optical Theorem.) It is now required to shew how the diameter of this circle is measured.

398. The cap lu (Fig. 62) covers a Ramsden eye-piece the eye-glass of which is divided into half-lenses, and the Dynameter is applied so that the bright circle is situated just beyond the field glass. Hence, as the axis of the eye-piece is directed towards this circle, when the half-lenses are separated by turning the micrometer-head, two images of the circle will be seen. (To certify that each is an image of the object-glass, it suffices to affix a patch of tin-foil of any shape upon the latter.) For effecting the measurement these images are brought into contact, as shewn at x in Fig. 62, first on one side and then on the other, and by means of a fixed index, the readings of the scale g (Figs. 61 and 62) are taken, for both positions, in scale-intervals and divisions of the micrometer. The intervals of the scale are one-fiftieth of an inch, and hundredth parts of an interval can be read by the micrometer, so that the difference of the readings can be obtained with great precision in parts of an inch. This difference is the measure of the diameter of the circle, because, as we have seen, the two images move in opposite directions. The ratio of the diameter of the object-glass or -mirror in inches to the diameter measured by the Dynameter is the magnifying power.

399. *Interpolations.* The interpolation which the astronomer has most frequently to perform is that of finding, when three values a_1, a_2, a_3 of any function, separated by a common interval h of time, are given, the value of the function at any

time t differing by less than half the interval from the epoch of the middle value a_2. Representing by X_t the required value, putting x for $\frac{t}{h}$, and supposing that $a_2 - a_1 = d_1$, $a_3 - a_2 = d_2$, and $d_2 - d_1 = e$, we shall have

$$X_t = a_2 + \frac{d_1 + d_2}{2} x + \frac{e}{2} x^2,$$

the values of x and x^2 and their coefficients being readily calculated from the data.

Occasionally it may be required to interpolate with greater exactness, for instance, by using five equidistant values of the function, as a_1, a_2, a_3, a_4, a_5. In that case if b_1, b_2, b_3, b_4 be the first differences $a_2 - a_1$, $a_3 - a_2$, &c., c_1, c_2, c_3 the second differences $b_2 - b_1$, &c., d_1, d_2 the third differences $c_2 - c_1$, &c., and e_1 the fourth difference $d_2 - d_1$, and if

$$X_t = a + bx + cx^2 + dx^3 + ex^4,$$

we shall have

$$e = \frac{e_1}{24},\ d = \frac{d_1 + d_2}{12},\ c = \frac{c_2}{2} - e,\ b = \frac{b_2 + b_3}{2} - d,\ a = a_3,$$

by means of which equations the values of a, b, c, d, e are easily derived from the data. In this interpolation the given time t should differ by less than half h from the epoch of a_3.

APPENDIX. (*See* Art. 397.)

It is proposed in this Appendix to demonstrate a general Proposition for determining the magnifying power and brightness of the image of an object seen by any optical combination of lenses, or mirrors, or both. The demonstration is restricted to the first order of approximation, and the thicknesses of the lenses are not taken into account, this degree of accuracy being sufficient for most practical purposes. The following is the enunciation of the general Proposition.

To find the magnifying power and brightness of the image of an object seen through any combination of lenses and mirrors.

The general determination of the magnifying power rests on the two following Theorems applicable to any one of the lenses or mirrors of the combination. The proofs of these Theorems are given in the common Treatises on Optics.

(1) The linear magnitude of the object is to the linear magnitude of the image, as the distance of the object from the lens or mirror producing the image, is to the distance of the image from the same.

(2) Each image, whether real or virtual, may be regarded as an object with respect to the next lens or mirror, taken according to the course of transmission of the light.

Let $n =$ the number of lenses and mirrors together.

$D =$ the distance of the object from the object-glass or object-mirror.

$d =$ the distance of the eye-glass from the eye.

$f_1, f_2 =$ the distances of the first image from the 1st and 2nd glasses or mirrors.

f_3, f_4 = the distances of the second image from the 2nd and 3rd glasses or mirrors.

....

f_{2n-3}, f_{2n-2} = the distances of the $(n-1)^{th}$ image from the $(n-1)^{th}$ and n^{th} glasses or mirrors.

$f_{2n-1} + d$ = the distance of the nth image from the eye.

Let L = the linear magnitude of the object.

Then by the two Theorems above enunciated,

$$L \cdot \frac{f_1}{D} = \text{the linear magnitude of the 1st image,}$$

$$L \cdot \frac{f_1}{D} \cdot \frac{f_3}{f_2} = \text{.............................. 2nd ......}$$

....

$$L \cdot \frac{f_1 \cdot f_3 \cdots f_{2n-1}}{D \cdot f_2 \cdot f_4 \cdots f_{2n-2}} = \text{.............................. } n^{th} \text{ ......}$$

Hence, dividing by $f_{2n-1} + d$, the distance of the nth image from the eye,

the angular magnitude of the final image

$$= \frac{L}{D} \cdot \frac{f_1 \cdot f_3 \cdots f_{2n-3} \cdot f_{2n-1}}{f_2 \cdot f_4 \cdots f_{2n-2} \cdot (f_{2n-1} + d)}.$$

Let c = the distance at which the object can be seen distinctly with the naked eye of any individual. Then the angular magnitude of the object is $\frac{L}{c}$, or $\frac{L}{D} \cdot \frac{D}{c}$. Dividing the former angular magnitude by this latter, we have for the advantage gained by the optical instrument, or the magnifying power, the following expression:

$$\frac{c}{D} \cdot \frac{f_1 \cdot f_3 \cdots f_{2n-3} \cdot f_{2n-1}}{f_2 \cdot f_4 \cdots f_{2n-2} \cdot (f_{2n-1} + d)}.$$

If the object be a heavenly body, necessarily $\frac{c}{D} = 1$.

Hence for any astronomical Telescope,

$$\text{magnifying power} = \frac{f_1 \cdot f_3 \cdots f_{2n-3} \cdot f_{2n-1}}{f_2 \cdot f_4 \cdots f_{2n-2} \cdot (f_{2n-1} + d)}.$$

Also if the final image be seen by pencils of nearly parallel rays, that is, if the observer be not short-sighted, the ratio of f_{2n-1} to $f_{2n-1} + d$ is nearly unity.

Another general expression for the magnifying power may be deduced from the foregoing, by means of a third general Theorem, which may be enunciated as follows:

(3) Supposing the pencils to fall very nearly perpendicularly on all the lenses and mirrors, the breadth of a given pencil, where it leaves any lens or mirror, is to its breadth, where it falls on the next lens or mirror, as the distance of the focus, virtual or real, of the intermediate portion of the pencil from the former, to its distance from the latter.

This Theorem, which depends on very simple geometrical considerations, is proved in the usual Treatises on Optics. The effect of the thicknesses of the lenses is not taken into account.

Let A = the breadth of a pencil falling on the object-glass or object-mirror.

Then by the above Theorem,

$A \cdot \dfrac{f_2}{f_1}$ = the breadth of the pencil falling on the 2nd glass or mirror.

$A \cdot \dfrac{f_2 \cdot f_4}{f_1 \cdot f_3}$ = the breadth of the pencil falling on the 3rd glass or mirror.

....

$A \cdot \dfrac{f_2 \cdot f_4 \cdots f_{2n-2}}{f_1 \cdot f_3 \cdots f_{2n-3}}$ = the breadth of the pencil falling on the nth glass or mirror.

Let e = the breadth of the pencil as it enters the eye.

Then $e = A \cdot \dfrac{f_2 \cdot f_4 \cdots f_{2n-2} \cdot (f_{2n-1} + d)}{f_1 \cdot f_3 \cdots f_{2n-3} \cdot f_{2n-1}}$.

Now referring to the general expression already obtained for the magnifying power, it is clear that we shall have,

$$\text{magnifying power} = \frac{c}{D} \times \frac{A}{e}.$$

This expression is of great practical utility. For a *Telescope* we have simply,

$$\text{magnifying power} = \frac{A}{e}.$$

In the Galilean Telescope the breadth A is determined by the aperture of the eye, which in this instance limits the size of the pencils. In other Telescopes the size of the pencils is generally limited by the object-glass or object-mirror, the diameter of which consequently determines the value of A.

If the breadth e be greater than the diameter of the pupil of the eye, part of the light of each pencil is lost, and the effect is the same for the same magnifying power as if the object-glass were of smaller diameter. If $e' =$ the diameter of the pupil of the eye, then $\frac{A}{e'} =$ the inferior limit of the magnifying power, for any practical purpose, of a telescope the diameter (A) of whose object-glass or object-mirror is given.

On the other hand, if the diameter (e) of the pencil entering the eye be very much smaller than the diameter (e') of the pupil of the eye, the image becomes obscure and ill-defined from the feebleness of the light. Consequently, to unite great magnifying power with distinctness of vision it is necessary to increase the size of the object-glass or object-mirror.

Suppose that the pencil of rays from *a single point* incident on an object-glass or -mirror has the breadth e when it crosses the axis of the telescope where the eye is usually situated. The pencils from an unlimited number of external points would all cross the axis at the same place and be of the same breadth. But the aggregate of the courses of these pencils between the object-glass and the position of the eye, is the same as if the object-glass were an illumined object from which rays are incident on the eye-piece. If the object-glass be exposed to the light of clouds, or the sky, and the eye be removed along the axis to some distance from the eye-piece, a bright round image of the object-glass will be seen formed in the front of the eye-glass. The diameter of this image is therefore very nearly the quantity e. Since the Dynameter, as shewn in Art. 397, measures this diameter, it also measures e. Hence, the diameter A of the object-glass being also measured, the ratio $\dfrac{A}{e}$ becomes known.

In the case of a *Microscope* the magnifying power is chiefly produced by the small distance D of the object from the object-glass. In consequence of this arrangement the object-glass must be of very short focal length and small diameter. Hence also e must be small, in order that $\dfrac{A}{e}$ may not be a small fraction. To compensate for the smallness of the breadth e of the pencils the object has to be strongly illumined by artificial means.

The brightness of an image produced by optical means depends greatly on the loss of light by transmission through lenses, or reflection at mirrors, and cannot be calculated with mathematical precision. Supposing, however, that no light is lost by transmission or reflection, it may be shewn as follows, that the brightness of the image is the same as the brightness of the object, if the pencils from the image fill the pupil of the eye.

The quantity of light which enters the naked eye from a given small portion of the surface of an object seen at the nearest distance of distinct vision, varies as $\dfrac{e'^2}{c^2}$. The quantity of light which enters the eye through the optical instrument from *an equal* portion of the image, varies as

$$\dfrac{A^2}{D^2} \times \dfrac{1}{(\text{mag}^r.\ \text{power})^2},$$

no light being lost by reflection or transmission. But the magnifying power is equal to $\dfrac{A}{e} \cdot \dfrac{c}{D}$. Hence, by substitution it will appear that the quantity of light through the instrument varies as $\dfrac{e^2}{c^2}$. We have, therefore,

$$\frac{\text{Brightness of object}}{\text{Brightness of image}} = \frac{e'^2}{c^2} \div \frac{e^2}{c^2} = \frac{e'^2}{e^2} = 1, \text{ if } e' = e;$$

which is the result it was required to obtain.

As it is known that an object seen with the naked eye appears equally bright at whatever distance it is viewed, it follows from the above proposition, that an object of *finite* dimensions would appear equally bright under whatever circumstances it is seen, provided no light were lost by transmission through the air or lenses, or by reflection at mirrors. Thus, if a *comet* appears brighter in a large telescope than in a small one, it is owing to the more effectual exclusion of stray light by the greater length of the telescope tube. The case, however, is different with respect to *a star*, which must be regarded as a luminous *point*, the brightness of the image of which, as seen in a telescope is, *cæteris paribus*, proportional to the aperture of the telescope.

THE END.

www.ingramcontent.com/pod-product-compliance
Lightning Source LLC
Chambersburg PA
CBHW020534300426
44111CB00008B/666